A Love of Discovery
Science Education
The Second Career of Robert Karplus

INNOVATIONS IN SCIENCE EDUCATION AND TECHNOLOGY

Series Editor:
Karen C. Cohen, Harvard University, Cambridge, Massachusetts

A Continuation Order Plan is available for this series. A continuation order will bring delivery of each new volume immediately upon publication. Volumes are billed only upon actual shipment. For further information please contact the publisher.

A Love of Discovery

Science Education
The Second Career of Robert Karplus

Edited by

Robert G. Fuller

University of Nebraska—Lincoln
Lincoln, Nebraska

Kluwer Academic/Plenum Publishers
New York, Boston, Dordrecht, London, Moscow

Library of Congress Cataloging-in-Publication Data

A love of discovery: science education, the second career of Robert Karplus/edited by
Robert G. Fuller.
    p.   cm.
Includes bibliographical references and index.

    1. Science—Study and teaching.   2. Karplus, Robert.   I. Karplus, Robert.   II. Fuller,
Robert G.

Q181.3 .L68 2001
507'.1—dc21

                                                                          2001038776

ISBN 978-0-306-46687-8   ISBN 978-94-007-0876-1 (eBook)
DOI 10.1007/978-94-007-0876-1

©2002 Kluwer Academic / Plenum Publishers, New York
233 Spring Street, New York, NY 10013

http://www.wkap.com

10 9 8 7 6 5 4 3 2 1

A C.I.P. record for this book is available from the Library of Congress

DEDICATION

To the memory of Robert Karplus

Robert Karplus
1927–1990
circa 1970

Preface

Early in the 1990s I was attending a conference in Washington, D.C. The conference was sponsored by the National Science Foundation and was directed at the difficulties in disseminating educational innovations. During that conference a videotape of a panel discussion among leading science educators was shown. One of the panelists was Alan J. Friedman, whom I had met years earlier at the Lawrence Hall of Science. During the videotaped discussion Alan said that the paradigm that he had always had for a scientist and educator was Bob Karplus. I turned to my neighbor and said something like; "Someone should collect the publications of Bob Karplus so people can read his work today."

A few years later the notice of the deadline to apply for a faculty development leave from the University of Nebraska-Lincoln crossed my desk just after I had spent several hours looking for a copy of the Karplus "seven countries" paper (see it reprinted in Chapter 6). The thought came, Well, why not me? So I made a phone call to my wife, Margaret, and I decided to apply for a semester of developmental leave to go to Berkeley to interview people and put a book together. That was the spring semester of 1999!

My original intent was to collect ALL of the published papers of Robert Karplus into one volume. I soon discovered that his research papers in physics alone would make a pretty large volume. Furthermore, his physics papers were, in general, published in well-known journals, easily accessible to physicists. On the other hand, his papers in education appeared in a wide variety of journals, many of which were not readily available in science libraries. Hence many of his articles were unknown to many physicists and physics educators.

As I continued to interview his co-workers during the spring of 1999 it became increasingly difficult for me to figure out how to organize the publications of Robert Karplus. I was drawn more and more to focus my attention on his contributions to science education. Finally, near the end of the semester in a telephone conversation with Rita Peterson it clicked in my mind. She said that she remembered Bob's infectious love of discovery, not just for himself, but for others as well.

In addition, many of the common ideas used to discuss science education today were new in the 1960s and 70s and, to me, grew out of the work of Robert Karplus. Clearly it was time for a collection of his important papers to be made available to science educators in the 21st century.

So here it is, my collection of some of the important papers of Robert Karplus as well as a pretty complete list of all his publications in the Appendix.

This book would not have been possible without the wonderful cooperation of all the co-workers of Robert Karplus who freely gave their time and talents to help me. Thanks to the physics department at the University of California, Berkeley who gave me visiting scholar status and an office near the office of Alan Portis. My work in Berkeley in 1999 would not have been as productive without the help of Alan Portis, Elizabeth Karplus and Margaret Fuller.

Robert G. Fuller
February, 2001

Contents

CHAPTER 7—TEACHING AND THE CONCEPTUAL STRUCTURE OF PHYSICS

CHAPTER 8—SCIENTIFIC LITERACY AND THE TEACHER DEVELOPMENT GAP: KARPLUS' CHALLENGE

CHAPTER 9—THE SCIENTIST'S SCIENCE EDUCATOR: HOW ONE YOUNG PHYSICISTS' LIFE WAS CHANGED BY THE EXAMPLE OF ROBERT KARPLUS

CHAPTER 10—THE FIRST CAREER OF ROBERT KARPLUS— THEORETICAL PHYSICS

CHAPTER 11—THE TRANSITION YEARS: WHAT A SHAME, WHAT A WASTE—TO GIVE ALL THAT UP TO WORK WITH CHILDREN AND TEACHERS

CHAPTER 12—THE SECOND CAREER—SCIENCE EDUCATION

APPENDICES

INDEX

Contributors

Robert G. Fuller, Professor of Physics, University of Nebraska—Lincoln, Department of Physics and Astronomy, Lincoln, Nebraska

Rita W. Peterson, Senior Lecturer in Education, University of California—Irvine, Department of Education, Irvine, California

Herbert D. Thier, Academic Administrator Emeritus, Lawrence Hall of Science, University of California, Berkeley, California

Anton E. Lawson, Professor of Biology, Arizona State University, Department of Biology, Tempe, Arizona

J. Myron Atkin, Professor of Education, Stanford University, School of Education, Stanford, California

Helen Adi Khoury, Associate Professor of Mathematics Education, Northern Illinois University, Department of Mathematical Sciences, DeKalb, Illinois

Jane Bowyer, Abbie Valley Professor of Education, Mills College, Education Department, Oakland, California

Alan Friedman, Director, New York Hall of Science, Flushing Meadows Corona Park, New York

A Love of Discovery

Science Education
The Second Career of Robert Karplus

CHAPTER 1

Introduction

Robert G. Fuller*

This book explores the work of Robert Karplus in science education. It is an attempt to support the claim that several of his contributions to science education are as important today as they were when he first made them nearly 30 years ago.

The book is organized into chapters that highlight the contributions of Robert Karplus to eight different aspects of science education. Each chapter is introduced by an essay written by a person who worked with Dr. Karplus. Each essay provides a context and perspective on the selection of reprints of publications by Robert Karplus that are contained in that chapter.

Over and over again as I interviewed the people who had worked with Professor Karplus, I was struck by the exceptional quality of the people with whom he worked. Not only was he an exceptional person himself, but he brought out exceptional qualities in the people who worked with him.

Before you begin the examination of the work of Professor Karplus, let me provide you with a brief overview of his career.

Robert Karplus attended Harvard University where he obtained bachelors, masters and Ph.D. degrees. He obtained his Ph.D. in chemical physics in 1948 when he was 21 years old. His Ph.D. thesis had included both experimental and theoretical work and his work was directed by E. Bright Wilson,

*University of Nebraska-Lincoln.

A Love of Discovery: Science Education—The Second Career of Robert Karplus,
Edited by Robert G. Fuller, Kluwer Academic / Plenum Publishers, New York, 2002.

Jr. After completing his Ph.D. he went to the Institute for Advanced Study at Princeton. When he was there he worked with Norman Kroll in the new field of quantum electrodynamics (QED). Together they published one of the first detailed calculations based on QED. Their joint paper, *Fourth Order Corrections in QED to the Magnetic Moment of the Electron*, was published in the Physical Review in 1950 and brought them both immediate recognition within the physics community.

In 1950, he returned to Harvard University where he remained an assistant professor of physics until 1954. That year he moved to the University of California, Berkeley (UCB) as an associate professor and he became a full professor of physics at UCB in 1958, at the age of 31. From 1948 to 1962 he published 50 research papers in physics, mostly in QED, but also on the Hall Effect and Van Allen radiation. He was the senior or only author on the first nineteen of those papers. He published with 32 different physicists, including two who later won Nobel prizes in physics. More than 90 percent of his co-authors went on to become fellows of the American Physical Society.

In 1948, Robert Karplus had married Elizabeth Frazier. Bob and Betty were the parents of seven children born between 1950 and 1962, three daughters and four sons. He made his first visits to his daughter's elementary school class in 1959–60. He probably did electrostatics demonstrations with a Wimshurst machine that he had inherited from his grandfather. Those visits piqued his interest and his career made a fairly rapid change from theoretical physics to science education. In the late 1950s, he joined in an elementary school science project with some other UCB faculty. He and Herb Thier started the Science Curriculum Improvement Study (SCIS) at the Lawrence Hall of Science (LHS) on the UCB campus in 1961. He published his first education paper with J. M. Atkin in 1962. It was entitled "Discovery or Invention?" and is included in Chapter 5 of this book.

Over the next decade, Dr. Karplus and Herb Thier and their co-workers developed a complete kindergarten through sixth grade science curriculum, the SCIS curriculum, which is still in use today. In the 1970s, he turned part of his attention to student reasoning beyond elementary school and his work in that area is represented by the ten articles reprinted in Chapter 6.

Dr. Karplus became the President of the American Association of Physics Teachers in 1977 and was awarded their highest prize, The Oersted Medal, in 1980 for his exceptional contributions to physics and physics education. His professional career was ended in 1982 when he suffered a cardiac arrest while jogging in Seattle in June of 1982 at the age of 55. He died in 1990.

Robert Karplus made many enduring contributions to science education. If you look at the national and state standards for science knowledge that have recently been developed in the United States, you will find ideas based on the work of Robert Karplus. I will highlight six of his important contributions.

When he became interested in how children learned science, he started to read the professional literature of education and psychology. He brought a scientific approach to his work. First, he thought, find the correct theory and then use that theory to develop the science materials for the children. It was during this time of his work that he studied the writings of psychologists and educational theorists. He found the concepts of Jean Piaget were the closest to his own understanding of scientific reasoning. He was one of the first professionals to take Piaget's work into account for science curriculum development. He realized, based on the work of Piaget, that children build, or construct, their own internal mental schemes of knowing sciences as they experience the world. Hence, Karplus reasoned, to be effective, a team of people who are developing a science curriculum for children must be aware of the reasoning patterns used by the children. He developed a series of films on student reasoning that are still available today.

Once Karplus and his co-workers had constructed an educational theory for their work, based on their understanding of Piaget's work, they knew that they needed an instructional strategy that would enable them to put their theory into classroom practice. They developed the learning cycle instructional strategy which has three phases, known as EXPLORATION, INVENTION and DISCOVERY (or APPLICATION) (see Chapter 4). This instructional strategy moved the learning of science from the study of a textbook to hands-on experiences. I think it is nearly impossible today to find a science curriculum or set of standards that does not have the words "hands-on" in it somewhere. And most of them have no idea that Robert Karplus was an important part of the invention of that method of teaching and learning science.

When Karplus and his co-workers began to develop materials for classroom use, they used a scientific process of curriculum development. First they would brainstorm content ideas and approaches that they thought would be appropriate and effective. Then they would develop their materials and take them into the schools for field testing by the development team and with schoolteachers. During these field tests they would make detailed observations of both children and teacher behaviors. These notes would then be brought back to team meetings and revisions in the materials would be made. In a process analogous to the cyclic processes used by scientists, theory → experiment → revision of theory, the Karplus team created science learning materials, develop → field test → revise. Many lessons were created and field-tested, only the very best lessons survived (see Chapter 3).

Perhaps because of the many hours Karplus spent in school classrooms during the development of the SCIS materials, he realized that school teachers needed is develop additional teaching skills to become effective science teachers. So he emphasized teacher development (see Chapter 8). He, along with others, created a series of teacher workshops on the theme of science teaching and the development of reasoning. The movies he made on student reasoning patterns became a part of the teacher development workshops.

From the beginning Robert Karplus thought of science as a subject for all students. In his first papers and presentations on elementary school science, he emphasized science for everyone. He would not be surprised by a national science standard that calls for science for all citizens. Karplus and his team tried SCIS lessons on students of all kinds.

Finally, throughout his life Robert Karplus was excited by the process of discovering new ideas. His career as a physicist had been fueled by his exploration of the new ideas of QED. His work in science education was driven by his fascination with the responses of children to nature. He enjoyed the act of discovery and wanted science curricula to make that joy available to others. He wanted children to know of joy of discovery. The title of a film for the SCIS materials was "Don't tell me, let me find out" (see Chapter 2).

The next eight chapters of this book are collections of the writings of Robert Karplus introduced by a person who worked with him on the ideas explored in those writings.

Chapter 2. A Love of Discovery (introduction by Rita Peterson, Univ. Cal. Irvine)

The infectious enthusiasm of Robert Karplus excited children and teachers when they worked with him. His excitement when children and adults discovered things for themselves drove him to seek to understand the development of concepts by children and adults. This process was carried over into his development of a scientific method of curriculum development, as described in the article that follows this essay.

Chapter 3. Exciting Science Lessons (introduction by Herb Thier, Lawrence Hall of Science)

Robert Karplus and Herb Thier and their co-workers developed many lessons. Only the best survived. There were 240 outstanding lessons in 12 SCIS books that illustrated the essential features of science. Two of their early favorites, Material Objects and Mr. O, are described in the articles that follow this essay.

Chapter 4. The Learning Cycle (introduction by Tony Lawson, Ariz. State Univ.)

This essay offers a discussion of the independent formulation of the learning cycle by Chester Lawson and Robert Karplus and their subsequent collaboration. This essay is followed by descriptions of the Karplus learning cycle in a SCIS handbook and an article on science teaching and the development of reasoning.

Chapter 5. Student Autonomy and Teacher Input (introduction by Mike Atkin, Stanford Univ.)

The balance between autonomy and input is essential in effective science teaching. This was explored by Atkin and Karplus in 1962 and became the basis of much of work of Robert Karplus. The development of Robert Karplus' thinking about this issue can by seen in the "Discovery or Invention?" article of 1962 and his response to the Oersted medal paper, "Autonomy and Input" of 1981.

Chapter 6. Central Role of Student Reasoning (introduction by Helen Adi Khoury, Northern Illinois Univ.)

Robert Karplus had a strong interest in students' thinking and reasoning and why that makes a difference in how teachers think about science teaching. In the 1970s, he, with a wide variety of co-authors, explored student reasoning beyond elementary school, as illustrated in the ten papers reprinted here.

Chapter 7. Teaching and the Conceptual Structure of Physics (introduction by Robert G. Fuller, University of Nebraska-Lincoln)

From his earliest days as a teacher Karplus was intrigued by the relationship between the formal structure of physics and how to teach it to students in physics courses. He found a way to relate his understanding of the concepts of physics to his understandings of the reasoning patterns those concepts demanded of the learners. These papers share his insights on this topic.

Chapter 8. Teacher Development (introduction by Jane Bowyer, Mills College)

Karplus and co-workers created a set of teacher workshops around the theme of science teaching and the development of reasoning. These workshop materials were expanded to college teaching and development of reasoning materials and continue on today as essentials in teacher and faculty development activities around the country. Following this essay is an article of teacher development and an excerpt from the workshop materials.

Chapter 9. The Scientist/Educator Paragon (by Alan Friedman, NY Hall of Science)

In a videotaped interview for the NSF, Alan Friedman said that when he thinks of a model for the scientist/educator, he thinks of Robert Karplus. In this chapter, Dr. Friedman explores and explains this observation. In addition, Dr. Karplus offered students some exceptional insights into the scientific process as illustrated by excerpts from his introductory textbook, *Introductory Physics: A Model Approach*. W. A. Benjamin, Inc., New York, © 1969.

Robert Karplus with his parents, circa 1948

From left to right: Isabell Goldstern Karplus (1900–1967), Hans Karplus (1898–1971), and Robert Karplus (1927–1990).

CHAPTER 2

A Love of Discovery

Rita W. Peterson*

Those who knew him or worked with him described Robert Karplus as brilliant, intellectually curious, and possessing a sense of openness, warmth and friendliness that he extended to everyone he met. Each of us writing in this volume has a different story to tell that focuses on some aspect of the lasting contributions of Robert Karplus to science education. In this brief essay, I will first try to capture a sense of the personal qualities that inspired the confidence, respect and affection of so many people who knew or worked with Bob Karplus; and second, I will describe that which I consider to be his major, lasting intellectual contributions to education.

KARPLUS: A SCIENTIST WHO SHARED HIS LOVE OF DISCOVERY

Robert Karplus loved discovery; he enjoyed every aspect of the process including the exploration that lead up to the act of discovery as well as the products of discovery—his own discoveries as well as those of others with whom he worked. He relished posing questions for which answers were unknown, and thrived on solving novel problems. His natural curiosity, enthusiasm for solving problems, and love of discovery were contagious and inspired those around

*University of California, Irvine.

A Love of Discovery: Science Education—The Second Career of Robert Karplus,
Edited by Robert G. Fuller, Kluwer Academic / Plenum Publishers, New York, 2002.

him. He shared this love with children and adults alike through the daily activities he engaged in with them during the two decades that I knew and talked with Bob Karplus.

These personal qualities of Bob Karplus were exceptional among the scientists who took time from their lives as research scientists to contribute to the national reform of the science education for the nation's elementary and secondary schools, during the 1960's and 1970's. Dr. Robert Karplus personified the qualities that were presented at that time in history as the idealized image of a scientist: someone whose natural curiosity led him or her to solve important problems and make rewarding discoveries. Bob as a scientist represented "the real thing." Other well-known scientists possessed these qualities, but the friendly out-going disposition of Robert Karplus made these qualities more accessible to many more hundreds of children, adolescents, teachers, graduate students, and science educators around the nation, through the SCIS program and through being seen on film by thousands of others, as he shared his joy of science and the processes of discovery. The following examples illustrate a few occasions when these exceptional qualities of Robert Karplus were immediately evident to me.

Working with Bob on the SCIS Program

The setting was the late 1960's. A team of nearly 50 scientists, science educators, educational psychologists, elementary teachers, equipment designers, and graphic designers worked together under Bob's leadership at the Lawrence Hall of Science (LHS) on the Berkeley campus of the University of California. We were developing the six-year elementary school science program called the Science Curriculum Improvement Study, or SCIS (pronounced SKISS) as it later came to be called by its acronym. Every member of the team felt privileged to work under Bob's leadership. Few if any of us were hired the usual way, through advertised positions; we all had met Karplus somehow, knew we wanted to work with him, and asked to join "the project."

On a typical day, Bob would appear suddenly in any one of our offices, and pose a new question or invite us to solve a science problem that he had just created to illustrate some particular scientific phenomenon. Smiling, eyes sparkling with enthusiasm, and eyebrows arched in expectation, he would pose the problem and wait for our response to this newest science challenge intended for children. Each new activity that Bob created led him to make the rounds to most of our offices, gathering individual and collective reactions and responses. This feedback provided useful information that guided the further development of all SCIS activities. Some might find this situation far from their ideal working condition; but those of us who worked with Karplus found it a stimulating and productive environment. We enjoyed his creative mind and boundless curiosity. It is not surprising that we were encouraged to adopt this interactive style of testing our own ideas with one another while we worked at SCIS; but it came as a surprise to find that many of us were imprinted for life, reflecting Bob's contagious love of exploration and discovery. We continued to behave that way long after we left the SCIS project.

Early afternoon on most days, small teams of three or four of us left our offices in the LHS which was perched high on a hill overlooking the rest of the University of California campus and drove down the narrow road to one of Berkeley's elementary schools where we piloted our newest ideas for SCIS activities. Bob usually carried the boxes of science materials and equipment that were to be used in the afternoon's activities, eager for what was about to happen next as we took turns introducing the newly designed science activities, materials, and pieces of equipment, and observed children's reactions to them. We worked our way through several classrooms at each school, and proceeded to a different school the next day, all the while absorbing children's reactions and questions to guide our refinement of each science activity. When I think about our development of SCIS today, a vivid image of Bob Karplus comes to mind; the image is of his keen sense of eagerness, interest and gentle playfulness as he talked with one student after another about discoveries they were making with each new probe or piece of equipment.

Making a Film About SCIS

Working at SCIS involved more than developing the K-6 science curriculum. Bob's sweeping vision included teacher/administrator workshops, films, newsletters, research, publications, and conferences. One day late in the 1960's, I was asked to accompany John Quick, a talented film-maker who was hired to make a film that depicted children's highly positive responses to the hands-on nature of the SCIS program. After an hour or so of "shooting" in classrooms full of children engaged in SCIS activities, John and I entered a classroom where the teacher had just opened a box containing six chameleons. She began to distribute the lively chameleons among the science groups of children whose terreria had previously been prepared for this surprise animal. Squeals of surprise and delight were recorded on film as each group of children watched the chameleon dart around their terrarium. John and I were pleased with the morning's film shoot.

By lunchtime John and I returned to LHS, and Bob asked John to play the film footage for the day. When we reached the chameleon sequence, Bob suddenly leaned forward with his Swiss Army Knife poised in one hand and an apple he had been slicing in the other, and when the sequence ended, asked John to re-run the last segment of film. As the three of us watched, Bob said, "Here it is, our title!" He had selected from the action-filled classroom sounds, a young boy's voice pleading with the teacher, "Please don't tell me; let me find out!" as the teacher started to explain how chameleons changed color in response to ... (the teacher never finished her statement). This moment of viewing the boy on film epitomized Bob's keen perception while watching children in the act of discovery, and also characterized the achievement of Bob's creative vision of a science program that would foster such discoveries in classrooms across the country. Shortly after the film, "Don't Tell Me; Let Me Find Out!" was released, it received the Golden Eagle Award which was the highest national distinction given to educational films at that time.

Making a Film About Piaget and the Development of Formal Reasoning

One of the finest visual records that captures the integration of Robert Karplus' boundless curiosity, great enthusiasm for solving novel problems, and love of discovery is found in a film called "Formal Thought" (which later was re-entitled and re-released as "Formal Reasoning Patterns" in its second edition). The film was produced by Jack Davidson and his crew in 1972. Some background is useful. Throughout the development of the SCIS program, Karplus was interested in the relationship between logical reasoning and understanding science concepts. Bob and I were both familiar with Piaget's research which suggested that children and adolescents were expected to pass through a series of stages (called Pre-operational, Concrete, and Formal) as they developed the ability to solve problems using abstract reasoning. Bob wanted to explore the connection between understanding science, being able to solve problems related to science, and logical abstract reasoning.

Soon after I joined the SCIS team, Karplus asked me if I would join him in making a film that depicted Piaget's description of the most advanced form of logical reasoning: Formal Thought. I was intrigued by the idea of making a film to show young adolescents solving problems that required abstract reasoning. Over the weekend I re-read *The Growth of Logical Thought from Childhood to Adolescence*, the book in which Piaget described a dozen or more interesting problems that could be solved through experimentation with equipment and use of mental constructions of hypothetical-deductive reasoning by children and adolescents. Bob and I chose three problems described by Piaget that involved interesting science materials and concepts. One was a problem that required proportional reasoning, using various weights on a balancing beam. A second problem required the separation of variables to test the effect of various weights on bending rods; and a third problem required combinatorial thinking about various combinations of colorless liquids that could be mixed to produce a specific color-change effect. To these Piagetian experiments, Bob added a fourth problem involving ratio, which he called "Mr. Tall and Mr. Short" and which I understood he had just created.

The next week I headed for the Berkeley lumberyard and hardware stores in search of materials that could be used to construct the equipment described by Piaget for three of the problems. As soon as the talented equipment designers at SCIS converted the materials into film-quality equipment, Bob and I were ready to explore the unknown. Neither of us had ever seen or heard students actually respond to the problems described by Piaget, even though we were familiar with Piaget's description of Swiss students' responses to the problems. As far as we knew, we were the first researchers in the United States to replicate Piaget's experiments. Our goal was to make a film that illustrated Piaget's clinical method of interviewing students, a process that entailed presenting problems to individual children and adolescents. Through his spontaneous interviews, Piaget described and interpreted students' unstated and previously unknown frames of reference, referred to as indicators of developmental stages in their logical or formal, abstract reasoning. Thus, it was critical that Bob and I gain some first-hand experience interviewing real students before we began filming ourselves conducting interviews, since there could be no safety-net provided by a detailed script.

Early the next morning Bob and I set up two card-tables in the foyer of the LHS where we attempted to entice students as they emerged from school busses, expecting to spend the day at the many exhibits at LHS. Our plan was to invite individual students to join us at our tables to solve new experimental science puzzles as each busload of students entered the foyer. One of the most memorable moments of those impromptu interviews occurred when a young adolescent girl with a distinctively European accent said, after she had responded to the equal arm balance problem, "You know, I have seen this very problem once before when I was a little girl in Switzerland. This very old man with white hair and twinkles in his eyes used to give us puzzles, and this was one of them!" I asked her if she knew the man's name and she responded quickly with reflective fondness, "Oh yes; I never forget his name. He is called Professor Piaget. He is a very important man in our city."

Looking back on our preparation for making the film, Bob Karplus and I must have looked like two peddlers trying to drum up business on opposite sides of a busy street, as we tried to entice students to participate in our science demonstrations. Nevertheless, after three short days of spontaneous conversations with students entering the LHS, we had discovered that our small sample of Californian youth responded very much as would have been predicted by Piaget's theory!

Following our brief three day preparation, Bob announced that we were now ready to conduct live interviews on film. I felt ready for the video stage next, followed by an in-house analysis to refine our spontaneous interviewing and analysis techniques; but that wasn't an option. We had to move fast to pull Berkeley students from their classes when the parental consent forms arrived, signed, granting permission to film their children. Filming began. In one of the Berkeley Unified School District's old wooden portable buildings, we met for the first time each of the students who had been recommended by their teachers, based on each student's "high potential and high self-confidence." Over the next four days this multi-racial group of bright young people who had agreed to be filmed, gallantly solved problems they had never seen before and responded to our probing questions with grace and goodwill, in front of a camera crew that was only a few feet away.

Viewers of the film find Karplus and myself taking turns interacting one-to-one with each student. The film reveals Bob's very compelling enthusiasm for solving problems and watching students discover new relationships. Playfully Bob challenges Vladimir, Jocelyn, Robert, and Monica on film: "Very good! You have found out how tall Mr. Short and Mr. Tall are in *small* paper clips. Now let's see if you can measure Mr. Small in *big* paper clips and then predict—without measuring—how tall Mr. Tall would be in *big* paper clips!" As students worked through this and other puzzles, Bob continues to encourage and challenge each student. Later in the film, viewers see Karplus present students with another problem and say, "Ah-ha! So now you have discovered how to balance two equal weights on each side of the balance beam. Now let's see what you can discover about balancing unequal weights!" Thousands of people by now have watched Robert Karplus on this film in the decades that have followed its release, and have witnessed his enthusiasm for the act of discovery itself, and his desire to share that enjoyment with the students we met on film.

Everyone who has watched the film has some favorite bits. One viewer's favorite is Bob's questioning of the girl who starts to explain how to control variables on the rods task, and Bob says "Can you show me?" The girl says something like, "Oh, you mean here?" and proceeds to show that she does not separate and control variables. . . . Another favorite is the boy near the end of the film who tells Bob he solved the equal arm balance task by making "an approximate ratio." Bob asks "Can you explain how you made an approximate ratio?" and the boy does so. The film is filled with good examples of Bob's keen interest in discovering how students reason.

KARPLUS: A SCIENTIFIC APPROACH TO CURRICULUM DEVELOPMENT

Robert Karplus approached the development of the Science Curriculum Improvement Study as a scientist approaches situations when scientific theories about natural phenomena do not exist; he conducted exploratory experiments in search of potentially predictable relationships while observing the phenomena of teaching and learning. During his transition from the field of theoretical physics to science education, Karplus explored the leading phychological, sociological and physiological theories of human development and learning that dominated the field of educational research in the 1960s and 70s. From the earliest days of the SCIS Project, Bob discussed these theories with others who were interested in the relevance of theory to classroom learning, teaching, and curriculum development. Still, Karplus found then-current theories lacking, as suggested by the following excerpt from his comments at the Cubberly Conference, entitled Confronting Curriculum Reform, held at Stanford University in 1969.

> As a relatively recent newcomer to the field of education, I am not committed to any particular theory in the social sciences. In fact, I believe that there is no satisfactory theory of instruction or of learning which leads unambiguously to a teaching-learning experience, once an educational objective has been specified. (This entire article by Karplus, "Some Thoughts on Science Curriculum Development," follows this essay.)

Bob's view that the field of education lacked a unified theory of instruction and learning led him to formulate an experimental approach to the development of a new science curriculum, whose success in the classroom could be predicted, based on extensive experimentation and field testing. His experimental approach included at least three distinct elements: 1) Combining the expertise of scientists, teachers and science educators in the creation of a curriculum, 2) A formal process for the initial testing of ideas for the new science curriculum, and 3) Extensive replication using the proposed science curriculum under the various conditions in which the curriculum was to be used, that is, by conducting field experimentation using SCIS Trial Editions at nationally distributed Trial Sites. The following examples illustrate Bob's experimental approach.

Combining Expertise

It is easy to imagine that many ideas of scientists were over the heads of children aged six to twelve years of age. Likewise, it is also easy to picture many excellent elementary teachers creating science activities that lacked scientific substance, accuracy, or relevance. Thus, an important feature of the model implemented by Bob Karplus involved pooling expertise through the formation of authoring teams. Teams usually consisted of a scientist, a science educator, and an elementary school teacher who collaborated on the development of all SCIS activities. When authoring teams brought their activities to the Berkeley classrooms, one member of the team introduced the science activity while a second team member videotaped the member teaching, and the third team member walked among the students along with the regular classroom teacher, watching and listening to children's reactions and responses to activities. After introducing a science activity in the classrooms of two or three Berkeley schools, the teams returned to LHS to pool their sources of evidence of each activity's success. Videotapes provided crisp—and sometimes disappointing—evidence of children's responses to activities, materials and challenges. Children's written or illustrated responses were examined for evidence of their understanding, and our direct observations of children brought clarity to how some activities could be improved. Revision and re-teaching revised activities in Berkeley's elementary school classrooms continued until a decision was made to include an activity in a Trial Edition, or to "archive it."

Testing Initial Ideas About Science Curriculum

When I first arrived at SCIS to begin working on the SCIS Project, George Moynihan took me on a tour of the offices where SCIS was produced by a staff of 50 people, each hired for his or her area of expertise. I met them all: receptionists, scientists, science educators, teachers, educational psychologists, editors and typists, illustrators, designers and inventors of equipment, film makers, and business operations staff. Near the end of the 2-hour tour, George took me into a vault-like room called The Archives, which housed shelved boxes of records, floor to ceiling, of all phases of SCIS operations. Pointing to the largest single collection in similar-looking binders, I asked George what was in them. He smiled and said, "Well, that's the collection of *every* good idea anyone has ever had for a science activity, but the idea never made it to the publication stage. The puzzled look on my face led George to add, "Well, Dr. Karplus will explain our process for testing ideas for science activities in the classroom." The size of the collection of archived science activities was astonishing, and I wondered how so many talented people (I had just met 50 of them) could produce so many ideas that "didn't make it" to the publication stage. Very shortly I discovered how Karplus had established an experimental environment and expectations for developing the new science curriculum.

At the day-to-day level, Bob demanded that every idea about the choice of science content, the language to be used with the teacher and the students, the instructional materials, and teaching methods, would be tested experimentally—in the reality of the classroom. The decision-rule for whether a single science lesson or a complete unit would "go to publication" was based upon multiple sources of replicable evidence, reviewed by teams of experts who produced the curriculum, and through extensive field testing with children and by teachers and school administrators who represented the potential consumers of SCIS. To be successful, every activity was expected to be scientifically accurate, educationally relevant to schools and society, and intrinsically interesting and understandable to children in the classroom. In this latter case, the "gold standard" was that children had to be able to explain a key concept in their own words. This approach for developing science activities for SCIS usually meant that any potentially successful activity had to be revised many times before it reached the multiple criteria for success that led to publication.

Replication: SCIS Trial Editions and Trial Sites

Once science activities were considered successful enough for sending to the next level, they were formatted into booklets called Trial Editions, which had the appearance of final illustrated publications, printed in sufficient numbers that they could be field tested extensively at each of five regional Trial Sites located on university campuses throughout the United States. Trial Sites were directed by university professors who were part of the SCIS coordinating team, and were located at Teachers College in New York City, Michigan State University at East Lansing, University of Oklahoma at Norman, University of California at Los Angeles, and University of Hawaii at Honolulu. The elementary schools surrounding the five Trial Sites were located in urban, suburban, and rural areas to represent the wide range of demographic variations that were then present in the nation's schools.

Dozens of teachers from elementary schools surrounding each Trial Site were introduced to the SCIS Trial Editions and the giant boxes of interesting equipment and other materials, when they attended summer workshops. In these workshops they learned about the SCIS program and philosophy, and the importance of discoveries made by children as the constructed new knowledge for themselves. Bob Karplus met all or most of these groups of teachers at each site and encouraged them personally to give honest feedback regarding the effectiveness of every science activity in every science unit. A part of my responsibility at SCIS was to visit Trial Sites, talk with the teachers, watch them teach SCIS activities, listen to their feedback, and carry back to Berkeley children's booklets with their responses to activities. This mountain of feedback was candid, insightful, and led to further revisions and refinements in activities and units that ultimately made it to a Final Edition of paperback Teacher Guides, accompanying Student Booklets, and motivating as well as durable science equipment and related material.

After nearly a decade, Bob's experimental approach to curriculum development led to the publication of 12 booklets containing approximately 240 science learning activities that were predictably successful in the wide variety of elementary school classrooms represented by the five nationally different Trial Sites as well as the schools in Berkeley. The final editions of the curriculum included six life science units and six physical science units. Today, subsequent editions of SCIS have continued to be published and advertised, almost 30 years later.

ROBERT KARPLUS: A SCIENTIST'S LEGACY

What is the legacy that Robert Karplus has given the world? Those of us who have written brief essays for this volume have a constellation of views that include Bob as a brilliant scientist who contributed to the field of theoretical

physics early in his career, and later as a scientist who pursued the improvement of science education for children, adolescents, young adults, and their
teachers and professors throughout the United States. Still later Robert
Karplus was embraced by educators in several European and Asian countries
as well. Future readers of this volume will find here a collection of observations that give only a glimpse of the gifted scientist and exceptional human
being that Dr. Robert Karplus was.

My final goal is to reflect on and suggest the legacy of Robert Karplus
by exploring connections between his exceptional personal qualities and his
intellectual contributions, first, to science education and then more generally
to the field of education as a whole.

Contributions to Science Education

Bob's personal qualities—his boundless curiosity, his unparalleled enthusiasm
for solving problems, and his love of discovery—were evident throughout the
development of the SCIS program. I have described several commonplace
instances where these personal qualities emerged on a daily basis when I
worked with him. But readers may find for themselves instances where the
personal qualities of Karplus generate the playful yet tantalizing questions
posed to children in the Student Booklets themselves, in randomly inserted
Brain Teasers where children are challenged to expand their thinking about
concepts they have already explored.

Readers also may find evidence of these same personal qualities in the
larger endeavor of Robert Karplus in his scientific experiment in curriculum
development. SCIS provides evidence of Bob's lasting impact as a scientist
upon science education, as one compares Bob's vision (expressed in various
journal publications, a few of which are included in this volume) with the SCIS
program booklets and the films about Piaget. My reasoning is this: Karplus
spoke at the Cubberly Conference about the absence of a single theory that
predicted various learning outcomes. Nevertheless, by the time the final editions of SCIS were published, they reflected a Piagetian influence; and the
end result of Bob's decade-long experiment developing a meaningful science
program for children provides clear evidence that children and teachers are
able to construct and reconstruct their ideas about the natural world. The fact
that 240 or so final activities "made it to publication" in the SCIS program is
substantial evidence that children and teachers can explain scientifically and
socially important science concepts in their own words when the concepts are
presented in the hands-on manner and problem-solving frame of reference
that Bob formulated and refined in the SCIS program. One even finds reflected in the titles of activities and science units themselves, such as "Serial
Ordering and Comparison of Objects" (in *Material Objects*); "Separating
Mixtures" and "Controlling Variables" (in *Subsystems and Variables*); and

"Variables of Paper Airplanes" (in *Energy Sources*), evidence that Karplus found Piaget's developmental theory relevant to curriculum reform. By way of illustration, one finds the Piagetian influence reflected in the Preface of every Teacher's Guide:

> The SCIS science curriculum is based on current theories of how children learn.
> ... A child's elementary school years form a period of transition; he moves away
> from a heavy dependence on concrete experiences and begins to use abstract
> ideas to interpret his observations. By investigating phenomena in the class-
> room, he continues the exploration of his environment he began in infancy, but
> in a more organized fashion. Ultimately, he will use modern scientific concepts
> to create his own view of the world.

Thus, we may all conclude that the natural disposition of Robert Karplus to experiment, as a scientist, led him to discover that Piaget's theory about the development of logical thinking from childhood to adolescence could be applied to the construction of a new science curriculum for children; and more importantly, that elementary school children and their teachers did find the activities meaningful to their understanding and interpretation of the world around them. Otherwise they would not have been able to describe these key concepts in science in their own words.

An examination of teachers' behavior also provides evidence of Bob's legacy to science education. Teachers' participation in the experimental phase of SCIS changed teachers' visions of teaching. When I visited and interviewed SCIS teachers, I still remember being astonished by their serious-yet-positive and global thought about the value of first-hand experience for advancing children's understanding of complex concepts, not just in science but also in other subjects like mathematics and social studies. These SCIS teachers generalized about the importance of hands-on activities in ways that other researchers and writers had not yet described. It struck me that the teachers shared a common vision of the value of their individual and collective contributions to children's long-term intellectual development, something I never heard expressed by other teachers at that time. SCIS teachers' behaviors were not the result of some effective marketing strategy by a publisher but rather represented a personal sea-change for teachers' thinking about the way children discover and understand the world around them when they are able to explain something in their own words, based on their hands-on experiences.

One final observation is of interest within the context of major increases in the numbers and percentages of the English Language Learner population in the nation. During the developmental and trial stages of SCIS, two elementary schools in a rural town south of the San Francisco Bay Area decided to assess the impact of SCIS on the advancement of spoken and written English among primary school children who spoke Spanish as a primary language. Over a period of two or three years the principals and psychologists assigned to the two schools had established a controlled experiment and reported finding significantly greater (in a statistical sense) advances in English speaking and writing among primary school children in the school where SCIS was being used. The results were reported in a SCIS Newletter around 1972, suggesting early evidence of the importance of hands-on experience with materials for improving second language acquisition.

When Robert Karplus was suddenly and prematurely forced to end his work at the University of California, Berkeley, all of us who had worked with Bob were devastated by the loss. Shortly thereafter Bob was honored by the initiation of a formal lecture called the Robert Karplus Lecture that was to be given annually at the national meeting of the National Association of Science Teachers. I was privileged to give one of these lectures.

Contribution to the Field of Education as a Whole

At the close of the decade of the 1990's, and the beginning of a new century and millennium, one finds among the dominant themes of scholarly educational literature, abundant references to the constructivist movement. Gradually over the past forty years American educators and researchers realized, and in many cases demonstrated, the fact that children react in instructional settings as though they hold certain beliefs or conceptions about naturally occurring events, and that children's prior experience or "constructed knowledge" influences their reactions to new instruction or activities introduced in the classroom. American scholars have demonstrated over the past 30 years that children appear to construct and reconstruct their ideas about natural events or phenomena after verifying first-hand that their previous conceptions no longer hold the same explanatory power.

Within this context, Robert Karplus was one of the earliest among American scholars and researchers to examine Piaget's ideas about the importance of concrete hands-on experiences for children's development of language, logical thinking, and ultimate construction of new knowledge. Primary sources of evidence to support this conclusion begin with Bob's collaborative empirical research with hundreds of children and adolescents in the United States and many foreign countries as they responded to Piagetian-like science tasks that Bob and others designed, and reported in publications that appear in many volumes elsewhere, describing students' responses to problems related to science. Additional evidence is seen in the formal and informal research conducted at the LHS under Bob's direction or support, where a body of new knowledge was generated about the growth and development of logical thinking during childhood and adolescence. This very early research began in the 1960s and continued through the 70s and 80s at LHS, and thereby, set the stage for and continued to give momentum to the constructivist movement in the United States today. One also must consider the films that Karplus made with others to educate teachers and researchers about the work of Jean Piaget.

Some of these films still are sold around the world today. Finally, I would point to evidence of secondary effects. To be considered are the hundreds of university faculty and their graduate students majoring in science education. Their research and publications continue to examine the wide range of factors that appear to influence the development of logical reasoning during critical stages in intellectual development. One has only to examine the publications in the *Journal of Research in Science Teaching*, and especially the bibliographies over the past 30 years, to get a sense of the very substantial influence that Robert Karplus' has had upon subsequent science educators' contributions to the constructivist movement.

CONCLUSION

In closing this essay, the preceding discussion of the legacy of Robert Karplus to education somehow diffuses the parting image I want to give readers. It is an image of Bob Karplus as a gifted scientist whose avid curiosity and love of discovery never overshadowed his interest in and affection for humanity. Bob's personal modesty led him to live his life in an unassuming and unpretentious manner, a characteristic that never detracted from his brilliance but caused people to gravitate to this man who generously shared his joy of discoveries in science.

BIBLIOGRAPHY

Karplus, R., 1969, Some thoughts on science curriculum development, in: *Confronting Curriculum Reform*, The Cubberly Conference, Stanford University, Elliott W. Eisner (ed.), Little, Brown and Company, Boston, MA, pp. 56–61.

Karplus, R., 1975, *Strategies in Curriculum Development: The SCIS Project, Strategies for Curriculum Development*. Jon Schaffarzick and David H. Hampton, eds., McCutchan Publishing Corporation, Berkeley, CA.

Karplus, R., and Peterson, R. W., 1970, Intellectual development beyond elementary school II: Ratio, a survey, *School Science and Mathematics*, 70(9):813–820.

Karplus, R., and Peterson, R., 1972, *Formal thought.* (second edition: Formal reasoning patterns). Excerpts from film produced by Davidson Films, Incorporated. San Luis Obispo, CA.

4 Strategies in Curriculum Development: The SCIS Project

Robert Karplus

The Science Curriculum Improvement Study has developed ungraded, sequential physical and life science programs for the elementary school—programs, which, in essence, turn the classroom into a laboratory. Each unit of these programs was carefully evaluated by SCIS staff as it progressed from early exploratory stages to the published edition. The units originated as scientists' ideas for investigations that might challenge children and that illustrate key scientific concepts. The ideas were then adapted to fit the elementary school, and the resulting units were used by teachers in regular classrooms. They were tested several times in elementary schools before they were published.

Central to these elementary school programs are current ideas of intellectual development. A child's elementary school years are a period of transition as he continues the exploration of the world he began in infancy, builds the abstractions with which he interprets that world, and develops confidence in his own ideas. Extensive laboratory experiences at this time enable him to relate scientific concepts to the real world in a meaningful way. As he matures, the continuous interplay of interpretations and observations frequently compels him to revise his ideas about his environment.

The teaching strategy is for the children to explore selected science materials. They are encouraged to investigate, to discuss what they observe, and to ask questions. The SCIS teacher has two functions: to be an observer who listens to the children and notices how well they are progressing in their investigations, and to be a guide who leads them to see the relationship of their findings to the key concepts of science.

Reprinted with permission from *Strategies for Curriculum Development*, Jon Schaffarzick and David H. Hampton (Eds.), pp. 69–88, McCuthon Publishing Corp., Berkeley, CA © 1975.

THE SCIS MATERIALS

Organization of Materials

The SCIS program consists of thirteen units, six for a physical science sequence, six for a life science sequence, and one for kindergarten. The unity of the physical science sequence derives from the fundamental concepts of change and interaction. The six basic units—*Material Objects, Interaction and Systems, Subsystems and Variables, Relative Position and Motion, Energy Sources*, and *Models: Electric and Magnetic Interactions*—introduce and develop scientific and process-oriented concepts considered necessary for scientific literacy (see under "Content," below). The units in the life science sequence focus on organism-environment interactions. The six basic units—*Organisms, Life Cycles, Populations, Environments, Communities*, and *Ecosystems*—make use of the scientific and process-oriented concepts, but add the special considerations appropriate to the study of life. The *Ecosystems* unit attempts a synthesis of the children's investigations in physical and life sciences. Each of the six units roughly corresponds to the sequence of first to sixth grade.

Format

The SCIS materials reach the classroom in the form of kits, which have been designed to simplify and make convenient the use, storage, and reuse of the required equipment and supplies. Each kit is packaged for a teacher and thirty-two children and contains all of the materials needed except standard classroom supplies, dry cells, and the living organisms, which are sent separately when requested by the teacher. For each unit there are a teacher's guide and, in most cases, a student manual, but no textbook.

Content

Central to the SCIS program is the view that changes take place because objects interact in reproducible ways under similar conditions. In the SCIS program, "interaction" refers to relations among objects or organisms that do something to one another, thereby bringing about a change. Students can observe change and use it as evidence of interaction. As they advance from a dependence on concrete experiences to the ability to think abstractly, they identify the conditions under which interaction occurs and predict its outcome. The four major scientific concepts the SCIS program uses to elaborate the interaction viewpoint are matter, energy, organism, and ecosystem. Students' experiences and investigations in the physical science sequence are based on matter and energy; organism and ecosystem provide the framework of the life science sequence. In addition to these scientific concepts, four process-oriented concepts—property, reference frame, system, and model—are used.

Cost

Each unit, packaged in a kit for a typical classroom, costs between $150 and $250. The initial cost of teaching a unit is approximately $5 per pupil, but this varies, depending on whether the kits are shared among teachers. In subsequent years the cost per pupil decreases, as the "permanent" equipment in each kit can be reused.

HISTORY OF THE PROJECT, 1958–1977

The chain of events leading to the SCIS project originated in the summer of 1958, when I perceived a relationship among three activities in which I had participated. These were meetings concerned with high school science teaching that had been stimulated by National

Science Foundation-supported curriculum development activities at that level, my "show and tell" science sessions with the elementary school classes of my two school-age children, and the oral examinations of many physics graduate students whose performance gave evidence of a highly compartmentalized and not a broadly integrated understanding of undergraduate physics. All of them indicated to me that there were serious educational problems and that a better utilization of the first nine school years for science education might lead to substantial improvement. This conclusion may seem naive. Still, with encouragement from the NSF, I embarked on what I expected to be a brief side activity in my career as professor of physics; it actually turned into a project of almost twenty years' duration that resulted in a complete professional redirection of my life. In recounting the actual project activities supported by the NSF with grants in excess of four million dollars, it is worthwhile to identify six phases, each about three years long.

Exploration, 1959–1963

The staff consisted of me, an assistant, and occasional consultants during some of the time. I taught and observed elementary school classes, met with teachers and science educators, designed science activities and tried them out, and read publications in science education, developmental psychology, history and philosophy of science, and public policy. Most influential were Bruner's *The Process of Education*, the writings of Jean Piaget, and Kuhn's *The Structure of Scientific Revolutions*. At the end of this period, the educational philosophy and conceptual structure of the Science Curriculum Improvement Study program had been outlined in very general terms. There remained the detailed and demanding tasks of refining this outline in the light of classroom experience and producing curriculum materials that would enable many elementary school teachers to implement the curriculum in their classrooms.

Berkeley Area Trials, 1963–1966

The development of curriculum materials for an articulated program lasting several years required a laboratory school situation in which children and teachers used SCIS materials consistently. Herbert D. Thier, who joined the study as assistant director, took responsibility for establishing three laboratory schools in public schools of the Berkeley area; he also contributed essentially to program development. Chester A. Lawson joined the staff to take responsibility for developing a life science teaching sequence. Sister Jacqueline Grennan of Webster College, Arthur W. Foshay of Teachers College, Columbia University, and John I. Goodlad of U.C.L.A. formed an advisory committee that helped guide the project. The project staff gradually expanded, with the addition of elementary and secondary school teachers, science educators, a psychologist, consulting scientists, an editor, technicians, and artist-designers. The staff prepared trial editions of teachers' guides, student manuals, and laboratory kits for use by local teachers. A newsletter was published, and the teachers' guides were offered for sale to interested individuals.

Preliminary Edition, 1966–1969

Growing public interest and the recommendation of the advisory committee led to the establishment of five trial centers associated with universities in five parts of the United States. These included rural, suburban, small-town, and inner-city schools. To supply these centers and other interested schools with the necessary printed and laboratory materials, publication of primary-grade units in the Preliminary Edition was arranged through D. C. Heath and Company. Development of additional units for the upper grades continued in the Berkeley area laboratory schools, with the staff expanded further to about forty persons. Dissemination activities, relying primarily on local leadership, were begun.

Final Edition, 1969–1972

Revision of the Preliminary Editions with feedback from the trial centers took place three years earlier than scheduled because of difficulties with the publisher and limitations of expenditures placed on the funds granted by the NSF. The final edition was published by Rand McNally & Company. A related project under Dr. Thier's direction adapted the materials for use by blind children.

Evaluation and Teacher Education Materials, 1972–1974

After the completion of the basic teaching materials, the project developed a set of evaluation supplements for classroom use by teachers who wish to identify their pupils' progress systematically. A kindergarten unit entitled *Beginnings* and teacher education materials, including films and a *Handbook*, were also completed. The project staff gradually transferred to other activities.

Public Information and Service, 1974–1977

Supervision of the publications and equipment kits will continue, to assure the necessary quality control and possible minor modifications necessitated by changing availability of supplies, identification of defects in design, and other new conditions. Dissemination will also continue.

DEVELOPMENT STRATEGIES

The overall approach to curriculum development by the SCIS has been pragmatic. As goals were set and problems were encountered, the staff devised procedures for solving the problems and attaining the goals. When the available time and resources appeared inadequate to reach a goal or the goal itself seemed intrinsically unattainable, it was modified accordingly. In looking back over our work and that of others during the last fifteen years, we find that much practical experience has accumulated. Yet it is my opinion that there has emerged no curriculum theory that promises to yield better results than a pragmatic approach that relies on highly qualified staff and applies available theories without dogmatism. The reason probably lies in the enormous complexity of the problem that arises from the diversity of institutions, personnel, local practices, and students.

Within the overall pragmatic approach, however, certain strategies were developed and used successfully by the SCIS. Nine of these, each relating to an aspect of curriculum development, are given below together with a brief theoretical justification. In a final section of this chapter, I describe the potentially more general value of these strategies and possible improvements.

Strategy Relating to Personnel

From its very early days the Science Curriculum Improvement Study pursued a strategy of bringing together individuals with backgrounds in the many different areas bearing on instruction in elementary science. The principal developers on the SCIS staff included university scientists, high school science teachers, and elementary school teachers. They were assisted by psychologists, science educators, editors, filmmakers, artists, designers, and technicians. Each development team worked together very closely, but certain primary responsibilities were divided. Thus, the scientists were concerned with providing leadership as regards the conceptual development, the secondary science teachers devised laboratory procedures and designed equipment, and the elementary school teachers outlined classroom procedures. Frequent discussions involving the entire staff helped to refine the content of the units and to establish priorities regarding general objectives.

Personnel recruitment was carried out very carefully, principally by the assistant director, through personal recommendations of scientists and science educators who were sympathetic to the SCIS philosophy. Every candidate for a continuing position was interviewed and then observed teaching a group of students; the teaching style was evaluated in light of the SCIS philosophy. Letters and personnel files were not accepted as providing sufficient information regarding prospective contributions.

Theory behind Personnel Strategy

The study depended on a long-term staff working throughout the year, primarily for five reasons: the slow pace at which children can accept truly new ideas makes summers inadequate for experimental teaching; the interaction of staff with pupils, which results in imaginative new teaching approaches, cannot be compressed into a few weeks' time; the pupils at each grade level must have prior SCIS experience to articulate the units, and that cannot be assured during voluntary summer schools; staff members who conceive a unit must be available to follow the classroom trials and interpret the feedback for revision; staff members working on different teaching units are expected to interact significantly. The three levels of personnel were clearly needed to provide gradation in scientific knowledge and sophistication that could ultimately communicate some modern scientific concepts from the university scientists to nonscientific elementary school teachers and pupils.

Strategy Relating to the Invention of Activities

One of the most important products of a curriculum development project is a set of thoroughly tested activities for pupils that contributes to the overall aims of the project. To invent activities that will hold the children's interest, stimulate them intellectually, have significant science content, fit the conceptual scheme of a unit, are safe, do not require costly equipment, have a reliable outcome, and can be readily understood and supervised by the teacher is a very demanding assignment. All project staff members were, therefore, always encouraged to propose ideas for such activities, and very considerable staff time went into considering, reviewing, laboratory testing, and further developing these ideas. Elementary school teachers and sometimes secretaries on the project staff served as the first "guinea pigs," using the new activity after it had been worked out by its inventors, since their lack of scientific training made them the most naive subjects readily available to the project. It was most important to provide a positive but also realistic supportive environment for the development of new activities. Unless a clear objection arose during internal discussions by members of the project, proposed activities were tried out in elementary school classrooms, where the developers could get feedback from the children rather than having to conform to the theoretical concerns of their colleagues.

From the above remarks it is clear that laboratory facilities and a design shop were important resources of the project.

Theory behind Invention Strategy

Open discussion and a supportive social environment, together with frank intellectual criticism, are the standard conditions for encouraging the expression of creativity. High quality of the personnel was, of course, also essential.

Strategy Relating to Classroom Trials

Classroom trials played an extremely important role in the work of the SCIS and were the ultimate way of acquiring the information on which final decisions regarding segments of the teaching program were based. Classroom experience and the reactions of children and teachers were more important than any theoretical considerations in this respect. Development staff participated in these trials as much as possible.

Classroom trials took place in three stages. The first, called "exploratory teaching," occurred very soon after an activity had been invented and again when a unit was revised. It was carried out completely by project staff members functioning as "guest instructors" in elementary school classrooms. At least two, and sometimes four or five, members of the development staff planned and participated in exploratory teaching. They took the necessary equipment, pages of instructions for students, and sometimes video or audio recording equipment to the classroom. One of the staff members actually conducted the class, while the others observed the sessions, sometimes interviewing children who were experimenting individually concerning their ideas, questions, or intentions. After every exploratory session, one or more of the participating staff members prepared a written report in which the activity, the children's reactions, and the operation of the apparatus were described fully. These reports were filed and later served as raw material for the redesign of equipment, revision of the activity, and preparation of teachers' guides and other publications. The activity was also discussed in a postmortem session in which every participant could express his or her views frankly.

The second stage of classroom trials was carried out by regular teachers in the three laboratory schools (one suburban, two middle-class urban) that had been established in the Berkeley area, after teachers' guides, student manuals, and apparatus kits for the teaching unit had been prepared by the project in the Trial Edition. These classes were visited regularly by the unit's developers and also by other staff members who observed the teaching, occasionally spoke to children if this did not interfere with the teacher's work, and conferred with the teacher concerning any questions the teacher might have. Reports of the observations were filed after the visits and ultimately served as source material for revision of the unit. Individual conferences, feedback meetings, and feedback questionnaires enabled teachers to communicate their reactions to the developers.

Regular teachers in the Berkeley area and in the five trial centers that had been established by the project in 1966 carried out the third stage of classroom trials. These teachers used the Preliminary Edition, published by D. C. Heath. Project staff observed the classes in Berkeley, and local coordinators observed classes in the trial centers. Project staff also visited the trial centers, observed classes, and met with teachers two or three times a year. The coordinators submitted quarterly reports of their experiences in teacher education, classroom observations, suggestions from teachers, and specific comments concerning the teaching activities. All of this information was used in the preparation of the final edition of SCIS units, published by Rand McNally beginning in 1970.

Theory behind Classroom Trial Strategy

The design of the classroom trials involved setting up three feedback loops through which the project staff could test curriculum ideas on an ever-larger scale: first with one class of children or possibly two, then with ten classes and ten teachers, and finally with fifty classes and fifty teachers in five different locations in the United States. The exploratory trials were used to test the children's reactions to proposed activities and equipment and to gather children's ideas to enrich the teaching activities. The Berkeley area trials were used to test the revised activities and to get a first response from teachers concerning the demands placed on them. Ghetto classes were purposely not included at this stage. The country-wide trials, finally, sampled reactions from a larger number of teachers of very diverse backgrounds, with activities that were known to function reliably with children.

Strategy Relating to Pedagogy

The pedagogical organization of all SCIS units is similar and employs the learning cycle of exploration, invention, and discovery. Exploration is the first stage of the learning cycle. Children learn through their own spontaneous behavior relative to objects and events, that is, by handling objects and experimenting with them. Children explore materials with minimal guidance in the form of instructions or specific questions. The materials are always carefully chosen to be easily used and to generate certain questions that the children have not asked before. Exploration affords the teacher informal opportunities for evaluation by observing what individual children actually do, listening to them describe their ideas to one another, and occasionally asking them questions about their intentions.

Invention is the second stage of the learning cycle. Children learn as the teacher provides a definition or term for a new concept, while illustrating with concrete examples from the children's prior explorations. The teacher is asked to be clear and explicit while explaining and to repeat if necessary. To give the children prompt opportunities for using the new concept, the teacher encourages them to identify examples in their everyday life or their recent science activities. These examples also give the teacher feedback concerning the children's understanding. The teacher's role during invention is more directive than during exploration or discovery.

Discovery is the third and last stage of the learning cycle. It refers to activities in which the child discerns a new application for the concept previously explained in an invention session. Several different discovery activities are suggested in the teachers' guides following each invention lesson. The child's discovery experiences enlarge and refine the concept's meaning and also enable the teacher to conduct further diagnostic observations concerning each child's understanding.

The learning cycle is applied in the SCIS program on three levels. On the lowest level, a learning cycle may occupy from two to four weeks of instruction, with one or two class periods for exploration, one for invention, and two to six for discovery; but these are not rigidly prescribed. On this level the structure of the learning cycle helps the teacher to define his role appropriately for each stage. It also prevents successive introduction of new ideas too rapidly for the children to digest, since several discovery activities usually occur between invention sessions. One SCIS unit may encompass two to four learning cycles.

On the next level, each SCIS unit represents a single, larger learning cycle revolving around a major concept such as interaction, community, or energy transfer. Finally, on the highest level, the entire SCIS curriculum can be viewed as a learning cycle extending over several years and revolving around the interaction concept. Thus, the learning cycle provides substantial guidance in the organization of the entire curriculum.

Theory behind Pedagogical Strategy

The most important learning theories are those of association learning, discovery learning, and equilibration learning. The learning cycle combines all three theories in an eclectic combination that does violence to none. Thus, exploration allows discovery learning, invention and discovery allow association learning, and the entire cycle leads to equilibration as the student is first forced to reconsider his conceptual approach during exploration of new phenomena, then hears the suggestion of a new conceptual scheme during invention, and finally, through his own actions and tests, consolidates the new scheme during discovery.

Strategy Relating to Content

Central to modern science and therefore also to the SCIS program is the view that changes take place because objects interact in reproducible ways under similar conditions, as described earlier in this chapter. This conceptual organization was developed by the university scientists on the SCIS staff. Because of its generality and power, it did not substantially restrict the choices of activities used for exploration and discovery. The entire staff, therefore, participated in inventing such activities and identifying the secondary concepts that could also be introduced. Classroom testing determined which of the secondary concepts were understood well enough by children so they could be used in discovery activities. The concepts of relative motion, solution, habitat, and predator were among those found acceptable for the program. Certain other concepts, such as equilibrium, state of a system, and periodic motion, were tried out but had to be eliminated because they were not sufficiently useful to children.

In other words, the overall conceptual viewpoint was well defined early in the history of the SCIS, but many details were determined through classroom testing during the development of the teaching units.

Theory behind Content Strategy

Only the university scientists on the project staff had the knowledge and self-confidence to propose the interaction concept as the central organizing theme. The curriculum elaborations contributed by the university scientists and high school science teachers reflect the project's pragmatic approach, according to which classroom trials and the reactions of children and teachers were given the greatest weight in decision making.

Strategy Relating to Intellectual Development

Many psychologists, but especially Piaget, have described the intellectual development of children during the elementary school years. The developmental stages of children in elementary school were taken into account in the program development in three distinct ways. First, the main thrust of activities had a perceptual emphasis in the very early grades, moved to a conceptual orientation with concrete referents in the intermediate grades, and then introduced some abstract referents for the upper grades. Thus, in the first level of the SCIS, the children observe, describe, and sort objects and organisms according to their properties. In the third-level units, the children deal with predators and prey, with solutions, and with systems and subsystems, all referring to real objects. On the fifth level, however, the children deal with communities of organisms and energy transfer, which refer to abstractions that cannot be directly seen, felt, or heard.

Second, the concepts introduced in the teaching program were of some help in advancing students from one developmental stage to the next, since they called attention to generalizable aspects of the children's observations. Examples are the concepts of properties and habitat (level one), interaction and systems (level two), food chain and variable (level three), producers and consumers (level five), scientific model and ecosystems (level six).

Third, the vast majority of activities could lead to student satisfaction regardless of the children's conceptual level, since each student was free to use the materials in the way he wished and could therefore adapt them to his immediate needs. In their work with electric circuits in the sixth-level physical science unit, for instance, some children got a sense of achievement from lighting their bulb as brightly as possible (preoperational), others investigated the materials that did or did not close the circuit (concrete operations), while still others were concerned with the flow of electricity through the circuit (formal operational). In other words, the teacher could identify preoperational students, concrete operational students, and beginning formal operational students together in the same class, and did not need to expect the same responses from all.

Theory behind Intellectual Development Strategy

Piaget's developmental theory was used to formulate the strategy. In my observation, this theory has not been taken seriously in any other curriculum project and therefore constitutes a unique characteristic of the SCIS program.

Strategy Relating to Dissemination

The dissemination strategy of the SCIS is based on internalization. Basic to this approach is the direct involvement in a leadership role of an individual or team from the community or school system that is adopting the program. It should be noted that this strategy emphasizes dissemination to meet local needs under local control. Accordingly, the implementation procedures have been exceedingly varied.

The following outline describes a typical implementation process:

1. Interest in the SCIS materials originates through reading of the literature, recommendation of nearby school systems, presentations at conventions, contact by the publisher's representative, or teacher education courses.
2. Discussions with the publisher's representative and/or SCIS staff lead to an overall implementation plan. Such a plan usually includes these elements: commitment from an administrator to obtain the necessary funds, arrange for teacher training, and provide intellectual as well as administrative leadership; identification of pilot schools and/or pilot teachers who will use the SCIS materials; selection of one or more individuals to inform themselves about the program in depth, carry on teacher education, and oversee the distribution and maintenance of the laboratory materials kits; a tentative schedule for full-scale implementation.
3. Leadership training for those responsible for educating the teachers takes place either through a study-visit to the Berkeley headquarters of the SCIS or in a Resource Personnel Workshop.
4. There is an in-service workshop conducted by the trained leader for teachers who will use the program.
5. SCIS units, including laboratory kits for the participating teachers, are selected and procured.

The participation of the SCIS staff in the dissemination activities is limited to providing information and serving in a consulting role. The initiative for implementation is expected to come from the community itself.

Theory behind Dissemination Strategy

The "internalization" approach is based on the experience in educational innovation that consistent and persistent on-site leadership is necessary to maintain a new approach. The local educational establishment is subject to so many unpredictable forces—economic, political, personnel, parental—that only local informed and pragmatic leadership can maintain the innovation against the pressures that would either restore the status quo or introduce a still newer innovation, after interest is lost in the present one. Dissemination initiated by the developer could never hope to counter all these forces in the many hundreds of communities using the new curriculum.

Strategy Relating to Teacher Education

Teacher education was accomplished through two principal means: teachers' guides for the conduct of ongoing classroom activities; participation in in-service workshops or preservice courses relating to inquiry-oriented science programs, especially SCIS. Four goals had to be achieved: familiarity with the science phenomena in the units under study; adoption of a child-centered, inquiry-oriented classroom environment; understanding of children's intellectual development; acquaintance with the nature of science and its logical and conceptual structure.

The teachers' guides concentrated on the first two of these goals. They were written in simple and direct terms, to encourage the teacher's use of the program even if he were not well prepared to do so and to help the teacher modify his techniques as necessary in the light of feedback from the children. In particular, the eagerness and interest of the children engaging in scientific activities were seen as a significant tool of teacher education. In this way, the use of the SCIS materials would provide an ongoing experience of teacher education.

All four of the goals were built into materials developed for teacher workshops. These materials include documentary films (completely unrehearsed) of classroom activities using the SCIS units, an *SCIS Teacher's Handbook* with background material relating to all four goals, and an *SCIS Omnibus* containing reprints of articles published by SCIS staff and collaborators, expressing their views at various times during the developmental process. Under the "internalization" concept of dissemination, these teacher education materials are resources for the local leadership personnel who plan and conduct in-service and preservice courses.

Theory behind Teacher Education Strategy

The approach was highly pragmatic, based on the results of classroom trials and teacher reaction to various written materials. It became clear that most teachers have little time to prepare for science teaching, that they are concerned primarily with their most immediate teaching needs, and that extensive changes in teaching style occur only as a result of social experiences involving leadership personnel, other teachers, and children.

Strategy Relating to Educational Objectives

The SCIS has not used explicitly stated educational performance objectives as a curriculum development tool. The considerations described in connection with other strategies were sufficient to select the content and form of the activities. Formative evaluation from repeated classroom trials concentrated on the teaching and learning processes and gave only secondary importance to performance outcomes. The SCIS program has the overall objective of helping students improve their scientific literacy, defined as a combination of basic science knowledge, investigative experience, and curiosity. This goal gives some guidance concerning the program activities, but leaves many options as well.

Educational objectives were used by the SCIS to communicate the possible thrust of activities to the teacher. Objectives are therefore included in each teacher's guide. They are, however, associated with parts of a teaching unit—about a month of instructional time—and not with individual chapters or activities that may occupy one or two class periods. They are, in addition, stated in general behavioral terms rather than in highly specific terms. An example from *Interaction and Systems* is "To detect evidence of interaction by using various senses." The conditions of the interaction and the kind of evidence are left unspecified, as are the reliability of the response, latency, and other details.

Educational objectives were also used in the construction of an evaluation program for use by teachers. Evaluation activities were related to some of the major objectives, though not to all.

Theory behind Educational Objective Strategy

Explicit educational objectives are intended to focus behavior on the stated outcomes and to permit rigorous evaluation. This was unnecessary since the public had no preconceived notion of what children should learn in science. Teachers and schools are, in fact, urged to develop their own objectives for using the SCIS program—objectives that might relate to science learnings, social outcomes, and/or other matters of local concern. Under these circumstances, explicit and detailed educational performance objectives tend to preempt local decision making and are counterproductive. The more general objectives included in the teachers' guides serve as helpful guidelines in the absence of local preferences.

General Value and Possible Improvements in Strategies

The pragmatic approach I have described, supplemented by certain theories, has been very satisfactory in the development of the SCIS curriculum materials. As the staff accumulated experience, the procedures underwent changes compared to the initially improvised strategies, so that the items above represent a significant refinement and not the zeroth approximation. Looking back at this point, I find it difficult to identify further significant and substantial areas of improvement that can be clearly seen. In other words, there are, certainly, alternative strategies, but it is not at all evident whether they will be as effective, more effective, or less effective than the strategies actually used by the SCIS project. Their relative effectiveness would surely depend on the educational level, subject matter, and target audience of the new curriculum materials to be developed.

Generalizable Strategies

I believe that the SCIS strategies relating to classroom trials, pedagogy, intellectual development, and dissemination can be applied directly and effectively in any curriculum development project. They do not depend on the science content of the SCIS program, and their applicability is not restricted to an audience of elementary school pupils. The classroom trial strategy has, in fact, become widely accepted as basic to successful curriculum development, and the comparative research design strategy involving predetermined objectives, an experimental group, and a control group has been discredited for this purpose.

The learning cycle in the pedagogical strategy has been used successfully at all educational levels, both in curriculum materials and for making "lesson plans" by an individual instructor. It appears to be a very valuable technique. Of course, the exploratory and discovery activities have to be appropriate for the learner and should provide challenges that involve manipulative, puzzle-solving, self-appraisal, or investigative activities. Likewise, the definition or explanation during invention has to be concrete or abstract and involve practical or hypothetical examples, depending on the learners. These details in using the strategy of the learning cycle have to be worked out as part of curriculum development.

In applying the strategy relating to intellectual development, the age of the learners has to be taken into account. Even though the traditional view has been that the elementary school years are the most important period for development, and that at the secondary level most students are capable of dealing with abstractions at the stage of formal thought, recent studies have shown that this is not true. (See the Proceedings of the Third Annual Meeting of the Jean Piaget Society, Temple University, Philadelphia, May 1973, for a summary.) It is likely that nonacademically oriented students suffer in school precisely because of the mistaken expectations in this respect.

The "internalization" strategy for dissemination does not directly involve the students and is therefore valid at the secondary level. Other approaches, such as those used by the Ford Foundation, that involved implantation of a new program with broad external financial support, appear to be less successful in that the innovation tends to become identified with the support and is discarded as the support ceases. Many ESEA Title III projects have encountered the same difficulty. In other words, it appears essential for long-term implementation that the new curriculum be funded through regular, ongoing budget categories of the institution. In higher education, dissemination has to contend with the self-image of many instructors as highly autonomous individuals. They are likely to accept new curricular materials only through a combination of institutional and individual approaches whose success has not yet been demonstrated. I have nothing to contribute there.

Strategies That Need Modification or Improvement

Several of the strategies employed by the Science Curriculum Improvement Study were specific to the area of elementary school science. These include strategies relating to personnel, invention of activities, content, teacher education, and educational objectives. All of these, of course, are generally relevant, but the specific modifications that are necessary depend on the area of curriculum development. I shall describe some considerations that appear important to me.

One important factor is the age of the learners and school level. The elementary school is a nonselective institution that accepts all children within a very broad range above and below average in physical and mental capacity. No prior learning is expected. Curricula for older students have to recognize their successful or deficient prior school experiences. These curricula may be aimed at students who are selected on the basis of criteria that are subject to influence by the developers or are determined by external agencies and require curriculum adaptation.

The overall goal of the educational experience is a second important factor. The SCIS program tried to enhance scientific literacy as part of the general education of the students. No specific science-related performance or skills are expected of the students by the public. Hence, the content and educational objectives could be treated in very general terms. In some courses of study, such as in the area of reading or in swimming instruction, the public has certain expectations of what children will learn. In other areas, such as type-writing, there are well-defined skills of speed and accuracy that can be appraised unambiguously. For all of these, the open and flexible approach to objectives is unsuitable and must be replaced by more detailed planning. At the same time, I would urge that any objectives originally devised be reviewed continuously in conjunction with classroom trials and that these objectives be considered subject to revision just as the curriculum materials themselves are.

The preparation of teaching materials and teacher education are closely related matters. For the SCIS, an important consideration was the nonscience orientation of most elementary school teachers. If the teachers have a closer professional identification with the curriculum area, which is likely to be the case at the secondary or college level, then the development strategies have to take this into account.

Publication and Distribution

I have not written earlier about publication and distribution because conditions external to the project determined the actions of the SCIS in this regard. These conditions included the publication policy of the NSF and current practices in the publishing industry. A complete account is much too long for this report.

As a matter of fact, the difficulties we encountered arose largely because the SCIS product was innovative, and no existing publisher had the capability to produce and market it. It was necessary, therefore, to arrange for cooperation among several firms, while at the same time controlling the overall cost, and this proved to be difficult. Because of many new products reaching the market, the publishing industry is changing. The experience of the SCIS dating back to the 1960s is, therefore, likely to be of limited value.

SUMMARY

I have described the history of the SCIS project and nine strategies that were employed consciously and consistently to guide the curriculum development activities. The overall approach was pragmatic, with classroom trials and field experience playing the key role in decision making. At the same time, however, the original program design and the responses to the classroom trials were shaped by a conceptual framework of science and an inquiry-oriented educational philosophy that determine the character of the SCIS program.

CHAPTER 3

Exciting Science Lessons

Herbert D. Thier*

When we first met in 1962 Bob Karplus was a theoretical physicist and I was an Assistant Superintendent of Schools. We both had a second love (at the time) called elementary science education and our chance meeting as consultants to an educational toy company redirected my career, and enhanced Bob's commitment to science education. It resulted in what became known as the Science Curriculum Improvement Study or SCIS which in Latin we later found out means to know. According to our editor the SCIS team developed "240 exciting science lessons." I never counted them and the fact that over 35 years after we started, the SCIS3+ (Their and Knott, 1998) revision of the original materials is still a viable commercial product attests to the fact that at least some of them are exciting. I think this is true because from the beginning (1963) SCIS (first published in 1970) was a pioneer in providing:

- a strong conceptual framework;
- a comprehensive equipment and materials kit; and
- an articulated teaching strategy (the SCIS Learning Cycle).

Furthermore, SCIS was the first elementary school science program to

---

* Lawrence Hall of Science.

A Love of Discovery: Science Education—The Second Career of Robert Karplus,
Edited by Robert G. Fuller, Kluwer Academic / Plenum Publishers, New York, 2002.

- use an ecological approach and live organisms to present the study of life science;
- establish Scientific Literacy as its overall goal for students; and
- document achievement of its goal of Scientific Literacy.

All of our distinguished colleagues who made up the SCIS team in Berkeley and our centers from New York to Hawaii contributed significantly to what became SCIS-the outcome of the vision and commitment of Robert Karplus.

Concepts and themes pioneered in the original version of SCIS now appear in many state and local guidelines for elementary science education, as well as the National Science Education Standards and Benchmarks for Science Literacy, Project 2061.

First stated by Bob in "One Physicist Looks at Science Education" (ASCD, 1964, pp. 12–14) the goal of SCIS has always been Scientific Literacy. Investigations are carried out daily by all of us—in fact, investigation is the way we learn. How much water does a plant require? Can a person jump across a stream more than 3 m (10 ft) wide? Will feeding a canary a special food cause it to sing more frequently? Our innate curiosity and search for truth compel us to investigate. When we investigate scientifically, the conclusions we draw are based on factual evidence, not hearsay or opinion. As students develop effective techniques for testing and observing everything around them, they learn the what, how, when, and why of things with which they interact and become *scientifically literate*. The commitment to literacy and understanding in no way diminished the central importance of content and process in SCIS. Rather than try to separate the two, SCIS stated "this philosophy of and approach to science teaching makes impossible the separation of process goals from content goals, or either one from concept development" (Karplus and Thier 1969). This commitment to the integral nature of science and how one learned it was the vision of the entire SCIS team under Bob's thoughtful leadership.

Prior to and early in the development of SCIS, Bob's research (see chapter 4) on how to effectively construct the interaction between a teacher and a classroom full of learners resulted in the commitment to guided student experiences rather than a textbook as the basis of the program. The concurrent development of the LEARNING CYCLE as the conceptual organizer for structuring the interaction between the teacher and a whole classroom of learners provided the approach needed to relate the guided student experiences to the ongoing intellectual and social development of each learner. Exploration, Invention, and Discovery are stages in a learning cycle; each stage can always lead to another. Exploratory sessions frequently include discovery activities for prior concepts while creating a need for the teachers introduction of the new concept. Invention sessions frequently lead to questions best answered by giving students opportunities to work on their own, thus to discover an application of the new concept. Discovery activities can provide opportunities to reintroduce concepts "invented" earlier, and they can motivate individuals to explore the next concept.

SCIS from the beginning used the Learning Cycle to organize all instruction for students. This emphasis prevented SCIS from becoming a mile wide and an inch deep and helped it accomplish in the 60's the following principle stated so clearly in 1996 by the National Standards for Science Education.

"Inquiry into authentic questions generated from student experiences is the central strategy for teaching science" (NSES, p. 31).

More important than the activities developed by SCIS was and is its approach to teachers as facilitators of learning and students as thoughtful individuals who through guided interaction with their world could further develop their reasoning about and understanding of their world. When we began in the 1960's little science was taught in the elementary schools. The science that was taught came primarily from fact-laden textbooks that taught students science was a complicated set of answers practiced by old men in white coats. On a snowy 1963 night in Maryland when Bob asked me to join him in SCIS, (funding at the time $17,000) as Assistant Director, he emphasized his concern about what students were learning was science and the perception of science held by many of their teachers. He also emphasized the need to understand the school as a social system if we really wanted the materials SCIS would develop to be used effectively.

I realized Bob's honesty, thoughtfulness and understanding of what it meant to work together when he said to me something like; one thing you should realize Herb is that I like the limelight and so you will have to shine by reflected light like the moon. I like the limelight also but something told me (more than proven) that orbiting with this sun was a good idea.

Above all else Bob was a learner and always asked what will we, the team, learn as a result of undertaking whatever new endeavor was being proposed. Bob had the unusual capacity to take on something new as a learner while using his accumulated background and knowledge to understand it. For example in the 1960's there was a great deal of controversy among researchers as to what theory or theories explained learning (so what else is new). Behaviorism was becoming popular as a result of the work of individuals like Ausubel, Gagne and Skinner. Others were committed to Piaget and his developmental approach to learning. Many projects and individuals chose one or the other with a commitment more related to religious belief than scientific thought. Bob on the other hand used his background as a theoretical physicist to consider them a set of competing theories. Instead of adopting one or another SCIS studied the work of many of the proponents and interacted directly with Ausubel, Gagne, Piaget and others in our efforts to refine and expand the learning cycle to encompass and reflect the qualities found in each approach. Later as many of you know and as discussed in Chapters 6 and 7 of this book, Bob continued his interest in learning and especially student reasoning. I am convinced that but for his untimely death our understanding of student learning in science would be much richer today because of the work Bob would have completed.

Many mornings I wake up wondering what science education would be like today if we all had the pleasure of enjoying the thoughtfulness and humanity of Bob longer. Some mornings I almost dial 254—

REFERENCES

Karplus, R., 1964, One physicist looks at science education, in *Intellectual Development Another Look*, Association for Supervision and Curriculum Development, Washington, 12–14.

Karplus, R., and Their, H. D., 1969, *A New Look at Elementary School Science*, Rand McNally, Chicago, IL.

National Academy of Sciences—National Resource Council, 1996, *National Science Education Standards*. National Academy Press, Washington, DC, p. 31.

Thier, H. D., and Knott, R., 1998, *SCIS3+*, Delta Education, Nashua, NH.

A Concept of Matter for the First Grade

Herbert D. Thier, Cynthia Ann Powell, and Robert Karplus

The development of the concept of matter with first graders should evolve from the level of understanding about matter usually held and verbalized by children of this age. The child's concept of matter at this age is usually specific and "use-oriented" for any piece of matter. For example, a flat stick about five inches long, a half inch wide with rounded ends is not so much a piece of wood as it is an ice cream or popsicle stick. Some first graders on seeing the stick have this relationship so clearly in their own minds that they appear to be tasting the ice cream. At the same time, the typical first grader is rather vague about the relationship between various different pieces of matter which have common properties, such as color, shape, hardness or roughness, or even those objects which are composed of essentially the same substance but have different uses. For example, the wooden ice cream stick and a wooden ball are two quite unrelated items to the typical first grader. Of course our general cultural description of pieces of matter is a use-oriented one (such as "paper clip" for wire bent in a certain way) and so we would expect the young child to have this orientation.

The scientifically useful concept of matter is more concerned with the properties of the substance which make up the individual pieces of matter rather than with the usual or customary use of that piece of matter. This concern with the properties of the substance

Herbert D. Thier, University of California, Berkeley, California.   Cynthia Ann Powell, University of Maryland, College Park, Maryland.   Robert Karplus, University of California, Berkeley, California.

A Concept of Matter for the First Grade, H. D. Thier, C. A. Powell, and R. Karplus, *Journal of Research in Science Teaching*, **1**, pp. 315–318, Copyright © 1963. Reprinted by permission of Wiley-Liss, Inc., a subsidiary of John Wiley & Sons, Inc.

out of which pieces of matter are composed leads to the development of a system of classification.

The development at this level concerns itself primarily with the classification of objects (pieces of matter) which are composed of substances with obvious properties. The choice of objects to be studied is restricted in order to postpone the handling of problems related to the differences between mixtures and compounds until the students have been exposed to the concepts of systems and interactions. The later development of the differences between compounds and mixtures eventually leads to an understanding of chemical reactions and the significance of the elements.

It is important to note that there is a limit to the understanding of the concept of matter as long as experience is restricted to the macroscopic, direct observational level. Any development beyond this level involves an awareness of atomic particles which cannot be simply derived from direct observation. Therefore the development of the concept of matter with young children will logically be restricted to the gross physical properties which can be directly observed; the understanding of the concept of elements is a goal which will probably not be achieved until the fourth or fifth grade.

The rest of this article is concerned with the description of a method by which first graders can be started along the path that will lead in future years to a scientific understanding of the concept of matter.

The first step is to reorient the child's interest and observation from the use to the properties of various pieces of matter. The term *object* is introduced to replace the word *thing* for pieces of matter. This introduction of the term *object* to denote pieces of matter is done in order to prevent the confusion caused by the rather general use of the word *thing* not only for pieces of matter but also for feelings or ideas as in "Love is a Many Splendored Thing."

The students carry out many operations with individual objects and groups of objects. At first the activity consists of observing and telling about the properties of certain objects such as whether they are hard or soft, rough or smooth, shiny or dull. Even this operation is really a comparison since the child decides what is rough, hard, or shiny, based upon his experiences with other objects and how the new object fits into his informally developed concept of these properties.

The next operation consists of having children work with groups of objects and then divide them into piles based upon some fairly obvious property which they choose. At first the task is a rather simple one with objects like parquetry blocks, so that the obvious means of division are shape and/or color. Later, less structured objects are used and the students learn to assort them into rough and smooth, shiny and dull, or light and dark piles. The objects used include both natural and man-made objects, therefore providing many opportunities for broadening the students' knowledge of their environment. These activities improve the students' ability to describe objects accurately by their properties.

Now the observation and description are focused on quantitative aspects, such as how heavy and how long, not only for what these ideas contribute to an understanding of matter, but also because of the inherent importance of including from the beginning a quantitative emphasis in the study of science. Measurements are introduced by the use of an arbitrary standard of length (such as a popsicle stick) to which various objects are compared in order to determine whether they are longer or shorter than the standard. The introduction of our system of feet and inches or pounds and ounces is postponed until after the idea that measurement consists of comparing an unknown to a standard is developed from the children's experience and understanding.

The beginning work with measurement also calls attention to the continuous nature of length or weight values. Activities such as the ordering of a group of sticks from the shortest to the longest and the ordering of a group of objects from the lightest to the heaviest are used to develop continuous scales for quantitative measurement. After ordering a group of objects, the students compare a "new" object to the ordered group and determines where it belongs on the scale. At this point, the need for a reproducible scale of measurement can be developed and the transfer over to the common scales of measurement can be carried out.

Once the students have learned to describe and group objects according to their gross properties, it is possible to generalize further their concept of matter and its properties. The additional understandings to be developed are the fact that one can change the shape or appearance of an object without changing the stuff (or material) out of which it is made, and that two or more different objects can be made of the same kind of stuff. The ideas that a number of objects can be composed of the same kind of stuff and that the same stuff can be used to form two or more different objects are significant steps in the achievement of an understanding of what matter is. At this point, the overall development of the concept of matter has reached a plateau for the first grader, since further understanding and refinements are more logically postponed until after the children have been introduced to the ideas of systems and interactions. (1)

The final section of this paper is devoted to a brief description of typical lessons used in the teaching program with a first grade class. This list is not intended to be all-inclusive but rather is intended to give an idea of the type of lessons and materials used in the program.

I. OBJECTS IN THE CLASSROOM

At first, the many objects found in the typical primary classroom, such as blocks, boxes, toys, etc., are used as a means for introducing the term *object* and for talking about the properties of an object rather than its use. Students are encouraged to tell about objects with the emphasis restricted to overall properties such as shape, color or feel.

II. PARQUETRY BLOCKS

The operation of grouping objects by properties is introduced by the use of parquetry blocks which are all of the same material and vary only in shape and color.

The first lessons involve grouping by both shape and color. Later either shape or color is used as criterion; this task is more difficult because it requires the students to disregard one property. Other objects such as buttons, nails, and pieces of unfinished wood (maple, cypress, walnut, and pine) can also be used at this point.

III. RECORD KEEPING

An important aspect of scientific work is the keeping of records. This is introduced by having the children make pictures or tracings of their groupings of objects. The children are also given pictures of sets of objects and draw loops around the groupings developed in their experiments.

IV. THE OBJECT HUNT

The operation of identifying and grouping objects is extended to less structured objects and to objects outside the classroom by the use of a field trip around the school or a park to collect objects from the environment. The properties of these objects are described and they are grouped on the basis of properties (such as roughness). This lesson also provides a good opportunity for extending the lessons to natural objects. The classroom did not provide enough of these.

V. DOWEL STICKS

An activity used to introduce measurement is the ordering of a group of dowel sticks by length. Children are given a group of dowel sticks with no two of the same length and are asked to arrange them in order from the shortest to the longest. Once they have carried out this operation, they are given another stick or some object of reasonable size and asked to determine where it would fit into the ordered group. They learn to describe the length of a "mystery" stick by describing where it would fit in relation to the sticks they have distributed by length. For example, the unknown stick is longer than the fourth but not as long as the fifth stick in any group.

VI. LENGTHS OF LICORICE

One introduction to measurement consists of offering the boys and girls pieces of licorice. If there are not enough whole long pieces to go around, the students are offered a finger's length of licorice. Awareness of the differences in the length of various student's fingers and of the difference in length between a student's finger and the teacher's finger leads to awareness of the need for a reproducible standard unit of length if everyone is to receive an equal length of licorice.

VII. OBJECTS GRAB BAG[2]

Another approach to the identifying and grouping of objects is the Objects Grab Bag game. A convenient set of objects to use is a set of parquetry blocks. Three or four children can play in one group. Each child draws five blocks from a large bag without looking into it. The children now match pairs of blocks that are alike in all respects. The game proceeds with each child in turn drawing a block from the bag and discarding one into the bag until someone has three matched pairs.

VIII. LEAVES

Leaves are collected and brought into the classroom and the operations involved in classifying them are carried out. Especially in the spring, leaves from a single tree can be collected and classified by size. Leaves from many trees can be collected and classified by shape and/or size. To some degree in the spring and especially in the fall leaves can also be classified by color. Seasonal variation in the properties of leaves can be discussed.

IX. CLAY AND CRACKERS

Operations with clay and crackers are used to develop the fact that one can change the shape or appearance of an object without changing the material or "stuff" out of which it is composed. A ball of clay is flattened into a pancake. Although the children see that the shape has changed, the stuff out of which the ball and the disk are made is the same. A cracker is crumbled. Again, although it appears to be quite different, the children are helped to realize that the crumbs are composed of the same material as the cracker. This idea is then extended to the case where two objects, a ball and a cylinder, are each formed from different portions of the same large piece of clay.

X. WEIGHT

A recipe or other scale on which the pointer moves vertically down as increasing weight is placed on the pan is used to introduce the measurement of weight. At first, a piece of paper is taped over the dial so that the student calibrates his own scale in "books." A book or other readily available standard is used and unknown objects are compared to it and separated as heavier or lighter. Sets of objects (for example, a set of cubes made of lead, brass, aluminum, wood, etc.) are ordered according to their weight, and "mystery" objects are placed properly into this sequence. Here it is helpful to use several similar scales side-by-side.

XI. CONSERVATION OF MATTER

The concept of conservation of matter is developed out of a reconsideration of the transformation introduced in Lesson IX. Now, attention is directed not only to the material composition of objects, but also to the amount (weight) of material present. The effects of adding material, of removing material, and of just changing the appearance of the objects placed on the scale are observed and compared. A scale permits more varied comparisons than does an equal arm balance.

REFERENCES

1. Karplus, Robert, "One Physicist Looks at Science Education," prepared for the 8th ASCD Curriculum Research Institute, Washington, D.C. (Department of Physics and Astronomy, University, of Maryland, April, 1963).
2. Powell, Cynthia Ann, and Robert Karplus, "Objects Grab Bag," Department of Physics and Astronomy, University of Maryland, May, 1963.

Meet Mr. O

Robert Karplus

Allow me to introduce Mr. O. Mr. O is an artificial observer. He knows where every object is at all times, but he always describes the location of everything relative to himself. He is the most important thing in the world. He does not wonder why; why events happen, or why objects appear the way they do. He "reports" in his very egocentric way only what can be observed and what happens.

Is Mr. O like a real person? No, of course not. Mr. O is an invention of the human mind. His characteristics are assigned to him for pedagogical reasons and not to make him resemble a person. In his "reports" he summarizes the knowledge of the person who uses the Mr. O concept, and he is not confined by perceptual limitations of his own.

If a Mr. O on a table is asked, "Where are you?" he can only answer, "I am right here," perhaps while pointing to himself. If he were asked, "Where is the table?" he would say, "Underneath my feet." For the pupils in a class, of course, Mr. O is on the table. Such a point of view, however, subordinates Mr. O to the table. For Mr. O, he himself is the central reference point, and the way he faces defines the reference directions. Since he accepts no external reference object, he has no way to describe where he is other than by pointing at himself and answering, "Right here."

As was pointed out above, Mr. O is an artificial observer. His characteristics are unlike common sense. Yet three examples will show that everyone's common sense outlook involves the use of the Mr. O concept in an unconscious way.[1]

The experiments in this article were carried out by the author at the Berkwood School, Berkeley, California, through the cooperation of the Director, Betty Halpern. Financial support for the project was provided by the National Science Foundation, Washington, D.C.

[1] Benjamin Lee Whorf. *Language, Thought, and Reality*. John Wiley and Sons, Inc., New York City. 1956.

Robert Karplus, Department of Physics University of California Berkeley, California.

Example 1.   Mother drives her daughter to school. The girl starts to climb into the back seat. "Don't move around so much. Sit still," mother says. The daughter obeys. That satisfies mother. But is daughter really not moving? That depends. For a Mr. O in the car, she is indeed sitting still. For a Mr. O on the sidewalk, however, the car, the mother, and daughter are moving past at perhaps a rate of 30 mph. Mother automatically uses both of these Mr. O's: one inside the car when she thinks about her daughter's behavior; one on the sidewalk when she thinks about the car as a whole.

Example 2.   In a bus, the situation is still more interesting. As the bus starts suddenly, the passengers seem to fall backwards. Do they really fall backwards? Not for a Mr. O on the road; for him, they are moving forward, but more slowly than the bus. For a Mr. O on the bus, of course, they do move backwards. Who is right?

Example 3.   In astronomy, everyone learns that the earth rotates on its axis and moves around the sun. Is this true? For an observer on the sun, it is. But for an observer on the earth, the earth does not move at all; it is fixed beneath his feet. Instead, the sun and moon move around the earth and show certain seasonal variations. Which idea is right?

Once the observer fixed on the earth was the only one considered consciously. Nowadays, man's thinking is more nimble, and he can conceive of the different observers that have been mentioned and many more. Each Mr. O is right for himself, from his point of view. The question, "What is *really* happening?" has no scientific meaning any more. Instead, the scientific mind thinks about a phenomenon from several points of view and then chooses the one that permits the simplest description and that leads to the best understanding. For this reason, everyone thinks about the solar system like an observer on the sun. Which observer to choose for studying the passengers on the bus is less clear.

The students, under the direction of the author, work with their own Mr. O's to get an understanding of the relationships of objects in a system.

Now, motion and change are fundamental aspects of natural phenomena. To communicate a clear and sensible description of motion and change, one must specify the imaginary observer who would make that same description. It is especially important that one be aware when one is using two or more different observers for different details of the same process or change, as mother was doing in the first example.

In brief, the concept of motion of an object is not an absolute one. It is meaningful only when referred to a certain environment, often called a reference frame, which may be selected consciously or unconsciously. In the common sense view, motion is usually seen in terms of the immediate environment of the object of interest. Thus, the daughter is seen relative to the automobile interior, the automobile relative to the road, the road relative to the surrounding countryside, the countryside relative to the whole earth, the earth relative to the sun, the sun relative to our galaxy, and our galaxy relative to the system of galaxies called the universe. Each environment can in turn become the object of study in a still larger environment. And with each environment that plays the role of a reference frame, a Mr. O can be associated so as to represent this function in our study.

MOTION AND CHANGE STUDIES

Because motion and change are so fundamental to science, the author has developed a series of lessons on that subject. One teaching objective was to make the children conscious of the reference frames they use. A second objective was to teach the children to use a particular reference frame more broadly than is done according to common sense. For example, it is

The students and Mr. O puppets observe the block as the wagon is pulled across the classroom. Is the block moving, or is it staying in the same place?

contrary to common sense to consider the behavior of the road relative to the automobile. Yet scientific understanding often requires thinking that transcends the ordinary patterns. Puppet Mr. O's served as concrete representations of abstract reference frames. The remainder of this article contains a sketch of procedures that have been used by the author with some success to teach elementary school children aspects of the relativity of position and motion.

The principal intellectual effort required of the children was the mental isolation of the experimental objects under study from the remainder of their environment. When a Mr. O was introduced, the positional relation between the objects and him had to be recognized. Often the cues inherent in the environment, such as the walls of the room, the trees, the desk, the edge of a paper, served as unconscious reference frames and interfered with the children's efforts to take Mr. O's point of view. At the beginning, indeed, the children treated Mr. O as an object (moving or standing still) relative to some conventional reference frame. With a little practice, however, the children between six and ten years of age learned to think in terms of Mr. O.

The Lessons

Lesson 1. *Misunderstandings that arise when the reference frame is not specified.* The teacher tells the story of Joe, his wagon, and his dog Spots who, when he is hitched to it, can pull Joe in the wagon. Joe likes his wagon very much. When he rides in it, he thinks just of Spots, the wagon, and himself. Joe goes for a ride one day in his wagon. Spots pulls them into a forest, and they get lost. Mother looks for Joe, and she calls him when she cannot find him, "Joe, Joe. Where are you?" Joe answers, "Here I am, in my wagon!" Was that right? Yes, it was exactly right. His mother replies, "You stay where you are. I'm coming after you. Don't move." Joe, who is a good boy, stays right where he is in the wagon, but Spots, who cannot understand words, keeps on walking. Did Joe obey his mother? Yes, he did. He did not move at all. After a while his mother calls again, "Joe, where are you? I told you to stay where you were!" And Joe answers, "I'm right here in my wagon where I was before . . . !" Eventually, the wagon hits a rock and tips over. Joe's daydream is shattered. He hears his mother calling him again, "Joe, where are you?" This time Joe answers, "I am next to the largest tree in the forest," and his mother soon finds him.

Children, who are often very literal, understand and enjoy the misunderstanding between Joe and his mother.

Lesson 2. *Explicit introduction of Mr. O.* A large wooden block and a wagon were used for this lesson. The teacher called the children's attention to the block standing on a desk and asked them to think just about the block—not about the desk, or the walls, or the floor. The pupils were reminded of this several times subsequently. In answer to questions, the children replied that the block could not move unless someone pushed it because it does not have legs or muscles. When the teacher waved the block in the air, the children variously commented that it was moving or being moved. Next, one child was asked to sit in the wagon and hold the block tightly on her knees. The teacher pulled the wagon across the room. Was the block moving now, or was it staying in the same place? The children in the class thought it had moved, while the child in the wagon insisted that she had held it still. Who was right? Everybody was right! And to help the class discuss the problem, the teacher introduced a puppet observer, Mr. O, with the properties described at the beginning of this article.

One special characteristic of Mr. O can now be explained. Mr. O does not wonder how or why objects look the way they do, nor is he concerned with what causes events to happen. He is interested only in what happens and the appearance and position of objects. This feature is introduced because some children tend to combine in their thinking the activity that causes the motion with the motion itself. To help them separate these two aspects of a process, Mr. O notices only the motion, the change in position. At this stage, the rules for using Mr. O are like the rules of a game that have to be learned. Eventually, the children will discover that the game helps in dealing with the reality.

The teacher gave a Mr. O to each child. The experiment with the block was repeated and discussed in terms of the observations of Mr. O. The Mr. O on the wagon considered the block stationary, while the Mr. O's on the desks observed that the block moved from one place to another. Who was right? Each one was right for himself. Mr. O served as a concrete representation of a reference frame.

Lesson 3. *Extension of use of Mr. O; viewing a process from two different reference frames.* The demonstration in lesson two was repeated in a modified way. Now one block

Figure 1.

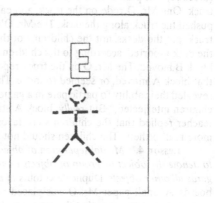

Figure 2.

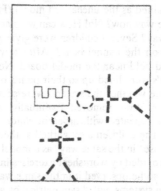

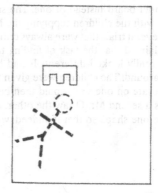

Figure 3.

(block A) was placed on a desk and a second block (block B) was loaded onto a large toy truck. One Mr. O rode on the truck, and each child had a Mr. O on his desk. The teacher pushed the truck along the desk. The Mr. O's were to observe only the two blocks—not the truck, not the desks, not the children, not the room. What do they report? The Mr. O's on the desk reported, according to the children, that block A stayed in the same place but that block B moved. The Mr. O on the truck reported that block B stayed in the same place but that block A moved, or seemed to move. This accomplishment on the part of the children revealed their ability to participate in a game even if the rules appeared to be strange. Some children interjected, "But *really*, block A stayed in the same place!" To this comment the teacher replied that the children were favoring the Mr. O on the desk whose report was more real to them. The children should not play favorites with the different Mr. O's.

Lesson 4. *Mental separation of objects from environment. The term "system" is used to denote the object or group of objects under observation by Mr. O. Temporarily, he disregards all other objects.* Duplicate cutouts of a colored paper "E" were placed on two flannel boards. A small paper Mr. O was placed near each one in such a way that each saw his system in the same way that the other saw his (Figure 1). The Mr. O's were observing only the system specified, *i.e.*, the paper "E"—not the flannel board, not the room, etc. Now the teacher picked up one shape and replaced it in a different orientation. The other flannel board remained as the unchanged model (Figure 2). Do the two Mr. O's see their systems in the same way now? No! How can we place the second Mr. O so he will see his system in the same way? Several children were given paper Mr. O's in order to try their ideas of his placement on the flannel board. After a while, this flannel board was picked up by the teacher and held near the model board. Now the model Mr. O and one of the children's trial Mr. O's were lined up so their points of view could be compared easily. It was simple to determine whether the systems were seen in the same way. If they were not, the incorrect trial Mr. O was removed. Eventually, only one trial Mr. O was left on the board. This activity was repeated with different children participants and other placings of the letter (Figure 3). The children also worked at their desks with puppet Mr. O's observing a block or a toy house "in the same way" as a model set up by the teacher or another pupil. Further drill was provided by worksheet exercises in which children drew a Mr. O near a picture of the system so he observed it in the same way as the model. Some children discovered that with certain systems, called symmetrical by adults, it is possible to have two or more distinct Mr. O's all of whom see the system in the same way.

Lesson 5. *Relation of objects in the system to each other.* This lesson began in the same way as Lesson 4, but two colored shapes, the letter "B" and the letter "E", were used on each flannel board instead of one. The situations shown in Figures 4 to 6 were set up successively with the children supplying the Mr. O in each case. In the last example shown, there were several trials that were always criticized by other children and eliminated. Finally, one girl exclaimed that the task of finding the correct position for Mr. O was impossible. The system really looked different. It had been changed. In later activities, the problem was turned around. The children were given two paper shapes that looked alike. They were asked to operate on one so it would then be impossible to find a Mr. O who saw it in the same way as a second Mr. O saw the other. A few children were baffled, but many tore or wrinkled the one shape so that it, indeed, would look different no matter how Mr. O was placed.

SUMMARY

Variations of these five lessons involving somewhat different activities were alternated. The children's ability to analyze the motion of objects was improved by the experience of the later work on isolating objects and systems in their thinking. One entertaining family

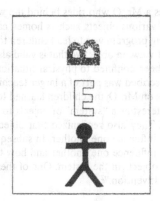

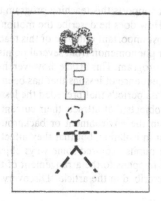

Figure 4.

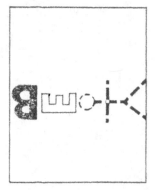

Figure 5.

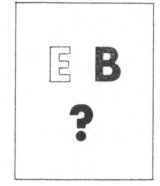

Figure 6.

activity that supports this teaching program uses a Mr. O who rides behind the windshield of the car. How does he describe the motion of various objects, such as homes, cars, etc.?

A most important outcome of this teaching program was to lead children to consider a process or phenomenon from several points of view, an objective that is valid also outside the science program. This article, however, has been confined to physical situations.

The sequence of lessons that has been described was part of a larger teaching experiment. In class periods that preceded the lessons on Mr. O, the children learned to describe a group of objects that attracts their current interest as a "system" of objects to be distinguished from the environment or background. They also recognized that objects do not usually exist in isolation, but that they affect or influence one another. In subsequent class periods, the pupils explored some ways objects influence one another and how this determines what happens to the arrangement of the objects in the system. One of these lessons has been described in the article "Discovery of Invention."[2]

[2] J. Myron Atkin and Robert Karplus. "Discovery or Invention." *The Science Teacher*, 29:45–51. September 1962.

CHAPTER 4

The Learning Cycle

Anton E. Lawson*†

Early approaches to science instruction in the United States consisted mainly of daily recitations from books and lectures. Use of the laboratory was unheard of prior to the mid 1800s. Physical materials and specimens, if used at all, were a means of verifying book or lecture information. But by the late 1800s, laboratory instruction became very popular because it was felt that first-hand observation and manipulation were useful in "disciplining" the mind. The idea of mental discipline stemmed from psychology and the then popular faculty theory. In general, faculty theory claimed that mental behavior was compartmentalized into several "faculties" such as logic, memorization, and observation. In theory, mental behavior could be enhanced by "exercising" these faculties and once the faculties were developed, they would function in all life situations. The theory was used to justify the use of abstract, meaningless, laborious tasks during instruction to exercise and strengthen the mind.

Largely through the research of psychologists such as E. L. Thorndike, faculty theory began to lose favor and emphasis in schools shifted from rote

* Department of Biology, Arizona State University, Tempe, AZ 85287-1501,
anton.lawson@asu.edu

† This material is based upon research partially supported by the National Science Foundation under grant No. DUE 9453610. Any opinions, findings, and conclusions or recommendations expressed in this publication are those of the author and do not necessarily reflect the views of the National Science Foundation.

A Love of Discovery: Science Education—The Second Career of Robert Karplus,
Edited by Robert G. Fuller, Kluwer Academic / Plenum Publishers, New York, 2002.

tasks to presenting meaningful information and developing positive attitudes and useful reasoning patterns. Indeed by 1898, organizations such as the National Education Association were making rather modern sounding recommendations such as: *"The high school work should confine itself to the elements of the subject . . . full illustration of principles, and methods of thought . . ."* (Hall and Committee, 1898).

The sentiment to teach scientific principles and reasoning patterns was even more apparent in the Central Association of Science and Mathematics Teachers' 1910 report on secondary education. The report identified major goals such as increasing student motivation and the selection of teaching materials that would teach *"the scientific spirit and method"* (Galloway, 1910). The committee suggested: (1) more emphasis on *"reasoning out"*; (2) more emphasis on developing a problem-raising and problem-solving attitude among students; (3) more applications of the subject matter to personal and social issues; and (4) less coverage of territory. Although the report criticized methods that emphasized memorization, no detailed teaching methods were advocated, except to suggest that problems or projects offered promise for better class discussions, more active student participation, and a better opportunity for *"research type"* learning.

John Dewey was an early and vocal advocate of science instruction that emphasized science as a method of inquiry. In an address to the National Education Association, Dewey (1916) argued that:

> science is primarily the method of intelligence at work in observation, in inquiry and experimental testing; that, fundamentally, what science means and stands for is simply the best ways yet found out by which human intelligence can do the work it should do, ways that are continuously improved by the very process of use.

But, according to Hurd (1961), it would take more than 40 years before this view would make its way into a large scale curriculum development movement. The movement Hurd referred to was the National Science Foundation sponsored curriculum development projects initiated in the late 1950s. These inquiry-oriented projects, such as the Biological Sciences Curriculum Study (BSCS), the Chemical Education Materials Study (Chem Study), the Science Curriculum Improvement Study (SCIS), the Elementary Science Study (ESS), the Physical Science Study Committee (PSSC Physics), and the Earth Science Curriculum Project (ESCP), sprang up largely as a reaction to the Soviet Union's perceived superiority in science and mathematics education as evidenced by their successful 1958 launch of Sputnik into outer space.

Although several of these alphabet soup projects, as they came to be called, developed some excellent inquiry-oriented activities, most of them, with the notable exception of the Karplus directed SCIS program and its learning cycle instructional method (see the Karplus, 1974 paper reprinted in this Chapter) failed to generate a systematic method of inquiry instruction. Rather, most projects only alluded to "discovery," "inquiry," or "problem solving" approaches, the steps of which were not always made clear to teachers, thus were sometimes difficult to implement.

ORIGINS OF THE LEARNING CYCLE IN THE SCIS PROGRAM

Origins of the learning cycle can be traced to the early work of the SCIS program on the Berkeley campus of the University of California during the early 1960s (*SCIS Newsletter*, No. 1, 1964. Reprinted in SCIS, 1973). To be more precise, its origin can be traced to a day in 1957 when a second grade student invited her father, Professor Robert Karplus, a physicist at Berkeley, to talk to her class about the family Wimshurst machine, a device for generating electrical charges. Professor Karplus found the visit enjoyable and so did the children. During the next few months other talks on electricity and magnetism to both elementary school and junior high school students followed. Soon Professor Karplus turned his thoughts to the possibility of developing a program for elementary school science.

With a grant from the National Science Foundation, Karplus prepared and taught three units entitled "Coordinates," "Force," and "What Am I?" during the 1959–60 school year. Although the experience proved interesting, analysis of the trial teaching revealed serious student misconceptions and other weaknesses. The experience prompted Karplus to raise a key question: "*How can we create a learning experience that achieves a secure connection between the pupil's intuitive attitudes and the concepts of the modern scientific point of view?*"

During the spring of 1960, Karplus continued to familiarize himself with the points of view children take toward natural phenomena as he taught lessons in a first, second and fourth grade twice a week. He also began to develop tentative answers to his question. Following that experience, Karplus was helped by a visit to the research institute of Jean Piaget, the Swiss psychologist and pioneer in the study of children's thinking.

When Karplus returned to the United States in the fall of 1961, he returned to the elementary classroom with a plan to stress learning based upon the pupils' own observations and experiences. However, he also planned to help them interpret their observations in a more analytical way than they would without assistance. During part of that school year, J. Myron Atkin, then a Professor of Education at the University of Illinois, visited Berkeley to share his views on teaching with Professor Karplus. Together Atkin and Karplus formulated a method of "guided discovery," which was implemented in subsequent trial lessons (see Atkin and Karplus, 1962—reprinted in Chapter 5).

The Atkin and Karplus guided discovery method was designed to be analogous to the way in which scientists invent and use new concepts to explain nature. In their 1962 paper, they offered the example of the ancients' observations and interpretation of the motions of the sun and planets. The geocentric theory of the solar system was taken to be a conceptual "invention" following initial observations. The heliocentric theory represents an alternative invention. With the help of these inventions, people attempted to "discover" other phenomena besides the ones that led them to propose the inventions in the first place, that could be understood using the invention. These attempts, if successful, led to a reinforcement and refinement of the invention. If they were unsuccessful, they revealed limits of the invention or, in some cases, led to its replacement.

Atkin and Karplus clearly distinguished between the teacher's initial explication of a new concept or conceptual system (called the invention phase) and its subsequent test or extension (called the discovery phase). They assumed that children are not generally capable of inventing (what some would now call constructing) the modern concepts of science, therefore, it becomes necessary for the teacher to explicate those concepts, but making sure that the students' previous observations can be interpreted (or reinterpreted) using the explicated concepts. Further, the teacher must follow this with opportunities for the children to discover that new observations can be interpreted using the concept. Atkin and Karplus likened the process, in some respects, to the Copernican teacher instructing students that the sun is at the center of the solar system while almost everyone else believes that the earth is at the center. Atkin and Karplus did not introduce the terms "exploration" or "learning cycle" in their 1962 paper, but as mentioned, the terms invention and discovery were clearly evident.

During the summer of 1962, Professor Karplus accepted an invitation to work with the Elementary Science Study. There it became clear to him that children need time to explore an experimental system at their own pace with their own preconceptions. Only after this initial "exploration" is it wise to discuss a more analytical point of view. Armed with this new insight, Karplus tried out the modified approach the following school year in several public school classes near the University of Maryland where the SCIS program was temporarily headquartered. A number of new staff members joined the effort at that time including Dr. Herbert Thier, then Assistant Superintendent of Schools in Falls Church, Virginia. In 1967, Karplus and Thier published a book in which the three phases of the instructional method are first explicitly stated: "*The plan of a unit may be seen, therefore, to consist of this sequence: preliminary exploration, invention, and discovery*" (Karplus and Thier, 1967, p. 40).

ORIGINS OF THE LEARNING CYCLE IN BIOLOGY EDUCATION

Origins of the learning cycle can be found in biology education as well. In 1953, the National Academy of Sciences convened a Conference on Biology

Education to examine past teaching practices and suggest alternatives. As a result of that conference, a project funded by the National Science Foundation under the direction of Professor Chester Lawson, a geneticist at Michigan State University, began in the fall of 1956. The result of that project, which involved the work of 30 high school and university biology teachers from throughout the country, was a sourcebook of over 150 laboratory and field activities appropriate for use in high school (Lawson and Paulson, 1958). Although no explicit statement of teaching method resulted from that work, it provoked Professor Lawson and others to begin a search for such a method. The project also served as the precursor to the well known Biological Science Curriculum Study.

Professor Lawson, like Professor Karplus, turned his attention to the history of science for insight into the process of conceptual invention. His 1958 book, *Language, Thought and the Human Mind*, carefully detailed the nature of scientific invention and identified a general pattern of thought he referred to as "Belief—Expectation—Test" (Lawson, 1958). This pattern can now be seen to be similar to Karplus and Atkins' pattern of invention and discovery as conceptual invention constitutes a belief which in turn leads to an expectation to be tested in the real world. If one discovers confirming evidence the invention is retained. If not, it is rejected in favor of another belief.

Following work on the biology sourcebook, Professor Lawson began a careful review of current psychological and neurological research in hopes of developing a comprehensive theory of human learning complete with a model of relevant neurological mechanisms and instructional implications. The theory that resulted from that work stipulated that learning involves: (1) attention directed to some undifferentiated "whole," (2) the differentiation of the whole through the identification of its parts, (3) the invention of a pattern by which the parts are interrelated, (4) testing the invented pattern to see if it applies, and (5) use of the new pattern in other similar contexts. Lawson's theory would not be published until 1967 (Lawson, 1967); however, his literature search uncovered the Atkin and Karplus (1962) paper to which he had this to say:

> If we substitute the term "initial unity" for system, "differentiation" for the identification of objects within the system, "pattern or relations" for invention, and "reinforcement" for discovery, we can see the relation of this teaching approach to our theory of learning. (p. 119)

Thus the same pattern of instruction had been independently "invented" by Atkin and Karplus and by Lawson. When Karplus, the physicist, needed a biologist to assist in developing the life science half of the SCIS program, he called Lawson. What began for Lawson as a two-week consultation in the summer of 1965 ended with a ten-year job as director of life sciences within the SCIS program.

The final product of the SCIS program in the mid 1970s was a K-6 life science and physical science curriculum based on learning cycles. In addition to the efforts of Karplus, Thier and Lawson, Jack Fishleder, Rita Peterson, Robert Knott, Carl Berger, and Marshall Montgomery made substantial contributions as staff members during the development years. Mary Budd Rowe, Stanford Davis, John Renner, Albert Carr and Glenn Berkheimer also made substantial contributions to the development effort as coordinators of trial teaching centers in five locations across the country.

CHANGES IN THE NAMES OF THE PHASES OF THE LEARNING CYCLE

Interestingly, the name learning cycle does not appear in any of the early SCIS publications, although the phases of exploration, invention, and discovery are clearly spelled out (cf., Karplus and Thier, 1967; SCIS, 1973; Jacobson and Kondo, 1968). First use of the name learning cycle appears to be in the Teacher's Guides to the SCIS program units beginning in about 1970 (e.g., SCIS, 1970).

Use of the name learning cycle with its phases of exploration, invention and discovery continued by Karplus and others through 1975 (e.g., Collea, Fuller, Karplus, Paldy and Renner, 1975). However in 1976, it was becoming apparent that many teachers were having difficulty understanding what the terms invention and discovery meant in the context of classroom lessons. So in a series of 1977 publications Karplus decided to refer to the learning cycle phases as exploration, concept introduction and concept application (e.g., Karplus, Lawson, Wollman, Appel, Bernoff, Howe, Rusch and Sullivan, 1977).

Still others have chosen to modify the names further. For example, Lawson (1995) refers to them as exploration, "term" introduction, and concept application. This modification was suggested primarily because of the belief that the names of the phases are intended to convey meanings to teachers (not necessarily to students). Teachers can introduce terms during the second phase of the learning cycle; but they cannot introduce concepts. Concepts must be "invented" by students.

THE LEARNING CYCLE IN THE BIOLOGICAL SCIENCE CURRICULUM STUDY PROGRAM

Tables 4.1 and 4.2 summarize the learning cycle as used in curricular materials, produced by the Biological Science Curriculum Study (BSCS). Notice that the tables divide exploration into engage and explore phases. And the tables refer to term introduction as the explain phase and to concept application as the elaborate phase. In addition to these three phases, BSCS includes another phase called evaluate. Of course students and teachers need to evaluate learning, so the addition of an evaluate phase is not really new. Thus, the

Table 4.1. The BSCS Version of Learning Cycle Instruction: Teacher's Role.

Phase of the instructional model	What the teacher does	
	That is Consistent with This Model	**That is Inconsistent with This Model**
Engage	• Creates interest • Generates curiosity • Raises questions • Elicits responses that uncover what the students know or think about the concept/topic	• Explains concepts • Provides definitions and answers • State conclusions • Provides closure • Lectures
Explore	• Encourages the students to work together without direct instruction from the teacher • Observes and listens to the students as they interact • Asks probing questions to redirect the students' investigations when necessary • Provides time for students to puzzle through problems • Acts as a consultant for students	• Provides answers • Tells or explains how to work through the problem • Provides closure • Tells the students that they are wrong • Gives information or facts that solve the problem • Leads the students step-by-step to a solution
Explain	• Encourages the students to explain concepts and definitions in their own words • Asks for justification (evidence) and clarification from students • Formally provides definitions, explanations, and new labels • Uses students' previous experiences as basis for explaining concepts	• Accepts explanations that have no justification • Neglects to solicit the students' explanations • Introduces unrelated concepts or skills
Elaborate	• Expects the students to use formal labels, definitions, and explanations provided previously • Encourages the students to apply or extend the concepts and skills in new situations • Refers the students to existing data and evidence and asks: What do you already know? Why do you think ...? (Strategies from Explore apply here also.)	• Provides definitive answers • Tells the students that they are wrong • Lectures • Leads students step-by-step to a solution • Explain how to work through the problem
Evaluate	• Observes the students as they apply new concepts and skills • Assesses students' knowledge and/or skills • Looks for evidence that the students have changed their thinking or behaviors • Allows students to assess their own learning and group-process skills • Asks open-ended questions, such as Why do you think ...? What evidence do you have? What do you know about x? How would you explain x?	• Tests vocabulary words terms, and isolated facts • Introduces new ideas or concepts • Creates ambiguity • Promotes open-ended discussion unrelated to the concept or skill

From Biological Sciences Curriculum Study, 1992, *Science & technology: Investigating human dimensions*, Dubuque, IA: Kendall/Hunt, p. 15 and Trowbridge and Bybee (1990).

Table 4.2. The BSCS Version of Learning Cycle Instruction: Student's Role.

Phase of the instructional model	What the student does	
	That is Consistent with This Model	**That is Inconsistent with This Model**
Engage	• Asks questions, such as Why did this happen? What do I already know about this? What can I find out about this? • Shows interest in the topic	• Asks for the "right" answer • Offers the "right" answer • Insists on answers or explanations • Seeks one solution
Explore	• Thinks freely, but within the limits of the activity • Tests predictions and hypotheses • Forms new predictions and hypotheses • Tries alternatives and discusses them with others • Records observations and ideas • Suspends judgment	• Lets others do the thinking and exploring (passive involvement) • Works quietly with little or no interaction with others (only appropriate when exploring ideas or feelings) • "Plays around" indiscriminately with no goal in mind • Stops with one solution
Explain	• Explains possible solutions or answers to others • Listens critically to others' explanations • Questions others' explanations • Listens to and tries to comprehend explanations the teacher offers • Refers to previous activities • Uses recorded observations in explanations	• Proposes explanations from "thin air" with no relationship to previous experiences • Brings up irrelevant experiences and examples • Accepts explanations without justification • Does not attend to other plausible explanations
Elaborate	• Applies new labels, definitions, explanations, and skills in new, but similar situations • Uses previous information to ask questions, propose solutions, make decisions, design experiments • Draws reasonable conclusions from evidence • Records observations and explanations • Checks for understanding among peers	• "Plays around" with no goal in mind • Ignores previous information or evidence • Draws conclusions from "thin air" • Uses in discussion only those labels that the teacher provided
Evaluate	• Answers open-ended questions by using observations, evidence, and previously accepted explanations • Demonstrates an understanding or knowledge of the concept or skill • Evaluates his or her own progress and knowledge • Asks related questions that would encourage future investigations	• Draws conclusions, not using evidence or previously accepted explanations • Offers only yes-or-no answers and memorized definitions or explanations as answers • Fails to express satisfactory explanations in his or her own words • Introduces new, irrelevant topics

From Biological Sciences Curriculum Study, 1992, *Science & technology: Investigating human dimensions*, Dubuque, IA: Kendall/Hunt, p. 15 and Trowbridge and Bybee (1990).

only substantive difference exists in use of and/or need for a separate phase of engagement. Clearly students need to be engaged for learning to take place. Good explorations do this as they provoke the raising of questions and the elicitation of connections with past experience. However, some teacher comments to "set the stage" for exploration are usually a good idea. In short, Tables 4.1 and 4.2 do an excellent job of identifying just what teachers and students should be doing, and not doing, during the phases of learning cycle instruction.

The learning cycle is a very flexible method for instruction. Certainly for young children and for anyone who lacks direct physical experiences with a particular set of phenomena, the exploration phase should involve that direct physical experience. This, however, does not imply that all explorations have to be conducted this way. Indeed, I had the pleasure of taking a history of science course in graduate school taught using learning cycles where the explorations consisted of slide presentations, lectures and discussions. The class explored various scientists' ideas and activities in this way and only later "invented" the concept of science. The key point is that one can change the learning format of the three phases of the learning cycle, but one cannot change the sequence of the phases or delete one of the phases. If the sequence is changed, or if a phase is deleted, one no longer has a learning cycle.

Thus, the main thesis is that a situation that allows students to examine the adequacy of prior beliefs (conceptions) forces them to argue about and attempt to test those beliefs. This in turn provokes disequilibrium when those beliefs are contradicted and provides an opportunity for self-regulation and the construction of more appropriate concepts. Along the way students become increasingly conscious of and skilled in using the reasoning patterns (forms of argumentation) required for concept construction. In other words, they learn how to learn. The central instructional principle is that correct use of the learning cycle accomplishes this end (see Karplus, 1977 reprinted in this Chapter).

CURRENT STATUS OF THE LEARNING CYCLE

Most educational policy makers and theorists agree that the educational system should help students: (1) construct sets of meaningful and useful concepts and conceptual systems, (2) develop skill in using the reasoning patterns essential for independent, creative and critical thought, and (3) gain confidence in the ability to apply knowledge to learn, solve problems, and make carefully reasoned decisions. Modern learning cycle theory (the primary postulates of which are listed in Table 4.3) argues that the most appropriate way, perhaps the only way, to accomplish these ends, is to teach in a way that allows students to reveal prior conceptions and test them in an atmosphere in which

Table 4.3. Postulates of Modern Learning Cycle Theory (after Lawson, 1995).

---

(1) Children and adolescents construct personal beliefs about natural phenomena, some of which differ from currently accepted scientific theory.

(2) These alternative beliefs (misconceptions) may be instruction resistant impediments to the construction of scientifically accepted beliefs (conceptions).

(3) The replacement of alternative beliefs requires students to move through a phase in which a mismatch exists between the implications of the alternative belief and the scientific conception and provokes a "cognitive conflict" or state of mental "disequilibrium."

(4) The improvement of reasoning patterns (procedural knowledge) arises from situations in which students state alternative beliefs and engage in verbal exchanges where arguments are advanced and evidence is sought to resolve the contradiction.

(5) Argumentation provides experiences from which particular forms of argumentation (i.e., patterns of reasoning) may be internalized.

(6) The learning cycle, a method of instruction consistent with the way people spontaneously construct knowledge, provides the opportunity for students to reveal alternative beliefs and the opportunity to argue and test them, thus become "disequilibrated" and acquire more adequate conceptions as well as more powerful and effective reasoning patterns.

---

ideas are openly generated, debated and tested, with the means of testing becoming an explicit focus of classroom attention.

A considerable amount of research has been conducted, much of it reviewed by Lawson, Abraham and Renner (1989), supporting the claim that correct use of the learning cycle in the science classroom at all levels of instruction is effective in helping students move toward the stated objectives. This is good news, as is the fact that a recent electronic literature search turned up nearly 400 reports of research and development efforts using the learning cycle since 1989. Clearly use of the learning cycle method is growing as more and more educators learn of its effectiveness. Indeed, research has shown that the learning cycle is effective in other fields as well. For example, Kral (1997) used learning cycles to teach 12th grade English and found significant improvements on the English portion of the American College Test (ACT). This comes as no surprise to those of us who teach science using learning cycles. Nor is it a surprise to find that research in the neurosciences is confirming the belief-expectation-test model of brain activity and human learning upon which the learning cycle is based (e.g., Kosslyn and Koenig, 1995).

Another piece of very good news are the position statements on teaching methods recently taken by organizations such as the National Science Foundation (1996), the American Association for the Advancement of Science (1989, 1990), and the National Research Council (1995), and by many state and local boards of education. These position statements clearly spell out that science teaching should be consistent with the nature of scientific inquiry—that science should be taught as science is practiced. These declarations lead teachers directly to learning cycle instruction.

Unfortunately, not all the news is good. Some educators still have not heard of the learning cycle or if they have, they have yet to see its importance and appreciate the successful curricular reforms that have resulted from its implementation. In the words of Sir Isaac Newton (Holten and Roller, 1958, p. 185), "*If I have seen farther than others, it is by standing upon the shoulders of Giants.*" One could argue that as a country we could see farther if we found the shoulders of giants such as Robert Karplus to stand on. The ones who suffer as a consequence of our often uncoordinated efforts to improve instruction are our children. The relatively poor performance of secondary school students in the United States is well documented (United States Department of Education, 1998). As a country we need to develop a more coordinated effort, one which builds on past successes, so that our curriculum development and implementation efforts produce the kind of science instruction our children and nation so desperately need and deserve.

REFERENCES

American Association for the Advancement of Science, 1989, *Science For All Americans*, Washington, DC.

American Association for the Advancement of Science, 1990, *The Liberal Art of Science*, Washington, DC.

Atkin, J.M. and Karplus, R., 1962, Discovery or invention? *The Science Teacher*, 29(5): 45.

Biological Sciences Curriculum Study, 1992, *Science & Technology: Investigating Human Dimensions*, Dubuque, IA: Kendall/Hunt, p. 15.

Collea, F.P., Fuller, R.G., Karplus, R., Paldy, L.G. and Renner, J.W., 1975, *Physics Teaching and the Development of Reasoning*. Stony Brook: American Association of Physics Teachers.

Dewey, J., 1916, Method in science teaching. *General Science Quarterly*, 1: 3.

Galloway, T.W., 1910, Report of the committee on fundamentals of the Central Association of Science and Mathematics Teachers. *School Science and Mathematics*, 10: 801–813.

Hall, E.H., and Committee, 1898, Memorandum concerning report of committee of sixty. *Addresses and Proceedings*. Washington, DC, National Education Association, 37: 964–965.

Holton, G.H., and Roller, D.H.D., 1958, *Foundations of Modern Physical Science*. Reading, MA: Addison-Wesley.

Hurd, P.D., 1961, *Biological Education in American Secondary School, 1890–1960*. Washington, DC: American Institute of Biological Sciences.

Jacobson, W., and Kondo, A., 1968, *SCIS Elementary Science Sourcebook*. Berkeley, CA: Science Curriculum Improvement Study.

Karplus, R., 1974, The learning cycle. In *The SCIS Teacher's Handbook*. Berkeley, CA: Regents of the University of California.

Karplus, R., 1977, Science teaching and the development of reasoning. *Journal of Research in Science Teaching*, 14(2): 169–175.

Karplus, R., Lawson, A.E., Wollman, W., Appel, M., Bernoff, R., Howe, A., Rusch, J.J., and Sullivan, F., 1977, *Science Teaching and the Development of Reasoning: A Workshop*. Berkeley: Regents of the University of California.

Karplus, R., and Thier, H.D., 1967, *A New Look at Elementary School Science*. Chicago: Rand McNally.

Kosslyn, S.M., and Koenig, O., 1995, *Wet Mind: The New Cognitive Neuroscience*. New York: The Free Press.

Kral, E.A., 1997, Scientific reasoning in a high school English course. *Skeptical Inquirer*, **21**(3): 34–39.

Lawson, A.E., 1995, *Science Teaching and the Development of Thinking*. Belmont, CA: Wadsworth.

Lawson, A.E., Abraham, M.R., and Renner, J.W., 1989, *A Theory of Instruction: Using the Learning Cycle to Teach Science Concepts and Thinking Skills*. NARST Monograph Number One, Cincinnati, OH: National Association for Research in Science Teaching.

Lawson, C.A., 1958, *Language, Thought, and the Human Mind*. East Lansing: Michigan State University Press.

Lawson, C.A., 1967, *Brain Mechanisms and Human Learning*. Boston: Houghton Mifflin.

Lawson, C.A., and Paulson, R.E. [Eds.], 1958, *Laboratory and Field Studies in Biology: A Source Book for Secondary Schools*. New York: Holt, Rinehart and Winston.

National Research Council, 1995, *National Science Education Standards*. Washington, DC: National Academy Press.

National Science Foundation, 1996, *Shaping the Future*. Washington, DC.

Science Curriculum Improvement Study, 1970, *Environments: Teacher's Guide*. Chicago: Rand McNally.

Science Curriculum Improvement Study, 1973, *SCIS Omnibus*. Berkeley, CA: Lawrence Hall of Science.

Trowbridge, L.W., and Bybee, R.W., 1990, *Becoming a Secondary Science Teacher*. Columbus, OH: Merrill Publishing Co.

United States Department of Education, 1998, *Policy Brief: What the Third International Mathematics and Science Study (TIMSS) Means for Systemic School Improvement*. Washington, DC: U.S. Government Printing Office.

The Learning Cycle

Suppose you are working with a class that is going to examine rocks and minerals, but has had no experience using magnifiers. You are planning activities that will give your pupils the necessary skill with magnifiers. Would you introduce magnifiers by

a. showing them to the class and demonstrating their use to look closely at a rock?
b. showing them to the class and then explaining their use by means of chalkboard diagrams of lenses and light rays before they are distributed?
c. distributing them to the children with the invitation to use them to look at objects on their desks, their hands, or at other things they would like to examine?

Compare your reactions with our comments on the alternatives:

a. This procedure is used very frequently to introduce unfamiliar pieces of apparatus; it provides the students with a model they can imitate in subsequent activities. This approach has the disadvantages—especially with young children—that your pupils may not be able to see the demonstration well because of the magnifier's small size, that they may not pay attention, and that they may not be able to relate what they see you do to their own actions later.
b. Since this approach is not directly relevant to the learner's needs, which relate to using the magnifier and not to the theory of how magnifiers are designed and used, we consider this plan to be unsuitable for learners of any age. We believe that they should be familiar with the operation of an instrument before becoming concerned with the theory on which it is based.
c. We usually take this approach in the SCIS program because it involves the children most directly. It also gives them an opportunity to explore various ways of using the magnifier so that they will later be able to adapt it to their needs in

Reprinted with permission from *SCIS Teacher's Handbook,* © 1974, The Regents of the University of California.

different situations. Furthermore, it enables you to identify children who have difficulty so you can help them individually, while giving others the satisfaction of having figured it out for themselves or of having shared their findings with their classmates. This approach requires more time than the one described in (a), but we consider the time well spent. Only when there is danger of damage to an instrument (e.g., breaking a thermometer) would we describe certain procedures to be followed and others to be avoided.

The stages in the learning cycle. The preferred approach in (c) is an example of the exploration stage in the learning cycle that is used to organize activities in the SCIS program. The learning cycle consists of three stages that we call exploration, invention, and discovery. During exploration children learn through their spontaneous reactions to a new situation. In this stage children explore new materials and/or ideas with minimal guidance or expectations of a specific achievement. (Note: For simplicity in explanation, we have based our example of exploration on the introduction of the magnifier; in the SCIS teacher's guides, the learning cycle is usually applied to the introduction of a concept, as it is in the discussion on the next page dealing with the teaching of the interaction concept.)

During the invention stage you define a new concept or explain a new procedure in order to expand your pupils' knowledge, skills, or reasoning. This step should always follow exploration. Since the magnifier is relatively simple to use, many children may "invent" how to use it themselves; for others you can provide the necessary instruction individually. The invention lessons in the SCIS program serve to introduce concepts, such as interaction or life cycle, that few children can phrase for themselves. You will "invent" these concepts in an activity involving the entire class or with small groups.

The last stage of the learning cycle is discovery, during which a child discerns new applications for the concept or skill he has learned recently. The children's investigations of rocks and minerals after they have partially mastered use of the magnifiers are discovery activities that enable them to practice and refine their skill. Discovery is most effective when there is wide variety in the examples and materials investigated, so that each child can test what he has just learned under many differing conditions. Note that the investigation of the rocks, which serves as a discovery activity in relation to the use of magnifiers, can also function as exploration to introduce a new learning cycle concerning the classification of rocks and minerals. In this way, a longer teaching unit can be built up from a sequence of learning cycles.

To summarize, the basic intent of the invention lessons is to introduce definitions of new terms and concepts that relate these immediately to objects and actions, not merely to other words. The exploration lessons provide a background for the new idea, and the discovery lessons permit its further application and extension.

The following example deals with activities related to the interaction concept. Identify the stage of the learning cycle that is represented by each of the following four scenes, or point out why it does not fit any of the stages; then see our comments below.

    a.   The children experiment with tumblers of water and vials of copper chloride after the teacher asks them to look for evidence of interaction in their investigations.

    b.   After student assistants have distributed trays with magnets, water-filled tumblers, colored candy balls, light bulbs, scissors, paper, crayons, batteries, wires, and paper clips, the teacher invites the pupils to find out what these objects can do.

    c.   While the teacher is demonstrating with a large rubber band that is stretched tightly between his thumbs, he asks "Are my thumbs and the rubber band interacting now?"

    d. While the teacher is demonstrating with a large magnet attracting a pair of scissors, he says "When two objects, like this magnet and the scissors, are doing something to each other, we say that the objects interact."

Comments:

    a. This is an example of discovery because the children carry out experiments with their attention focused on evidence of interaction and not completely undirected. Note, however, that the teacher's instruction is still divergent in that the evidence of interaction may take many different forms. Presumably, the concepts of interaction and evidence of interaction were "invented" during an earlier science session.

    b. Since the children are investigating materials with no specific focus from the teacher, this is exploration.

    c. Here, the teacher is gathering feedback in a quick group activity. He is trying to determine whether most pupils respond knowledgeably to the term "interacting." In spite of the teacher's leadership, this is not an invention because the teacher is not explaining or defining.

    d. Along with the demonstration, the teacher is defining the term "interact"; this is an example of invention, therefore.

Many experimental and longitudinal studies suggest that the amount of new information children can incorporate into their understanding is dependent both on the knowledge they have already and on the way that knowledge is organized. This finding highlights the importance of the exploration lessons, where the children are left largely on their own to explore and discover. These lessons enrich their experience, enable them to reorganize their knowledge, and provide opportunities for you to appraise to what extent they are accommodating their ideas to those presented in past invention lessons.

The SCIS program aims to nurture the children's ability to discover new relationships and to think imaginatively, at the same time as it facilitates the transition from preoperational to concrete or formal operational thought (see Chapter 3). Accordingly, it also includes invention and discovery lessons. The invention lesson provides guided practice in using new labels and categories. It is clearly teacher-directed and should provide an opportunity for each child to stretch his already acquired association of meanings for the objects in his world to include new meanings. The child accommodates his thought to that of the teacher as he imitates the teacher's classification or designations. Such momentary accommodation may have little effect on the child's ability to use the new concept, however, unless he can also try it out independently in new situations.

The discovery lessons leave the children somewhat on their own in order to test and eventually assimilate the new information. This kind of lesson also provides opportunities for children to make observations, perhaps paying attention to aspects of their world that have been highlighted for them in previous activities or perhaps focusing their attention on aspects uniquely and personally of interest to them.

To assist young children in making the transition from preoperational to operational thinking as expeditiously as possible, keep the different functions of the three stages of the learning cycle clearly in mind.

Learning theories. You are probably familiar with several psychological theories of learning that have been used in the development of educational materials. Best known is the theory of "learning-by-association," which views the student's behavior as a response to well-planned stimuli. With repetition, practice, correction of errors, and suitable rewards or punishment, the learner is expected to master the desired behavior. This is the theory behind rote learning of much of past education and has led to programmed instruction in more recent times. As usually applied, this theory leaves no room for spontaneous or creative expressions by the student.

A sharp contrast is provided by the theory of "learning-by-discovery," which claims that everything of which an individual is capable is latent within him. Given a sufficiently rich environment, the learner is expected to discover the properties of objects, the conditions under which interaction takes place, the principles governing energy exchange, and the concepts necessary for understanding life. In its extreme forms this theory allows no direct input that might limit or focus the student's natural interests.

Still another theory, based on the work of Jean Piaget, may be called "learning-by-reasoning." According to this theory, the student is brought to understand relationships through logical reasons provided by himself, the teacher, or classmates. Drill and practice, according to this view, are less important in learning than analysis and interpretation of a few problem situations.

We believe that all three of these learning theories have merit and that no one of them presents a complete picture that applies in all situations and to all individuals. The learning cycle of the SCIS program, therefore, reflects the three theories in a complementary way, using each of them for its strongest points but giving greatest weight to "learning-by-reasoning." The process of self-regulation to which Piaget ascribes intellectual development (see Chapter 3) is facilitated especially well by all three stages of the learning cycle taken in sequence. (The relationship of learning theories and the learning cycle is described in the *SCIS Omnibus*, page 126.)

For further investigation of the learning cycle, we suggest that you look for examples of exploration, invention, and discovery in an SCIS teacher's guide, some of the SCIS films, and your colleagues' teaching approaches. Discuss your conclusions with your friends and/or science consultant.

Three Guidelines for Elementary School Science

Robert Karplus, Director, SCIS

I wish I could present to you a comprehensive theory of instruction that might serve as a basis for curricula in science and other areas. Unfortunately, I can't do this; I can only describe some guidelines that have been useful to the Science Curriculum Improvement Study. The teaching objective is to give the students sufficient knowledge and experience so they will have an understanding of the natural environment and an appreciation for original scientific work being carried out by others.

Guideline 1. Two aspects of the teaching program should be distinguished from one another: the experiential (student experience with a wide variety of phenomena, including their acting on the materials involved) and the conceptual (introduction of the student to the approach which modern scientists find useful in thinking about the phenomena they study). A key problem in planning instruction is how to relate these two aspects to one another, a matter to which I shall return in my third guideline.

Let me list a few examples of what may be done to give the students experience. The observation of magnetized or electrically charged objects interacting without physical contact, of chameleons eating crickets in a terrarium, of trajectories that can be controlled through the launching conditions, and of seeds germinating under certain conditions are useful in helping to form a picture of the broad range of physical and biological phenomena. Also necessary are more elementary experiences with the change in appearance of a liquid sample as it is transferred among differently shaped containers, with the "feel" of specimens of high and low density, with the "disappearance" of solids when they dissolve in liquids, and with the details of surface structure that become visible when a wood or mineral specimen is examined with a magnifying glass. In all these areas it is essential that students experience these directly, and that they have an opportunity to act on the materials and thereby influence what happens.

Reprinted with permission from *SCIS Newsletter No. 20*, © 1974, The Regents of the University of California.

Three Guidelines For Elementary School Science

Supplementing the above is the conceptual approach, which leads to general principles interrelating the phenomena I have just described. Being a physicist, I began my educational activities ten years ago with the notion that force was the fundamental explanatory concept, since force is the cause of motion, and motion is a part of all change. Now I believe that this approach, which is also taken by most physics texts, is not valid. The reason is that observable motion accompanies only a small fraction of phenomena. Thermal, chemical, electrical, optical, and acoustic phenomena do not involve observable motion, hence the force concept is not of direct value in dealing with them. Instead, the broader concept of interaction does apply as an explanatory concept for all these areas, and this concept therefore plays the central role in the SCIS program. The conceptual framework is described in greater detail in each teacher's guide, where supporting concepts such as matter, energy, organism, ecosystem, system, evidence, reference frame, and scientific model are identified.

Guideline 2. Major theories of intellectual development and learning should be drawn upon in curriculum construction, even though they are in conflict with one another. I find it useful to distinguish three major types of theories.

The "learning-by-conditioning" theory views the learner's behavior as a response to a well-planned stimulus. With repetition, practice, and suitable reinforcement, the learner will exhibit the desired behavior. Note that in this theory there is no room for spontaneous or creative expressions by the student. Everything of educational value reflects the inputs accumulated during the teaching program.

The "learning-by-discovery" theory claims that everything of which an individual is capable is latent within him. The educational program must give him opportunities to express these latent tendencies, but should not provide any input that might inhibit or redirect his natural inclinations. Given a sufficiently rich environment, the learner will discover the properties of objects, conditions under which phenomena take place, and general principles relating the isolated incidents and observations in his experiments and investigations.

The "learning-by-equilibration" theory, associated with Jean Piaget, views the individual as capable of mental operations which function in a self-sustaining feedback loop (equilibrium) as he acts on his environment and receives stimuli in return. When the feedback loop is disturbed by events that don't fit the scheme (disequilibrium), changes in the mental operations ultimately lead to more powerful mental operations that can cope successfully with a larger class of events (equilibration).

Guideline 3. By putting together guideline 1 (experience and concepts) and guideline 2 (theories of development and learning), SCIS arrived at a prescription for designing instructional units. We conceive of a learning cycle with three phases: *exploration*, referring to self-directed, unstructured investigation; *invention*, referring to the introduction of a new integrating concept by teacher or by learner; and *discovery*, referring to applications of the same new concept in a variety of situations, partly self-directed, partly guided. Each SCIS teacher's guide describes in detail how the learning cycle relates to the activities of a unit.

Note that the learner is active during the exploration and discovery phases, which occupy most of the teaching time. Experiential input is provided during these phases. He is least active during the invention phase, which should occupy only a brief interval between the other two. Conceptual input is provided during this phase.

Note also how this prescription utilizes the three learning theories. Exploration is in accord with "learning-by-discovery" and "learning-by-equilibration." It allows the learner to impose his ideas and preconceptions on the subject matter to be investigated. If he comes up with a successful new idea, more power to him. If, as is often the case, his preconceptions lead to confusion, the teacher learns about these difficulties. At the same time, the exploration should create some disequilibrium, since not all students can cope with the materials with equal success.

Invention is in accord with the "learning-by-equilibration" theory, as the new idea introduced at that time suggests a way for the learner to resolve his disequilibrium.

Discovery, finally, is in accord with "learning-by-equilibration," and also with the "learning-by-conditioning" view that repetition and practice are necessary for learning. It is essential, however, that the repetition and practice occur largely through self-directed activities by the learner, so that he will actually resolve his disequilibrium by interacting with the experimental materials and by establishing a new feedback pattern for his actions and observations. At this time, the same concepts are applied repeatedly in a wide variety of activities.

These three guidelines are useful to a teacher planning his lessons, as well as to the curriculum developer creating new course materials. After formulating his objectives, the teacher selects the experiential and conceptual inputs that have to be provided for the students; he identifies the elements leading to disequilibrium and conceptual reorganization; and he decides whether highly self-directed exploration, partly guided discovery, or more structured invention is the most appropriate mode for the activity. Over a period of time, the teacher will prepare a program that is balanced in these respects, and that provides for substantial self-direction even at the expense of limiting the course's "coverage."

This article was adapted from papers published in Curriculum Theory Network, Winter, 1969–70, *and presented to Section 1 (physics) of the 41st Conference of the Australian and New Zealand Association for the Advancement of Science, Adelaide, South Australia, August 1969.*

Science Teaching and the Development of Reasoning*

Robert Karplus

In your interactions with secondary school students learning science, you have probably become aware of large differences in student ability to understand science concepts, conduct investigations, and/or solve specific problems. Some students are extremely capable, while others demonstrate peculiar and inappropriate reasoning strategies. Sometimes, even after your best efforts, they seem unable to grasp ideas that to you are eminently clear. Often students are able to follow problem solutions but are at a loss when required to transfer those strategies to slightly different problems. You wonder why students are able to respond successfully to examination questions and then, after a month or so, forget almost all of what they learned.

Teachers' understanding of these situations and of student differences can be significantly aided by the developmental theory of Jean Piaget. For several years, Wollman and

*This position paper was prepared for the 1976 national meeting of the National Association for Research in Science Teaching. It was sponsored by the Educational Resources Information Center of the National Institute of Education and The Ohio State University. This publication was prepared pursuant to a contract with the National Institute of Education, U.S. Department of Health, Education and Welfare. Contractors undertaking such projects under Government sponsorship are encouraged to express freely their judgment in professional and technical matters. Points of view or opinions do not, therefore, necessarily represent official National Institute of Education position or policy, nor of the National Association of Research in Science Teaching or the view of this *Journal*.

David P. Butts, Editor

Robert Karplus, Lawrence Hall of Science, University of California, Berkley, California 94720.

Lawson have worked with me investigating the relation of this theory to science teaching at the secondary school level. Other researchers with whom we have been in touch and whose work has influenced our thinking include Arnold Arons (University of Washington, Seattle), Kenneth Lovell (Leeds, England), Eric Lunzer (Nottingham, England), John W. Renner (Norman, Oklahoma), Michael Shayer (London), and Antonio Suarez (Zurich, Switzerland).

In addition to researching students' reasoning, it is essential to communicate the important features of Piaget's theory to secondary school teachers so that they might apply it in their classrooms and provide larger field tests of its applicability than our small research group could undertake. This paper focuses on an interpretation of Piaget's theory derived from recent findings applicable to adolescents.

PIAGET'S THEORY AND REASONING PATTERNS

Piaget has characterized human intellectual development in terms of four stages (Inhelder & Piaget, 1958). The first two, called sensory-motor and preoperational, are usually completed when a child is seven or eight years old. Following these are two stages of logical operations, called concrete thought and formal thought, which are relevant to secondary school students.

Piaget has ascribed the process whereby individuals advance from one stage to the next to four contributing factors: maturation, experience with the physical environment, social transmission, and "equilibration." The last item designates an internal mental process in which new experiences are combined with prior expectations and generate new logical operations.

To make the stage concept useful, one has to describe the reasoning of an individual whose development has reached each of the stages. This description has been stated by Piaget in terms of the mental operations the individual uses when facing certain problems. To avoid confusion with other uses of the term "operation" in science, it is useful to employ the phrase "reasoning patterns." Two examples of behavior based on reasoning patterns are (1) serial ordering a set of sticks according to their length and (2) investigating the effect of fertilizer on clover by setting up several test plantings that are treated alike in all respects except in the amount of fertilizer applied to them.

From the research of Piaget and others, certain rules have been formulated for identifying reasoning patterns as belonging to concrete or to formal thought. In general, reasoning that makes use of direct experience, concrete objects, and familiar actions is classified as a concrete reasoning pattern, such as example (1) above. Reasoning that is based on abstractions and that transcends experience is classified as a formal reasoning pattern, such as example (2) above. Here is a more extensive list of clues that are helpful in classifying reasoning patterns (Karplus et al., 1977, Module 2).

When using concrete reasoning patterns, the individual:

C1   Applies classifications and generalizations based on observable criteria (e.g., consistently distinguishes between acids and bases according to the color of litmus paper; all dogs are animals, but not all animals are dogs).

C2   Applies conservation logic—a quantity remains the same if nothing is added or taken away, two equal quantities give equal results if they are subjected to equal changes (e.g., when all the water in a beaker is poured into an empty graduated cylinder, the amount originally in the beaker is equal to the amount ultimately in the cylinder).

C3 Applies serial ordering and establishes a one-to-one correspondence between two observable sets (e.g., small animals have a fast heart beat while large animals have a slow heart beat).

By using these patterns, the individual can reason and solve problems beyond his ability in the preoperational stage. Yet there are many limitations if concrete reasoning patterns are compared to formal ones.

When using formal reasoning patterns, the individual:

F1 Applies multiple classification, conservation logic, serial ordering, and other reasoning patterns to concepts, abstract properties, axioms, and theories (e.g., distinguishes between oxidation and reduction reactions, uses the energy conservation principle, arranges lower and higher plants in an evolutionary sequence, makes inferences from the theory according to which the earth's crust consists of rigid plates).

F2 Applies combinatorial reasoning, considering all conceivable combinations (e.g., systematically enumerates the genotypes and phenotypes with respect to characteristics governed by two or more genes).

F3 States and interprets functional relationships in mathematical form (e.g., the rate of diffusion of a molecule through a semipermeable membrane is inversely proportional to the square root of its molecular weight).

F4 Recognizes the necessity of an experimental design that controls all variables but the one being investigated (e.g., sets up the clover experiment mentioned above).

F5 Reflects upon his own reasoning to look for inconsistencies or contradictions with other known information.

In Table I, the most important differences between concrete and formal reasoning patterns are summarized.

Table I. Concrete and Formal Reasoning Patterns.

	CONCRETE	FORMAL
(a)	Needs reference to familiar actions, objects, and observable properties.	Can reason with concepts, relationships, abstract properties, axioms, and theories; uses symbols to express ideas.
(b)	Uses reasoning patterns C1–C3, but not patterns F1–F5.	Uses reasoning patterns F1–F5 as well as C1–C3.
(c)	Needs step-by-step instruction in a lengthy procedure.	Can plan a lengthy procedure given certain overall goals and resources.
(d)	Is not aware of his own reasoning, inconsistencies among various statements he makes, or contradictions with other known facts.	Is aware and critical of his own reasoning; actively seeks checks on the validity of his conclusions by appealing to other known information.

PRESENT STATUS OF THE THEORY OF FORMAL THOUGHT

While each of the two lists in the previous section has a certain theme, this enumeration of formal reasoning patterns does not communicate the unity originally proposed by Piaget. Piaget conceived of formal reasoning patterns as dealing with logical propositions and having the organizational structure of an algebraic group called the INRC group. Recent workers, however, have not found evidence to support Piaget's proposals in these respects. Neimark (1975) has summarized the present status by concluding that there *is* more advanced intellectual functioning than concrete thought, but that such reasoning is not used as reliably and universally as Piaget's writings imply. Lunzer (in press) has expressed similar views. In fact, Piaget (1972) has adopted a more flexible position in the last few years.

One example of a study that leads to difficulty for the highly unified view of formal thought is an unpublished survey of student reasoning in seven countries carried out in 1974 (Karplus et al., 1975). Several thousand eighth- and ninth-grade students were presented with a task in proportional reasoning and a second task requiring control of variables (reasoning patterns F3 and F4). The results indicated that United States students succeeded more frequently on control of variables, while Austrian students succeeded more frequently on proportional reasoning. For British students, performance on the two tasks was more closely similar than for either of the other two samples mentioned. Using the British results as a guide, one might claim that the two tasks are about equally difficult. The lack of correspondence in Austria and the United States can then not be explained by asserting that eighth-graders in one of these countries are more or less advanced toward formal thought than students in the other. The researchers concluded that there were differential effects of instruction that did not generalize directly from one formal reasoning pattern to another.

In spite of the fact that the unity of formal reasoning patterns appears to be elusive, some research studies suggest strongly that there is a bond relating them. Consider, for instance, the factor structure of 10 Piagetian tasks reported by Lawson and Nordland (1976) after interviewing 96 seventh-graders in an urban school. They identified two principal components that accounted for 55% of the variance in their data. Four tasks—equilibrium in the balance, separation of variables, conservation of displaced volume (both cylinders and clay)—loaded primarily on the first component. Three other tasks (conservation of number, solid amount, and liquid amount) loaded exclusively on the second component. The last three tasks (conservation of length, area, and weight) loaded on both components. The authors concluded that the first and second components represent formal and concrete thought more comprehensively than these are represented in any single reasoning pattern. Perhaps, then, there is a unity after all!

One shortcoming of the Lawson-Nordland study is its emphasis on conservation tasks. Lawson is now planning a more comprehensive investigation that includes a wider variety of reasoning patterns.

APPLICATIONS TO SCIENCE TEACHING

The science teacher who is interested in applying Piaget's theory can benefit but must be cautious to avoid the difficulties with the theory that are even more prominent in a classroom than they are in a research study. Teachers need to concentrate on identifying their students' reasoning patterns and should not expect that each student's entire behavior can be classified neatly as reflecting either concrete or formal thought. Most important is the teacher's willingness to accept the conclusion, documented in recent studies, that a large fraction of students will use concrete reasoning patterns extensively (Karplus & Peterson, 1970; Karplus et al., 1975).

By becoming aware of reasoning patterns needed to understand a particular science course, a teacher can both identify the conceptual emphasis and demands of the subject matter and help students develop more advanced reasoning patterns than they use currently.

Here are nine concepts that are usually included in secondary school science courses: density, temperature, cell, gene, environment, chemical bond, periodic system of elements, acid-base, and ideal gas. What reasoning patterns must a student use to understand these? (Karplus et al., 1977, Module 7).

Look first at density. Density must be understood in terms of other concepts—mass and volume—rather than in terms of direct experience. Furthermore, the ratio relationship must be applied to mass and volume. Both of these mental steps make use of formal reasoning patterns, items F1 and F3 in the earlier list. For this reason, density may be called a "formal" concept.

Temperature can be defined in terms of sensations (warm/cold) or thermometer readings. When this is done, temperature may be called a "concrete" concept, because it is based on observable criteria and thus requires concrete reasoning patterns (item C1 in the list) for understanding. Temperature, however, can also be defined as a measure of the average molecular kinetic energy. If this is done, temperature becomes a "formal" concept whose understanding derives from other concepts (molecule, kinetic energy), the kinetic molecular theory, and mathematical relationships (items F1 and F3).

This example illustrates that a concept with several meanings may be either "concrete" or "formal," depending on the meaning used. To identify the reasoning required of the students in a course, the teacher must be clear about the meaning of the concepts that are introduced. Special care must be taken to use a concept always with the meaning that was explained to the students, and not to expect that the introduction of temperature or another concept as "concrete" concept can be extended automatically to an application of the concept's "formal" significance.

All the other concepts listed above can be defined as "formal" concepts. Still, "cell," "environment," and "acid-base" can also be given definitions as concepts in terms of familiar actions and examples. A cell can be observed when tissue is examined through a microscope; the environment and environmental factors such as heat, moisture, and light can be observed easily; and acid-base can be distinguished by the use of a chemical indicator, interaction with washing soda, or—in safe cases—by tasting.

The remaining concepts—"gene," "chemical bond," "periodic system," and "ideal gas"—all require formal reasoning patterns for their understanding. They can only be defined in terms of other concepts, abstract properties, theories, and mathematical relationships. There is no way of defining them as "concrete" concepts.

THE FORMATION OF REASONING PATTERNS BY SELF-REGULATION

Problems that a secondary school science teacher is likely to encounter were mentioned at the beginning of this article. Some of these problems can be ascribed to the fact that many students use concrete reasoning patterns, yet that subject matter often requires formal thought. Unless science courses are to become highly selective and admit only students who use formal reasoning patterns with ease, the formation of formal reasoning patterns should be made an important course objective (at least as important as the covering of a certain body of subject matter!).

Let us, therefore, return to the process of intellectual development. Rather than using Piaget's term "equilibration" for the essential but hard-to-define fourth contribution, one may employ the term "self-regulation," which has fewer science connotations and emphasizes the active role played by the individual (Karplus et al., 1977, Module 5).

The key to the formation of new reasoning patterns is an individual's responding to his or her inadequacy in using the present reasoning patterns to cope with a demand. An analogy in physical actions is your response to driving an unfamiliar car with a brake of different stiffness from that in your car. You first use your accustomed foot pressure, discover that it is unsatisfactory, and then try variations until the car responds smoothly. Your first encounter with an unsuspected power brake can lead to near-disaster!

A child using a concrete reasoning pattern in a pizza parlor may decide that the eight-inch pizza costing $1.25 is too small and may order a 16-inch size without looking at the price, in expectation that it costs $2.50, "Because it's twice as big." Imagine the dismay when the giant pizza arrives, together with a check for about $5.00! Here is a surprise that may trigger the search for a more successful reasoning pattern to cope with the pizza size/price problem, a mathematical relationship requiring a formal reasoning pattern.

Others have been gathering evidence (Lawson, Blake, & Nordland, 1975; Lawson & Wollman, 1975; Lawson & Wollman, 1976) that the learning cycle, which was introduced as part of the Science Curriculum Improvement Study (1970–1974) to facilitate concept development at the elementary school level, is also effective with older students and the introduction of formal concepts (Shoemacker, 1967; Sticht, 1971). The learning cycle consists of three instructional phases that combine experience with social transmission and encourage self-regulation (SCIS, 1974). These three phases are *exploration, concept introduction,* and *concept application.*

During *exploration,* the students gain experience with the environment—they learn through their own actions and reactions in a new situation. In this phase, they explore new materials and new ideas with minimal guidance or expectation of specific accomplishments. The new experience should raise questions or complexities that they cannot resolve with their accustomed patterns of reasoning, as in the pizza example. As a result, mental disequilibrium will occur and the students will be ready for self-regulation.

The second phase, *concept introduction,* provides social transmission—it starts with the definition of a new concept or principle that helps the students apply a new pattern of reasoning to their experiences. In the pizza problem, the relation of area to diameter would be the key idea, but might be first illustrated by means of the area and side of a square rather than a circle. The concept may be introduced by the teacher, a textbook, a film, or another medium. This step, which aids self-regulation, should always follow exploration and relate to the exploration activities.

Concept introduction is especially effective when it involves the formal definition of a concept whose concrete definition is already understood by the students. Since, for instance, a square can easily be subdivided into unit squares, determining the area of a square need only make use of concrete reasoning patterns. This illustration would help lead the students toward conceptualizing and approximating the area of a circle, a step that requires a formal reasoning pattern because a circle cannot be subdivided completely into unit squares by a finite number of steps.

In the last phase of the learning cycle, *concept application,* familiarization takes place as students apply the new concept and/or reasoning pattern to additional situations. In the pizza area example, a valuable application activity might involve the construction of sets of similar rectangles, ellipses, and other figures out of cardboard and investigating the relationship of their diameter to their weight. In this phase, physical experience with materials and social interactions with teacher and peers play a role.

Concept application is necessary to extend the range of applicability of the new concept. This phase provides additional time and experiences for self-regulation. Furthermore, concept application activities aid the students whose conceptual reorganization takes place more slowly than average or who did not adequately relate the teacher's original explanation to their experiences. Individual conferences with these students to identify and resolve their difficulties are especially valuable.

CONCLUSION

This has been a very simplified introduction to a complicated area of research that holds a great deal of promise for the improvement of secondary science teaching. It is most important that Piaget's ideas can and should be used actively for instructional improvement, and should not be interpreted as implying that education must wait until development has occurred spontaneously. Piaget (1973) has described the interaction of education and development in these words: "Thus education is . . . a necessary formative condition toward natural development itself." Of course, the theory will not solve all educational problems, but it can help in those aspects of concept development and understanding which make science courses especially difficult for many students.

REFERENCES

Inhelder, B., & Piaget, J. *The growth of logical thinking from childhood to adolescence.* New York: Basic Books, Inc., 1958.

Karplus, R., & Peterson, Rita W. Intellectual development beyond elementary school II: Ratio, a survey. *School Science and Mathematics,* 1970, **70,** 813–820.

Karplus, R., Karplus, E. F., Formisano, M., & Paulsen, A. C. *Proportional reasoning and control of variables in seven countries.* Berkeley, Calif.: Lawrence Hall of Science 1975.

Karplus, R., Lawson, A. E., Wollman, W. T., Appel, M., Bernoff, R., Howe, A., Rusch, J. J., & Sullivan, F. *Workshop on science teaching and the development of reasoning.* Berkeley, Calif.: Lawrence Hall of Science, 1977, Modules 2, 5 and 7. (Arrangements for the distribution of the workshop materials are being made.)

Lawson, A. E., Blake, A. J. D., & Nordland, F. H. Training effects and generalization of the ability to control variables in high school biology students. *Science Education,* 1975, **59,** 387–396.

Lawson, A. E., & Nordland, F. H. The factor structure of some Piagetian tasks. *Journal of Research in Science Teaching,* 1976, **13,** 461–466.

Lawson, A. E., & Wollman, W. T. Teaching proportions and intellectual development. Unpublished manuscript, University of California, Berkeley, 1975.

Lawson, A. E., & Wollman, W. T. Encouraging the transition from concrete to formal cognitive functioning—An experiment. *Journal of Research in Science Teaching,* 1976, **13,** 413–430.

Lunzer, E. A. Formal reasoning: A re-appraisal. *Proceedings 3rd Annual Symposium of the Jean Piaget Society,* Temple University, Philadelphia (in press).

Neimark, E. D. Intellectual development during adolescence. In F. D. Horowitz (Ed.), *Review of child development research* (Vol. 4). Chicago: The University of Chicago Press, 1975.

Piaget, J. Intellectual evolution from adolescence to adulthood. *Human Development,* 1972, **15,** 1–12.

Piaget, J. *To understand is to invent.* New York: Grossman, 1973.

Science curriculum improvement study. Chicago: Rand McNally, 1970–1974.

SCIS teacher's Handbook. Berkeley, Calif.: Lawrence Hall of Science, 1974.

Shoemacker, H. A. The Functional context of instruction. Professional paper 35–67, Human Research Office, George Washington University, 1967.

Sticht, T. G. Comments on Kenneth Lovell's paper: Does learning recapitulate ontogeny? In D. R. Green, M. P. Ford, and G. B. Flamer (Eds.), *Measurement and Piaget.* New York: McGraw-Hill, 1971.

CHAPTER 5

Student Autonomy and Teacher Input

J. Myron Atkin*

The Congress of the United States created the National Science Foundation (NSF) in 1950. It stipulated that the Foundation would have two missions: to support basic research in science and to improve science education. From the inception, the first of these missions was pursued enthusiastically, but not the second. While there was considerable interest within the Foundation for support of education in science at the Ph.D. level, there was little enthusiasm for becoming involved in undergraduate science curriculum—and there was none for education below the college level. In fact, funds to address issues of improving pre-college science education were not made available until 1956. Under a new "course content improvement program," a grant was awarded to the Physical Sciences Study Committee, a consortium of science professors and some teachers in the Cambridge, Massachusetts area, to design a new course for high school physics. Support for several other secondary-school curriculum projects followed quickly. Nevertheless, there was continuing and pronounced reluctance, as well as active opposition, to a move into developing science curriculum for elementary-schools.

*Stanford University.

A Love of Discovery: Science Education—The Second Career of Robert Karplus,
Edited by Robert G. Fuller, Kluwer Academic / Plenum Publishers, New York, 2002.

To shorten an interesting but tangential story, the NSF finally moved into the elementary-school science curriculum-development field in 1960, but only tentatively. Two modest, "experimental" grants were made. One went to Bob Karplus at the University of California at Berkeley. The other was awarded to me and a professor of astronomy at the University of Illinois. Bob had begun working with students in primary grades on concepts of force and motion. An astronomer-colleague and I had conceived of a project on astronomical understandings for students in grades 5 through 9.

Bob and I first met each other at one of the regular meetings convened by NSF of its principal investigators in the course content improvement program. As the only two people at the table with NSF-supported curriculum-development projects at the elementary-school level, we decided to join forces. I arranged to spend the autumn semester of the 1961–62 academic year on a sabbatical leave to work with Bob directly at his school sites in California. It turned out to be a lively time personally and professionally. It also afforded us the opportunity to frame the conceptual work that provides the subject of this chapter.

Some of Bob's first "subjects" had been and still were his own children. (At least the first three, as I recall, had entered elementary school by then.) He was in the regular habit of discussing science topics with them seriously in a range of family settings. It was quite an experience to watch Bob and his wife, Betty, engage the children in conversations about their views of the natural world. Probing and puzzling questions, and attempts to grapple with them, seemed to be a regular part of the dinnertime ritual. Bob and Betty were intent on learning the kinds of explanations that Bev, Peg, and the other children were able to devise to account for phenomena like the phases of the moon, magnetic attraction, or the processes of digestion. Though an already distinguished physicist, teaching clearly was a passion for Bob, a task at which both Betty and he excelled and that they clearly enjoyed—and whatever Bob learned from the experiences at home was regularly transferred to the elementary schools in which he and I had begun to work collaboratively. Even beyond teaching, however, Bob was avidly trying to comprehend the underlying principles that might guide children's thinking in science. This quest, it turns out, was never far from the center of his interests for the rest of his career.

For purposes of this chapter, it should be emphasized that both of us were part of an NSF-supported group of curriculum developers that emphasized course *content* improvement. Our starting point was to identify science concepts of considerable explanatory power: the ideas in science at the particular grade level of the student that had the greatest potential for deepening an understanding of the subject. This general orientation was a hallmark of all NSF-supported curriculum work at the time. In each of the course content improvement projects, the aim was to cut through the maze of disparate facts to identify the ideas that provided conceptual coherence to the subject, whether it be physics, or chemistry, or earth sciences, or any other facet of science. Consequently, it was essential for NSF that an outstanding scientist, a person who knew the subject deeply and well, be the principal investigator. (I, a professor of education, could be a co-director of an NSF curriculum project, but only if the other co-director was a scientist.) And there was little doubt that outstanding scientists were more likely to be involved in matters of education if financial support came from the NSF rather than from any other government agency.

This strong emphasis on content identification, however, was not the dominant one in elementary-school science at the time. The point of view that prevailed and that was attracting the most rapt and serious attention in the field was one that most valued individual investigation and discovery by the students. It was a period in which it was commonplace to talk, somewhat separately, about science *content* and about science *process*. By using appropriate processes, students could discover the major concepts that underlie the various fields of science, it was believed. While there is no necessary separation of process and content—indeed Bob and I claimed that it is impossible to isolate one from the other—it was not uncommon at the time, especially at the elementary-school level, to assume that it was the processes of science that should receive priority. And several of the active curriculum developers, distinguished scientists among them, believed that if more could not be done at the elementary level than engage students in rigorous inquiry, that was probably a satisfactory outcome. The specific science that was chosen for children to pursue was a lesser matter. Often it seemed to be selected casually.

As one example, there was a project in the late 1950s called Inquiry Training, based at my own institution at the time, the University of Illinois. I had a modest connection with it during its first year. Children were shown some intriguing phenomena, then asked to explain them. For example, a bimetallic strip was held in a flame, and the students observed how it bent. By asking questions, they were to come up with a scientific explanation. The exercise was a variation on "Twenty Questions." The objective for the student was to devise a line of questioning. The "training" centered on honing one's question-asking skills. Which questions are most productive? Why? Is there an optimal questioning strategy for a particular line of inquiry?

Entire curricula were developed in which the content to be learned was secondary. No less an organization than the American Association for the Advancement of Science launched a project called *Science—A Process Approach*. Young children were asked, for example, to watch corn popping. The objective, it must be emphasized, was not to understand the science associated with the phenomenon, but rather to develop children's skills at observation and hypothesis formation. What did they see? Were there differences among the individual kernels? What conjectures might the students advance to explain what they saw? However important such processes might be—and of course they are part of many scientific investigations—the belief was that children could learn such procedures in a somewhat isolated fashion *prior to* delving into the scientific principles associated with the phenomena. They could then apply these methods to any scientific investigation. Process outcomes alone would justify a curriculum at the elementary-school level.

Bob and I were skeptical on several grounds. First, committed as we were to helping even young children understand the underlying science, we saw no necessary conflict between learning science and learning something about how science is done. In fact, the ideas would be understood at a deeper level, we believed, if students came to know how they gained acceptance among scientists. Second, we entertained profound doubts about whether children could discover many, if not most, of the fundamental ideas of science. Both of us in our work with children had seen them incorporate their observations and conjectures about events in the world into their existing conceptions of how things work, even if those conceptions were at variance with current scientific understanding of the subject. Adults do the same, of course. When we see something new, we usually try to figure out how it may be an instance of something we already know. It is a sign of considerable intelligence, in fact, to be able to relate the strange and puzzling to the familiar.

It is self-evident to a six-year-old that an object moves only as long as a force is exerted. Stop exerting a force, and the object stops. Exert a steady force, and the object moves at a steady speed. The child has functioned perfectly well in accepting this view of the world. Almost all adults do, too. Of course, Galileo and Newton demonstrated centuries ago that these ideas are false. A twelve-year-old asked to draw a diagram to account for the retrograde motion of Mars as seen from earth often develops the idea of an epicycle. Almost always, the student devises an epicycle if asked to assume that Mars moves around earth. I always considered such description to be reasonable, and sometimes brilliant. It, too, is wrong. And students have even greater difficulty in accounting for retrograde motion if asked to assume that the sun is at the center of a system that includes both Mars and earth. "Discovery," even very creative discovery, does not necessarily comport with scientific understanding that matches current knowledge of the subject.

In the field of education, John Dewey put process elements of science and underlying conceptions of discovery at the center of his educational philosophy. His seminal work, *How We Think*, published in 1910, advanced the view that scientific methods of problem solving should be at the core of education in a democracy. His ideas were interpreted to mean, however, that students should first and foremost develop the abilities to define problems, hypothesize, test their hypotheses, collect data, and then draw and apply conclusions. They were also understood as suggesting that these abilities could be developed within virtually any investigation in which children might engage.

In 1938, in his *Experience and Education*, Dewey found it necessary to enter a caveat to those who professed to be developing programs based on his ideas and who, in his view, were misinterpreting the kind of progressive education he was advocating. In this short volume, he emphasized the importance of the *substance* of the investigation. For example, he said, "[T]he experimental method of science attaches more importance, not less, to ideas as ideas than do other methods. There is no such thing as experiment in the scientific sense unless action is directed by some leading idea." Then, two paragraphs later, "[E]xperiences in order to be educative must lead out into an expanding world of subject-matter, a subject-matter of facts or information and of ideas."

It seems a bit puzzling that 20 years later, in the late 1950s and early 1960s, there were educational theorists and researchers who failed to absorb a key aspect of science: concepts and processes are interconnected. Facts that are not developed within a compelling and powerful conceptual framework tend to be forgotten. Scientific "methods" that are not directed toward understanding foundational concepts are naïve about how science is actually conducted.

As he and I discussed such matters in the autumn of 1961, we began to distinguish between "invention" and "discovery." Children could profitably spend their time in science classes trying to discover important relationships, we postulated, but those inquiries would be more meaningful if associated with big ideas of considerable explanatory power, the development of which, after all, was the great triumph of modern science. There are countless facts, but scientists have identified certain over-arching concepts that give coherence to these facts, and thereby greatly simplify our understanding of the world. Those who understand these broad concepts could conduct serious investigations that deepen their meaning and extend their applicability. But it was unlikely that children could "discover" a heliocentric solar system, or the universal law of gravitation, or the theory of plate tectonics, or the means by which genetic information is transmitted. If the big idea were "invented" for them, we suggested, they could then engage in worthwhile and important discoveries that would help them to understand the concept at more profound levels. These ideas were expanded in the article Bob and I wrote for *The Science Teacher* that was published in 1962 and that is reprinted in this chapter.

It may be of more than passing interest that while Bob and I were writing the article he had a conversation about our work with a colleague of his at Berkeley at the time, Thomas Kuhn. Kuhn was deeply interested, talked with us at length, and gave us a copy of a manuscript he coincidentally had just completed, titled "The Nature of Scientific Revolutions." The book that was published from this manuscript several months later, now titled *The Structure of Scientific Revolutions*, became one of the seminal works in history and philosophy of science of the 20th Century.

Kuhn took the now-famous view that major advances in science often are somewhat discrete events associated with the creation of major conceptual frameworks that he called "paradigms." Copernicus' heliocentric solar system is such a paradigm, as is the universal law of gravitation that likened the falling apple to the falling moon. These transformations in human thinking, what Kuhn called "paradigm shifts," are deep, new insights into the nature of the world, and they occur infrequently. While much of science is an inductive process, what Kuhn called "normal science," occasionally there are breakthroughs that completely alter how humans understand their world. These new paradigms usually coexist with prior conceptions for a while, but gradually they overwhelm them. The replacement of the old with the new is not solely based on observed facts, however. To his dying day, Tycho rejected the Copernican paradigm, despite the fact that Tycho had the best observatory at the time and the most complete data. He simply could not detect parallax in the near stars, a necessary test of the theory that earth revolves around the sun against a relatively stationary background of stars; so Tycho continued to refine the Ptolemaic epicycles. Parallax was not detected, in fact, until the 1830s, by which time the heliocentric view was widely accepted—but on other grounds, primarily the universal law of gravitation.

The Kuhnian view of the advancement of scientific thought provided Bob and me with a powerful foundation for our views about discovery and invention. The most we could expect of students when studying any phenomenon is that they be Ptolemys with respect to solar-system astronomy and Aristotles with respect to understanding motion (and that would be high expectation, indeed!). Since we wanted to teach post-Aristotelian laws of motion and post-Ptolemaic views of the solar system to children, these conceptual frameworks had to be "invented" for them. Our debt to Kuhn was strong, and our reference to Kuhn's manuscript in "Discovery or Invention" may have been the first printed citation of this famous work.

One final word: Bob went on for many years to probe the nature of children's thought. He (and Betty) became deeply conversant with the work on children's conceptual development that was advanced by Jean Piaget and his colleagues in Switzerland. Throughout his exploration of psychological and other theories to explain how children learn science, however, Bob never took his eye off the target of helping people to understand how scientists see the world and what they see. In this realm, he was able to make contributions to science education of which few other people were capable. Deeply knowledgeable about science, he was able to observe and work with children with the mind of an investigator and theoretician. He was able to help us fathom how they think, the relationship between this set of mental and social processes and understanding the world of modern science, and the implications of all this for curriculum.

Discovery or Invention?

J. Myron Atkin and Robert Karplus

More recently, the discussion of the role of discovery in teaching has intensified. Many authors have stressed the great educational benefits to be derived if pupils discover concepts for themselves.[1] Other authors have warned that discovery teaching is so time consuming and inefficient that it should not, in general, replace expository teaching.[2]

There is a way in which autonomous recognition of relationships by the pupils; *i.e.*, "discovery" can and should be combined with expository introduction of concepts in an efficient program. This will produce understanding rather than rote verbalization. The approach can be described more clearly, if a historical example[3] is given of how a particular scientific concept is developed.

In ancient times the sun and the planets were observed by man. These observations gave rise to various conceptual interpretations. There were the mythological interpretations,

NOTE: The experiments in this article were carried out by the authors at the Berkwood School, Berkeley, California, through the cooperation of the Director, Betty Halpern. Financial support for the project was provided by the National Science Foundation, Washington, D.C.

[1] Jerome Bruner. *The Process of Education.* Harvard University Press, Cambridge, Massachusetts. 1960.
[2] David Ausubel. *Learning by Discovery: Rationale Mystique.* Bureau of Educational Research, University of Illinois, Urbana, Illinois. 1961.
[3] The historical development and its analysis in this illustration have been greatly oversimplified. For a fuller and more profound discussion see: Thomas Kuhn, "The Nature of Scientific Revolutions," University of Chicago Press, Chicago, Illinois. (In process.)

J. Myron Atkin, Professor of Education, University of Illinois, Urbana, Illinois.   Robert Karplus, Professor of Physics, University of California, Berkeley, California.

the interpretation as "celestial matter" with certain properties, and eventually the modern interpretation of planets orbiting around the sun. With the help of each of these concepts, man could attempt to understand other phenomena beside the ones that had led him to suggest the interpretation originally. These attempts, if successful, led to a reinforcement and refinement of the concept; if they failed, they revealed limits of the usefulness of the concept or even stimulated a search for a new concept. Of the three interpretations we have mentioned, the final and currently accepted one has turned out to be much more powerful than its predecessors.

In the development of a concept, it is useful to distinguish the original introduction of a new concept, which can be called invention, from the subsequent verification or extension of the concept's usefulness, which can be called discovery. Of course, this distinction is not completely clear-cut, because the inventor must recognize that the new concept is applicable to the phenomena he is trying to interpret; otherwise he would discard the invention immediately. Return therefore to the example for determining how the distinction can be applied. Assume that the deities, the celestial matter, and the solar system were inventions. In the mythological framework, one could then discover that the deities intervened in human affairs in certain ways and refine one's idea of the characteristics of the gods. In the framework based on the existence of celestial matter, one could discover that celestial objects move in cycles and epicycles. Finally, in the framework of the solar system, one could discover additional planets.

Undoubtedly, an invention is not complete and static, but it is the germ of a concept that is developed to greater significance by the subsequent discoveries. When an invention is made, its full significance is not evident. Still, the concept must be introduced and the invention must be made, if it is to grow in meaning.

Applying this distinction between discovery and invention to science teaching, acknowledge the fact that the pupil has experience both before he enters school and also outside the school environment during the school years. He therefore makes observations all the time, and he invents concepts that interpret the observations as well. He also makes discoveries that enable him to refine his concepts. Most of the discoveries and inventions reveal a type of natural philosophy—a "common-sense" orientation popular in the culture at a given point in history.

Yet, the objective of the science program is to teach children to look at natural phenomena from the distinctive vantage point of modern science. And in the mid-twentieth century, this vantage point differs from the culturally prevalent view. In a small way the situation is analogous to that of a Copernican teacher instructing his students that the sun is at the center of the solar system while almost everyone else in the society *knows* that the earth is at the center of the universe.

In general, no results are evident if a teaching program is based on the expectation that children can invent the modern scientific concepts, because their spontaneously invented concepts, some of which even exist at the time the child enters school, present too much of a block. After all, concepts were developed to interpret their experience; why should they change these concepts on their own? Indeed, it does not seem crucial to teach the children to invent concepts, because they can and do invent concepts readily. The educational problem, rather, is to teach the children to carry out their creative thinking with some intellectual discipline. And the development and refinement of modern scientific concepts in the light of observations would seem to be one excellent vehicle for achieving this goal.

If the children are not able to *invent* the modern scientific concepts, it is necessary for the teacher to *introduce* the modern scientific concepts. During this introduction, the teacher must make clear which previous observations of the children can be interpreted (or perhaps reinterpreted) by using a concept. Further, he must follow the introduction with opportunities for the children to discover that new observations can also be interpreted by using a concept. This type of discovery is made possible by the availability of a concept to the children, because their perception is oriented by the teacher's formulation of the new idea. This type of discovery can be extremely valuable to solidify learning and motivate the children; it is essential, if a concept is to be used with increasing refinement and precision. Categorically, the teacher must not present the concept in a complete, definitive, and authoritarian way, for concepts are never final.

As an example of this teaching approach, a thirty-minute lesson will be described in which second graders discovered the usefulness of the magnetic field concept after the teacher had invented it for them. In thirteen previous sessions, a class of fifteen pupils had discussed the selection of systems by the specification of the objects in the system, and the existence of interactions among the objects, and had been introduced to the notion of the free energy of the system.

Two new concepts were introduced to the class in the lesson: the interaction-at-a-distance, and the magnetic field. These ideas were developed through a series of experiments.

Experiment 1. Two boys pulled on a rope in opposite directions. The pupils identified the system of interest as consisting of Bruce, James, and the rope. Interactions between objects in the system included the one between James and the rope and the one between Bruce and the rope. Bruce and James were considered not to interact with each other. *Invention of interaction-at-a-distance:* Next, the teacher pointed out that Bruce and James were really the important objects in the system; that the whole class could think that there was an interaction between the two boys (at a signal, Bruce yanked James with the rope); but that it was not a *direct interaction*, it was a *distant interaction*. The new term was stressed. The rope made the distant interaction possible. The teacher further asked Bruce and James to interact strongly, then weakly, then strongly again.

Experiment 2. The teacher produced two wooden balls that were held together by a strip of rubber tacked to the balls. Five objects were identified in the system: the two balls, the rubber strip, and the two thumbtacks. The pupils identified the direct interactions ball-rubber, ball-thumbtack, and rubber-thumbtack. They identified the distant interactions ball-ball, thumbtack-thumbtack, and end of rubber-end of rubber. The pupils called the interaction weak when the balls were close together, strong when they were far apart, and medium when they were somewhat separated. The strip of rubber made the distant interaction possible.

Experiment 3. The teacher put his hand on the head of one boy. The pupils correctly identified the direct interaction, teacher-boy, and the distant acoustic interactions, teacher-all pupils.

Experiment 4. The teacher repeated Experiment 2 with a long brass spring. The distant interaction between the ends of the spring and the strength of the interaction were identified. The spring made possible the distant interaction between its own ends.

Experiment 5. The teacher produced two large U-magnets mounted to attract one another on roller skates so that they could move easily. The pupils identified the distant interaction between the two magnets in the system. They also determined that the strength of the interaction decreased as the magnets were separated. The interaction was sufficiently strong for the magnets to roll toward one another at a separation of four inches. *Invention of magnetic field:* In response to the teacher's question, "Which do you like better, direct interactions or distant interactions?" The pupils expressed a strong preference for direct interactions. The teacher now told the pupils that most people prefer to think in terms of direct interactions. In the earlier experiments there had been something between the two objects that made possible the distant interaction between them. Was there something now between the two magnets that made possible the distant interaction between them? There was nothing visible. (Curiously enough, no pupil suggested that the air was involved.) Even

The students experiment with two large magnets mounted on roller skates so as to move easily. Identity of distant reaction between the two magnets in the system is established.

though the magetic field was not yet mentioned by name, the children were given the crucial idea of a mediator (an "it") for the distant magnetic interaction. This step constituted what we have called the "invention."

Experiment 6. Discovery of the significance of the magnetic field:

a. Three children came to the demonstration table to find "it" by feeling with their fingers. They did not find "it."

b. Two children came to the demonstration table to find "it" with a wooden ruler. They did not find "it."

c. One child came to find "it" with a nail held at the end of a piece of wire. The nail responded to something!

d. All pupils wanted to explore with the nail. As others had an opportunity to do so, the teacher verbally confirmed the fact that the nail indeed seemed to have responded to something. When the teacher was going to name it, he was

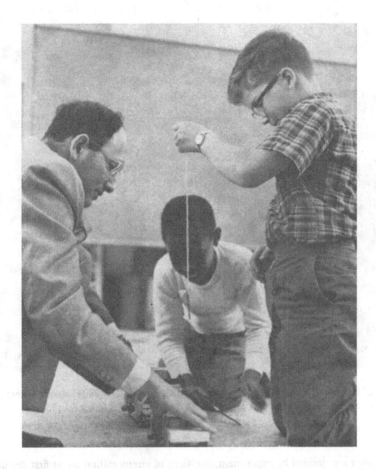

As a nail is held above the magnets mounted on the roller skates, the students note that a reaction was produced in the nail from something.

interrupted by one pupil who said "I know what 'it' is called. 'It' is a magnetic!" The teacher agreed to the "magnetic," because it occurred in a magnetic interaction, but proposed the name "magnetic field." Thereafter, the discussion was carried out in terms of exploring to find the magnetic field—the "it" which made possible the distant interaction between the two magnets.

e. Several more pupils explored and found the magnetic field with the nail.

f. As a final step of this experiment, the teacher invited the pupils to find the magnetic field with other objects. Screws, paper, paper clips, a screw driver, jewelry, and coins were used. Some of these objects responded to the magnetic field, some did not.

It is necessary to point out here that the appeal to the children's intuitive preference that was significant in the invention of the field concept was not at all unscientific, but was the necessary first step in the adoption of a new concept. While questions of scientific

During a classroom discussion of equilibrium and the approach to equilibrium of a system placed in a new environment, the students are guided to discovery by Professor Karplus (r.).

observation are decided by experiment, questions of interpretation are at first decided by preference in the light of past experience and later by the usefulness of the interpretation in generating discoveries. The magnetic field is a useful *invention*, but it is not essential to describe magnetic interactions.[4] Without the invention of the magnetic field, the subsequent explorations with the nail, etc., would have resulted in the discovery of additional distant interactions between the magnets and the nail or the other objects.

There is one feature of the preparatory Experiments 1–4 which should be emphasized. These experiments had been carried out earlier in a somewhat different way, but the distant interactions and the strength of interaction were newly introduced in this lesson. The sequence in which these experiments are carried out is not important. The pupils are not led step-by-step to the magnetic field concept. Rather, they are; "led in a circle around the magnetic field concept so they may then converge on the center of the circle from several directions." This strategy offers more promise of success.

Now, the lesson described must be placed in a science course. It is essential that the discussion of magnetic fields not be terminated and wrapped-up with the discovery

[4] J. A. Wheeler and Richard P. Feynman. "Classical Electrodynamics in Terms of Direct Interparticle Action." *Reviews of Modern Physics*, 21:425. 1949.

described. Instead, this discovery should in itself be part of a strategy of attack on another more profound concept. In teaching the second grade, the concept of energy is the next higher order of understanding. Springs, rubber bands, dry cells, candles, and air, all these had been introduced earlier. Now magnets are seen as systems in which energy can be stored. The energy concept, in turn, is part of the strategy being developed for teaching about interactions among the objects in a system, a still higher order concept.

The pedagogical point to be stressed in conclusion is that this type of discovery teaching appears to be strongly motivating and rewarding. Yet, the teaching seems also to be reasonably efficient even when compared with a more verbal expository approach. The pupils come to the point where they know they will discover something, and they know what their discovery will mean. Hence, perhaps they did not invent the new concepts, but they did make discoveries.

Response by the Oersted Medalist: Autonomy and Input

Robert Karplus

I believe that people become vitally interested in their studies and learn best when direction and guidance from a source of authority are combined with ample opportunities for the students to direct and control their own learning. The former may be called instructional input, the latter student autonomy. Hence the phrase "autonomy and input" characterizes a teaching program that balances these two factors. Yet one must also consider the relation between these two elements. They may be uncoordinated as in a physics course with lectures (input) and concurrent but independent open laboratories (student autonomy). Alternatively, the course might provide for more interaction between the two elements when the lectures (input) respond to questions or issues raised during previous investigations (autonomy) and follow-up investigations provide for applications of the new material.

Thank you for your generous words. I will begin by also expressing my appreciation to the University of California, which has encouraged my work in the teaching of physics and other sciences through the Department of Physics, the Lawrence Hall of Science, and the Group in Science and Mathematics Education. I am further happy to acknowledge the wise counsel of my wife Elizabeth as well as the stimulation provided by our seven children, who were the often eager and occasionally reluctant subjects of my teaching inventions. Financial support for much of the work came from the National Science Foundation's Course Content Improvement Program and its successors in the present Science Education Directorate.

Robert Karplus, Department of Physics and Lawrence Hall of Science, University of California, Berkeley, California 94720.

Reprinted with permission from Karplus, R., "Response by the Oersted Medallist: Autonomy and Input," *American Journal of Physics* **49**(9), pp. 811–814. Copyright, American Institute of Physics [for *American Journal of Physics*], 1981.

Richard E. Paulsen was the officer of the Foundation who reviewed my initial proposals carefully and ultimately recommended approval of the development program of the Science Curriculum Improvement Study as designed with Herbert D. Thier and Chester A. Lawson.

When I look back over my work in teaching and learning during the past 20 years, I find a pervasive idea, which has potential applicability to the teaching of any subject that requires understanding of concepts and principles. I will come to this idea through two examples.

Suppose you were a high school physics teacher or that you were assisting a high school teacher beginning a course's section on uniformly accelerated motion. How would you begin?

(a) Presenting a film that shows carts rolling down an inclined plane and golf balls falling freely (slow-motion photography), together with stroboscopic photographs, timing tapes, and graphs of speed and distance versus time?

(b) Arranging for a laboratory period in which your students would use level or inclined air tracks, springs, weights, falling golf balls, timing tapes, flexible ramps, polaroid cameras with strobe lights, and stop watches to observe accelerated motion under various conditions of their own choosing?

(c) Discussing with your students their own experiences with accelerated motion in automobiles, buses, airplanes, or elevators, and analyzing observations?

(d) Presenting an explanation beginning with Newton's second law and deriving from it the relation between the constant acceleration, velocity, and distance moved by a particle subject to a constant force?

(e) Providing a laboratory where students carefully measure the distance intervals on a timer tape produced by a glider sliding down a slightly inclined air track, then graph the measurements and relate the slope of the graph to the angle of inclination of the track?

Certainly the preparation of your students and the materials available to you will influence your choice. Compare my comments regarding each of the five options with your ideas.

(a) Films are popular ways of introducing a new topic. In this case, the film's content is similar to what students might observe in a laboratory of type (b) or (e) if they worked very conscientiously and carefully with good equipment. I would recommend the film be used *after* a laboratory period if laboratory materials are available. Films are less effective in raising questions, provoking inquiry, or presenting contradictions than are first-hand laboratory experiences. Since paying attention to the film preempts their initiative, few students watching a film for the first time think critically about what they observe. Additionally, seeing a picture of an object or process does not carry the impact of handling the object or influencing the process oneself.

(b) I would highly recommend an approach of this kind, where the students have a great deal of freedom to use their own judgment, try out their own ideas, and learn from their own mistakes as they gain practical experience with the objects they will study theoretically later. The teacher can evaluate the students' reasoning and later provide more direction or extend the autonomous investigations as needed.

(c) Even though this procedure appeals to the students' direct experience, the occurrence of uniformly accelerated motion is hard to isolate from the everyday events listed. A jet accelerating to take-off provides the best sample I can think of. Yet it is very difficult for the passenger in the jet to be aware of the distance covered. While this discussion may be appropriate when motion in general is taken up, I consider it unsuitable at this time.

(d) This rather theoretical approach, often in the form of a lecture, would be completely inappropriate as the introduction to a new topic, because it takes for granted that the students have a good grasp of the distance, speed, acceleration, and force concepts. Furthermore, it depends on algebraic relationships that can only be understood through relatively abstract reasoning patterns.

(e) This type of laboratory prevents the students from asking their own questions and taking responsibility for satisfying their own curiosity. The reason for graphing the distances will not be clear at this time either. Such a laboratory would be more appropriate later in the instructional sequence, but even then it might better focus on sources of error or other deviations from uniform acceleration rather than on merely verifying the linear relationship of speed and elapsed time.

Two of the options you just considered were laboratory activities. I explained why I thought they were substantially different from one another and why (b) was preferable to (e) to introduce the topic of uniformly accelerated motion.

To compare various laboratory alternatives in more detail, put yourself into the position of teaching about periodic motion and the pendulum to the same group of students at some time after the introduction of dynamics. Rank the following four alternatives in terms of how they will encourage students to conceptualize the phenomena, properties, and regularities of periodic and simple harmonic motion.

(1) Provide the students with a mass on a string. Indicate the variables of the system and suggest that they verify the square-root relationship between the length of the string and the period of oscillation.

(2) Provide the students with a mass on a string. Supply a list of possible variables of the system (angle of swing, mass, length of string, acceleration of gravity, the period of oscillation, etc.). Supply a list of possible relationships between variables. (The period of oscillation is directly proportional to the mass, the period is directly proportional to the length of string, etc.) Ask the students to identify the relationship fitting the data best.

(3) Provide the students with a variety of periodic systems, e.g., a cork floating on water, a baseball bat swinging by a hole in its handle, a clock pendulum, a mass on a string, a uniform metal rod with pivot holes in it. Ask the students to identify variables of these systems and to search for quantitative relationships between the variables.

(4) Provide the students with a mass on a string. Indicate that for small angles of oscillation there is a relationship between the length of the string and the period of oscillation. Challenge them to discover the relationship based upon their data and then compute the length of string required for a 10-sec period.

Since the students will be expected to recognize periodic motion and identify relevant variables and relationships, alternatives (2) and (3) are superior to (1) and (4). Furthermore, (3) is somewhat more open and therefore will encourage students to investigate aspects of periodic motion that relate to their own preconceptions without subordinating their reasoning to what they think the instructor expects them to say. Conversely, alternative (1) encourages students to follow directions and perhaps even to select their data carefully to make sure they fit the expected outcomes. Their strategy will be aimed more at getting a good grade rather than at understanding the periodic phenomena.

In both examples—uniformly accelerated motion and periodic motion—the options I described ranged from directed and informational to open ended and exploratory. Much control of the content of the open-ended activities lies with the students, while content control of the directed or expository activities lies with the instructor, film, or laboratory manual. I like to use the terms *autonomy* for student control and *input* for the content of teacher controlled activities.

Input was my original concern as a physics teacher, as it appears to be for most of my colleagues. Then, about 1960, the observations and theories of Jean Piaget called to my attention the potential value of student autonomy in learning at all levels. Subsequently I recognized that the coordination of autonomy and input was a crucial factor in the planning of physics courses, individual laboratory assignments, lectures, and other learning situations.

In an autonomous situation, students learn through their own actions and reactions to a new situation. They have to form explicit or implicit hypotheses, then test these through planned trials and careful observation of many factors since they do not know in advance what effects will be important. Students are likely to be guided by their misconceptions, if any, and they will get data that call these misconceptions into question. Finally, students get a sense of control and responsibility that will encourage them to take more intellectual initiative in their studies and investigations in the future.

Please do not identify autonomy exclusively with laboratory activities. Theoretical investigations involving thought experiments can also be carried out autonomously and lead to thought-provoking results. The following problem provides an example appropriate to introductory mechanics: "A car of mass m is driving at the speed v_0. Find the distance it has traveled after a time interval T under various conditions of constancy of one or more physical variables. If, for instance, you take the speed as constant, the distance will be $v_0 T$. Get as many different results as you can. If several conditions lead to the same result, explain the equality in terms of applicable physical principles."

As another example of autonomy, consider the early sections of this article. Asking you to consider the five ways to begin teaching and the four versions of laboratory activities was my appeal to your teaching experience and educational judgment intended to stimulate autonomous reflection.

After these examples, I introduced the notions of autonomy and input in an educational setting. My proposing these terms to emphasize the input–autonomy contrast and their possible coordination was an example of educational input in this article. The contrasting and complementary roles of autonomy and input form the theme I had chosen for this presentation. Other examples of input may be suggested in conjunction with the high-autonomy laboratories I have described. For instance, appropriate input after the laboratory (b) dealing with accelerated motion could present the concept of acceleration in terms of the second differences of displacements during equal time intervals. After the periodic motion laboratory (3), appropriate input could present the notions of frequency and amplitude.

Note that I have not only distinguished between input and autonomy in teaching and learning, I have also suggested that activities with these two emphases should be carried out in close conjunction with one another. In other words, input introduces the concepts and relationships of the subject matter, to complement the data base that is acquired by students through autonomous activities. The concepts of vectors and force, Newton's laws of motion, and the laws of energy and momentum conservation are all appropriate instructional input in an introductory mechanics course. Students should be encouraged to speculate about the significance and interpretation of the observations they make in a laboratory or the procedures they need to solve their problem assignments. However, expecting them to introduce the complex ideas of Newtonian or other branches of physics is unrealistic. Yet they

Table I. Approaches to teaching.

Approach	Procedures	Goals
High input	Lectures	Accuracy
	Textbooks	Completeness
	Cookbook labs	
	Drill, practice	
	Associative	
High autonomy	Rich environment	Originality
	Open laboratories	Self-realization
	Libraries	
	Bull sessions	
	Discovery	
Interaction	Discussions	Consistency
	Explanations, reasons	
	Investigations	
	Open-ended problems	
	Feedback loop	

need to weigh evidence, formulate hypotheses, select procedures for laboratory investigations or problem solutions, and evaluate their own results to form a basis for understanding the new ideas.

What I am suggesting is a teaching program in which there is a great deal of interaction and feedback between students and teacher. The students make observations and raise questions to which the teacher provides feedback. The teacher asks questions, suggests investigations, defines terms, and gives explanations to which students provide feedback by responding in appropriate or inappropriate ways. The introduction of force, energy, momentum, or other new concepts makes use of the students' prior experience and leads, in turn, to new questions and theoretical or practical investigations. Similarly, each investigation applies previously introduced concepts and leads to new questions that require the definition or identification of still other concepts.

To summarize my remarks, I present in Table I three types of teaching programs in idealized form. The first, high input, is characterized by lectures, textbook reading, cookbook labs, and a great deal of drill and practice with the concepts and techniques that have been introduced. Student work is evaluated according to its accuracy and completeness—have all problems been solved? Have all factors been taken into account? Are all answers correct (i.e., matching the teacher's answers)? The psychologist of learning calls such an approach associative, in that certain words, formulas, steps, and results are associated with certain other words, principles, and procedures. Most introductory physics courses take this approach.

The second approach, high autonomy, is characterized in this idealization by the students' use of a rich learning environment that may include open laboratories, libraries with readings and media materials, and bull sessions for comparing ideas or charting new approaches. Student work is evaluated according to its originality and contribution to a student's own understanding. The psychologist of learning calls this a discovery format, in that the students' are discovering certain facts, ideas, and relationships new to them, without regard to whether they are, in fact, new to the physics community or even consistent with other physical principles the students may not know. Some independent study courses may take this approach, but I doubt that it is widespread.

The third approach, interaction, follows the procedures I have described above. It is characterized by discussions and explanations making reference to the students'

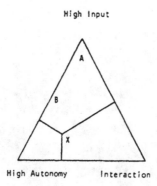

Figure 1. Isometric description of teaching programs.

observations and preconceptions as well as to text readings and other sources. Open-ended problems are used to stimulate thought and evaluate the consistency of students' understanding. Learning psychologists rarely consider this approach, which has been called developmental as it is concerned with the development of students' prior understanding to new levels of insight. Individual student research projects leading to a thesis often make use of this approach with its linked feedback loops centered on the student and the instructor.

As I have pointed out, the three teaching approaches have been idealized. I would expect any actual teaching program to incorporate aspects of all three in varying degrees, depending on the purposes of instruction, the students' background, and the time, staff, and materials available.

One way of representing the relationship of a real course to the three types is to represent the course by a point X in an equilateral triangle with vertices labeled by the three types (Figure 1). The point's isometric coordinates then describe the degree to which the course includes input, autonomy, and interaction. The introductory course to which I referred would cluster in region A near the high-input vertex, while a traditional course with a concurrent open laboratory could be placed in region B along the input–autonomy edge. For the course to move off this edge, some ongoing evaluation of student learning and appropriate feedback would have to be included, as may happen in individualized instruction programs. Perhaps you have taught or know of a course that could be represented by a point near the center of the triangle, in which input, autonomy, and interaction are balanced. Yet perhaps such courses are too expensive in their demand for staff time and learning materials.

Nevertheless, faculty groups at several colleges and in many secondary schools are working on special programs that avoid the input vertex. Most of these groups include representatives from several disciplines, in an effort to provide their students with an approach that is consistent from one course to the next. I invite you to inform yourself about these programs[1-3] and to apply some of their ideas to your own teaching.

[1] ADAPT, *Multidisciplinary Piagetian-based Programs for College Freshman* (University of Nebraska, Lincoln, 1977).

[2] F. Collea, R. Fuller, R. Karplus, L. Paldy, and J. Renner, *Workshop on Physics Teaching and the Development of Reasoning* (American Association of Physics Teachers, Stony Brook, NY, 1975).

[3] R. Karplus, A. E. Lawson, W. T. Wollman, M. Appel, R. Bernoff, A. Howe, J. Rusch, and F. Sullivan. *Workshop on Science Teaching and the Development of Reasoning* (Lawrence Hall of Science, Berkeley, CA, 1977).

Elizabeth Frazier and Robert Karplus
December 1948

CHAPTER 6

Central Role of Students' Reasoning

Helen Adi Khoury*

In this chapter, the focus of Robert Karplus's work on students' reasoning will be revisited. The emphasis on exploring and understanding students' reasoning in multiplicative conceptual domains, beyond the elementary school, is central in the mathematics and science education-related work of Robert Karplus. About 30 years ago, at a time when the emphasis in mathematics and science education research was on students' achievement test scores and related statistical group averages, Robert Karplus launched a sequence of studies focusing on identifying and classifying students' reasoning as they attempted to solve a variety of well-designed problems (or at times they were referred to as puzzles) that purposefully embedded key mathematical concepts such as proportions, control of variables, probability, correlations and others. It is the quality of students' thinking and how they had reasoned that Karplus and his colleagues focused on, rather than the correct/incorrect answers students may have given.

It is the intellectual development of the individual student that was valued and was at the center of the research, teaching, and teacher development efforts of Robert Karplus and his colleagues. More than two decades

*Department of Mathematical Sciences, Northern Illinois University, DeKalb, Illinois 60115.

A Love of Discovery: Science Education—The Second Career of Robert Karplus,
Edited by Robert G. Fuller, Kluwer Academic / Plenum Publishers, New York, 2002.

later, the calls for reform, in education (e.g., National Council of Teachers of Mathematics, 1989, 2000) placed students at the center of teaching and of research. Currently, many mathematics and science educators are now focusing on understanding students' multiplicative thinking, beyond the elementary school level, and they are using qualitative research methodologies to do so. Robert Karplus and his colleagues assumed these challenges early on in the 1970's, and their pioneering work which extended over a span of more than ten years set an example of how a student-centered constructivist perspective may be applied in both teaching, teacher development, and research situations.

With a deep knowledge-base of the work of Jean Piaget on the development of thought among young adolescents, coupled with a strong desire and commitment to help understand and improve students' reasoning in science and mathematics, Robert Karplus and his colleagues believed both in the merits of qualitative-based research and in the significance of including classroom teachers in the process of research to improve students' learning. In a paper presented in 1980, Elizabeth Karplus described the significance and the challenges of this avant-guarde action-research perspective. She said, "For the practicing teacher, every classroom can be a research laboratory in which new data about individual students' learning . . . be analyzed daily. This research must be carried out under conditions that would stun a scientist or a mathematician: the vast number of variables, the lack of controls, the many subjects to be observed at the same time, the variations among the observers' knowledge and sensitivities, are combined with the 'noise' of random careless errors and slips . . . that occur frequently." To help teachers observe and assess their students' reasoning, and to help teachers learn how to act upon their observations in order to enhance their students' mathematical and scientific reasoning, R. Karplus and his colleagues (1977) developed and conducted a sequence of five inservice teacher-development programs titled, "Science teaching and the development of reasoning."

Robert Karplus and his colleagues explored students' reasoning both in the United States and in six other countries (see Karplus, Karplus, Formisano, and Paulsen, 1977). In comparing the reasoning levels of about 2,000 students from seven different countries, the commonalities among students' reasoning patterns were of interest much more than the differences. It is also interesting to note that R. Karplus's interest in the global individual student was not limited to the cognitive aspect only. He valued, respected, and was sensitive about the opinions and attitudes of the students with whom he worked. In a research report prepared in the late 70's, the following question: "What are students' opinions concerning -our- (sic) research in their classrooms?" was raised and analyzed (Karplus, R., and Karplus, B. 1979). One of the conclusions confirmed that, ". . . it is fortunate that many young people responded positively to such an approach—an approach where students are required to explain or justify their answers and conclusions—even in a test situation" (Karplus and Karplus, 1979, p. 6).

In most of the Intellectual Development studies of Robert Karplus, the problems were administered in both group or individual interview settings. A common observation was consistently made by Karplus and his co-authors in favor of the one-on-one interview settings in order to get a sharper reflection of students' true reasoning patterns. It was also reported that by probing students' thinking during the interviews, students were able to operate in general at higher intellectual levels than what they would have performed individually in group paper-and-pencil settings without any interaction. The implication was often made by Robert Karplus that one-on-one individual interviewing of students during problem solving is the preferred approach not only as a methodological tool for research purposes, but also as an educational tool to advance students' thinking.

In the sections that follow, an introduction and discussion of some of the Karplus studies on students' intellectual development beyond the elementary school will be presented. These studies focused on students' reasoning in multiplicative situations. In these studies, students' control of variables, proportional, probabilistic, correlational and logical reasoning were explored by identifying reasoning patterns on related tasks. These concepts were selected by Karplus and colleagues because of the conceptual interrelationship among them and because higher levels of reasoning on these concepts may reflect formal abstract thinking beyond the elementary school level. A brief discussion of the possible impact of the findings of these studies on our current teaching, research, and teacher development practices will also be included.

CONTROL OF VARIABLES

To identify and to control variables in problem-solving situations is an important form of thinking in mathematics as well as in science. With the recent calls to promote algebraic thinking among all young adolescents in our mathematics classrooms (Silver, 1997), it becomes very important for teachers to understand how students identify, assign labels to, and control variables in various problem-solving situations. Based on Piaget's work, Karplus and colleagues designed interesting problems using physical objects to study this capability. These and similar other problems were used later on by many teachers, researchers, and teacher educators (e.g. Monk, 1992; Hines, 1998). For further details on students' reasoning patterns in control of variables situations, please read Karplus, Karplus, Formisano, and Paulsen (1977) which is reprinted in this chapter.

PROPORTIONAL REASONING

Proportional reasoning is essential for students' meaningful understanding of many mathematics and science concepts, beyond the elementary school. Many researchers and teachers have been studying students' problem-solving strategies and levels of thinking in proportion situations for at least the last 30 years. During the last 10 years, several research studies and books were published on students' development of multiplicative reasoning in general, and on proportional and rational number reasoning in specific (e.g., Behr, Harel, Post, and Lesh, 1992; Harel and Confrey, 1994). Robert Karplus and his colleagues (Karplus, Adi, and Lawson, 1980; Karplus and Karplus, 1970, 1972; Karplus, Karplus, Formisano, and Paulson, 1977; Karplus, Karplus, and Wollman, 1974; Karplus and Peterson, 1970; 1972,1974, 1977, 1980) were the pioneers among this group of researchers and teachers.

Many problems were designed and used by Karplus and colleagues to investigate proportional reasoning. Also several variations of the same problem were at times used. For example, the well-known Mr. Tall/Mr. Short problem is such a problem. Robert Karplus and his colleagues designed and used Mr. Tall/Mr. Short (Paper Clips problem, Mrs. Jones and Jim problem) to assess and to enhance students' levels of proportional thinking. Since then, this problem, or one of its modifications, has been used by many other researchers, teachers, teacher educators, and test developers during the last 20 years (e.g., Billings,1998; Hart, Brown, Kerslake, Kuchemann, and Ruddock, 1985; Khoury, 1997; Thrun, 1999). At least four different levels of proportional reasoning were identified. These levels included the illogical, the additive, the transitional, and the multiplicative ratio levels. Please read the reprinted studies for further details on the identified proportional reasoning patterns and on the role of the contextual problem variables as well as the role of students' cognitive styles in proportional thinking.

PROBABILISTIC AND CORRELATIONAL REASONING

Patterns of probabilistic and correlational reasoning were identified in at least two of the studies reprinted in this chapter (see Adi, Karplus, Lawson, and Pulos, 1978; Karplus, Adi, and Lawson, 1980). Again the levels of thinking varied from the illogical, the additive to the multiplicative. It was noticed that many adolescent students ignored significant given quantitative information in making their decisions about the probabilistic or the correlational situations.

As students looked for confirming evidence for a high correlation between two attributes, many tended to look only for cases were both attributes were present. Only students who were thinking multiplicatively in correlational situations tended to include cases where both attributes were absent and then tended to compare both of these groups multiplicatively to the remaining cases.

More recently, the teaching of probability and statistics concepts has been recommended into the mathematics school curriculum by the National Council of Teachers of Mathematics, but still many teachers tend to exclude teaching these concepts in their classrooms. Recently published research studies on students' understanding of probability and correlations are still isolated and not well-integrated. The reported findings of these recent studies are very similar to the contributions of R. Karplus and colleagues in this domain, and at times the earlier work is not acknowledged. We have not seen yet recent studies that have significantly emerged beyond the earlier findings on students' probabilistic and correlational reasoning patterns.

LOGICAL REASONING

Two of the articles reprinted in this chapter refer to students' logical reasoning (Adi, Karplus, and Lawson, 1980; Karplus and Karplus, 1970). In the first article in the Intellectual Development studies by Karplus and Karplus (1970), the authors analyzed students' logical reasoning (application of a transitive relationship) on the Islands Problem. In this study, they reported that intellectual development in abstract reasoning at the high school level reached a "plateau," and that this plateau was "disappointingly" low. Following this study, Robert Karplus published many other studies in the Intellectual Development series to explore this observation further in other conceptual domains. In most of the later studies in the Intellectual Development studies sequence, an attempt was made to include students from the 6th grade level through the early college level.

In 1980, a study was published by Karplus and colleagues (Adi, Karplus, and Lawson, 1980, which is reprinted in this chapter) that compared the conditional logical abilities (both performance and reasoning) of about 500 adolescents and young adults on the Four-Card problem. This study is interesting for several reasons:

First, it was a study that provided lots of evidence to show that students may give correct answers in logical situations for the invalid reasoning. Thus, the implication was that as educators, teachers, and researchers, a focus was needed on students' justifications and reasoning patterns and not only on their correct/incorrect answers as was the practice in mathematics education at the time.

Second, this qualitative-based research was written for the eyes of the mathematics education researchers to read, at a time when the application of quantitative parametric statistics was the acceptable research paradigm to use. In the concluding remarks, it was affirmatively pointed out that "Although no statistical hypotheses were tested in the present study, many such hypotheses could be generated based on the present report." The process of constructing new research hypotheses based on observations, and linked to a significant theoretical or educational framework, was highly valued by Karplus and his colleagues.

Third, various logical reasoning patterns were identified that helped many teachers and mathematics educators understand some of the cognitive dilemmas students experience as they try to prove mathematical statements. It seems that many young adolescents do not discriminate between the roles and meanings of a hypothetical statement versus a declarative descriptive statement. They perceive them as having them the same meaning. For example, when asked to check for the validity of a hypothetical statement of the form *If p, then q*, many students do not appreciate what the question is asking them to do, because to them they think that *If p, then q*, means the same thing as asserting *p and q*. They tend to accept the hypothetical statement as a true statement confirming both propositions. To them, they assume that a hypothetical statement is like a rule, and rules are written in society to be accepted as true. This finding explains why we observe in geometry classrooms, for example, students who during the process of proving a statement, they go ahead and apply the very same statement as if it were true, the very relationship that they were trying originally to prove.

It needs to be noted that in exploring the logical reasoning of students, Karplus and his colleagues (1980) were not attempting to suggest that formal thought culminates in propositional logic or in the INRC group as was suggested earlier by Jean Piaget. Instead, in exploring students' reasoning patterns in key mathematical and scientific concepts, such as in logical, proportional, probabilistic, and correlational situations, Karplus and his colleagues were able to show, in their intellectual development studies, that formal levels of thinking develop much slower than what was anticipated by Piaget's theory. They also showed how these formal reasoning levels across concepts did not co-develop similarly, at the empirical level. They identified many intervening variables that may have an impact on students' development of reasoning. By increasing the awareness levels of many educators nationally and internationally to the development of students' reasoning and its relation to classroom teaching, Robert Karplus was delighted to have witnessed, about 20 years ago, that at least some secondary school and college teachers at that time were starting to "provide their students with more opportunities to construct knowledge—and to reflect at their own thinking (sic)—rather than merely to accept and learn hypotheses and conclusions formulated by others" (Karplus, 1978, p. 8). Looking back, the impact of the work of Robert Karplus on students' reasoning and the related messages we received from his work did carry us to our current research and reform-based mathematics and science classrooms of today.

REFERENCES

Adi, H., Karplus, R., Lawson, A., and Pulos, S., 1978, Intellectual development beyond elementary school VI: Correlational reasoning. *School Science and Mathematics*, 78(8):675–683.

Adi, H., Karplus, R., and Lawson, A., 1980, Conditional logic abilities on the four-card problem: Assessment of behavioral and reasoning performances. *Educational Studies in Mathematics*, 11:479–496.

Behr, M., Harel, G., Post, T., and Lesh, R., 1992, Rational number, ratio, and proportion. In *Handbook of Research in Mathematics Teaching and Learning*, D. Grouws, ed., Macmillan Publishing Co., New York.

Billings, E., 1998, *Qualitative-based Reasoning of Preservice Elementary School Teachers in Proportional Situations*, Unpublished doctoral dissertation, Northern Illinois University, DeKalb, Illinois.

Harel, G., and Confrey, J. (Eds.), 1994, *The development of multiplicative reasoning in the learning of mathematics*, SUNY Press, New York.

Hart, K., Brown, M., Kerslake, D., Kuchemann, D., and Ruddock, G., 1985, *Chelsea Diagnostic Mathematics Tests: Ratio and Proportion*, NFER-NELSON, Berkshire, Great Britain.

Hines, E., 1998, *Analysis of Processes Used by Middle School Students to Interpret Functions Embedded in Dynamic Physical Models and Represented in Tables, Equations, and Graphs*, Unpublished doctoral dissertation, Northern Illinois University, DeKalb, Illinois.

Karplus, E., 1980, *Classroom Research: Valuing Students Errors*, A paper presented at the 4th annual MERGA conference in Hobart, Australia.

Karplus, E., and Karplus, R., 1970, Intellectual development beyond elementary school I: Deductive logic, *School Science and Mathematics*, 70(5):398–406.

Karplus, E., Karplus, R., and Wollman, W., 1974, Intellectual development beyond elementary school IV: Ratio, the influence of cognitive style, *School Science and Mathematics*, 74(6):476–482.

Karplus, R., May, 1978, *Final Report: Advancing Education Through Science-Oriented Programs (AESOP); Pre-college Components*, NSF funded projects to Lawrence Hall of Science, University of California, Berkeley, California.

Karplus, R., Adi, H., and Lawson, A., 1980, Intellectual development beyond elementary school VIII: Proportional, probabilistic, and correlational reasoning, *School Science and Mathematics*, 80(8):673–683.

Karplus, R., and Karplus, B., 1979, *Student Attitudes Toward Research in the Classroom*, Unpublished research report, AESOP Project: Lawrence Hall of Science, University of California, Berkeley.

Karplus, R., and Karplus, E., 1972, Intellectual development beyond elementary school III: Ratio, a longitudinal study. *School Science and Mathematics*, 72(8):735–742.

Karplus, R., Karplus, E., Formisano, M., and Paulson, A. C., 1977, A survey of proportional reasoning and control of variables in seven countries. *Journal of Research in Science Teaching*, 14(5):411–417.

Karplus, R., and Peterson, R., 1970, Intellectual development beyond elementary school II: Ratio, a survey. *School Science and Mathematics*, 70(9):813–820.

Khoury, H., 1997, *Measuring Up: A Middle School Experience*, An Eisenhower professional development project, Department of Mathematical Sciences, Northern Illinois University, DeKalb, Illinois.

Kurtz, B., and Karplus, R., 1979, Intellectual development beyond elementary school VII: Teaching for proportional reasoning. *School Science and Mathematics*, 79(5):387–398.

Monk, S., 1992, Students' understanding of a function given by a physical model, in G. Harel, and E. Dubinsky. eds., The concept of a function: Aspects of epistemology and pedagogy. *MAA Notes*, 25:175–194.

National Council of Teachers of Mathematics, 1989, *Curriculum and Evaluation Standards for School Mathematics*, NCTM, Reston, VA.

National Council of Teachers of Mathematics, 2000, *Principles and Standards for School Mathematics*, NCTM, Reston, VA.

Silver, E., 1977, Algebra for all: Increasing students' access to algebraic ideas, not just algebra courses. *Mathematics Teaching in the Middle School*, 2(4):204–207.

Thrun, J., 1999, *College Students' Rational Number as Operator Strategies: A Focus on Students' Coordination of Units and Distributivity of Operators in Problem-solving Situations*, Unpublished doctoral dissertation, Northern Illinois University, DeKalb, IL.

Wollman, W., and Karplus, R., 1974, Intellectual development beyond elementary school V: Using ratio in differing tasks. *School Science and Mathematics*, 74(11):593–613.

Intellectual Development Beyond Elementary School I: Deductive Logic*

Elizabeth F. Karplus and Robert Karplus

The Islands Puzzle (Figure 1) was created as a tool to assess abstract reasoning ability. The "clues" given as part of the puzzle must be analyzed and used to draw certain conclusions, as explained in Figure 1. Note that both an answer and an explanation of the answer in terms of the clues are required. In this way our study is similar to those of Piaget.[1] Furthermore, we have found it most useful to examine and categorize the explanations, just as is done by Piaget. Our procedure deviates from that of Piaget, however, in that the puzzle is administered to a group of subjects and that the individual responses are written. It is therefore not possible to investigate a particular subject's thinking beyond the level of the standard questions. Furthermore, we have not conducted any longitudinal studies in which the development of a single individual is observed over several years.

Because the answers are written, the puzzle has been presented only to fifth graders and older persons. Phillips,[2] who has recently reported on a classroom presentation of

* Part I of a series of articles.
[1] Inhelder, B. and Piaget, J. *The Growth of Logical Thinking from Childhood to Adolescence.* New York: Basic Books, 1958.
[2] Phillips, Darrell G. *Individual Interview versus Classroom Group Presentation of Piaget-type Tasks,* paper presented to the National Science Teachers Association, Dallas, Texas, March, 1969.

Elizabeth F. Karplus and Robert Karplus, Science Curriculum Improvement Study, Lawrence Hall of Science, University of California, Berkeley.

"Intellectual Development Beyond Elementary School I. Deductive Logic" was taken from *School Science and Mathematics,* 1970, vol 70, pgs. 398–406.

Figure 1. Islands Puzzle.

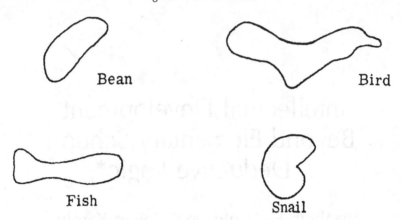

Bean

Bird

Fish

Snail

Introduction: The puzzle is about four islands in the ocean. People have been traveling among these islands by boat for many years, but recently an airline started in business. Listen carefully to the clues I give you about possible plane trips. The trips may be direct or they may include stops on one of the islands. When I say a trip is possible, it can be made in both directions between the islands.

This is a map with the four islands, called Bean Island, Bird Island, Fish Island, and Snail Island. You may make notes or marks on your map to help you remember the clues. Raise your hand if you have questions about the clues.

First clue: People can go by plane between Bean and Fish Island.

Second clue: People cannot go by plane between Bird and Snail Islands. Use these two clues to answer question 1.

Question 1: Can people go by plane between Bean and Bird Islands?

Yes ____ No ____ Can't tell from the two clues. ____

Explain your answer.

Third clue: People can go by plane between Bean and Bird Islands. Use all three clues to answer questions 2 and 3. Don't change your answer to question 1.

Question 2: Can people go by plane between Fish and Bird Islands?

Yes ____ No ____ Can't tell from the three clues. ____

Explain your answer.

Question 3: Can people go by plane between Fish and Snail Islands?

Yes ____ No ____ Can't tell from the three clues. ____

Explain your answer.

Piaget-type tasks, was able to work with fourth, fifth, and sixth graders and found considerable agreement between individual and group responses.

METHOD OF PRESENTATION

For all groups studied, the task was administered by the authors or a specially trained assistant. The subjects were shown a map of the four islands (Fig. 1), which we designated by the indicated names for the children, by letters for the high school students and adults. Then the introduction was presented; it was emphasized that all plane routes operated in both directions, and that nonstop trips or trips with stops or plane-changes were equivalent. No "practice questions" or other teaching was provided. The results of a pilot study where practice was provided are reported in an Appendix to this paper.

Most of the groups had answer sheets with the map and the three question items, but the introduction and clues were always presented orally and clarified with the aid of a blackboard map. The clues were repeated as often as requested by the subjects. Nevertheless, it was clear from a few of the answers that some subjects were confused by the clues and did not remember the details correctly. The completion of the puzzle required between ten and fifteen minutes.

At this point the reader may wish to attack the Islands Puzzle and record his answers and explanations on a piece of paper.

CATEGORIZATION OF RESPONSES

One subjective aspect of this study is our categorization of a subject's explanation of his answers. As we have pointed out, the fact that the task was presented to groups prevented detailed exploration of the reasoning of individuals. Some of the responses are very brief and therefore ambiguous; these may represent inadequate reasoning or the subject's judgment that the answer is so obvious it could be explained briefly. Nevertheless, we have established a sequence of six categories which comes very close to the sequence Piaget and his collaborators have found applicable to the tasks investigated by them.[1] A description of the categories follows.

Category N: no explanation or statement "I can't explain."

Category I (pre-logical): an explanation which makes no reference to the clues and/or introduces new information. Subcategories are the mere repetition of the answer to be explained (to #2, "Yes, because there are flights"), appeal to the diagram itself (to #2, "No, because it is the diagonal" or to #3, "Yes, because it is close"), and fanciful stories (to #1, "No, because there is a strong air pocket that no one can survive" or to #2, "No, because the plane can run out of gas and go down in the water").

Category IIa (transition to concrete models): direct appeal to or repetition of clues (#1, "No, because you did not say so" or to #1, "Can't tell because you didn't say"). Since all three questions require inferences, a direct appeal to the clues does not provide a logical justification.

Category IIb (concrete models): the clues are used to construct models which are then used to make the predictions. The most common model provides for the presence or absence of airport facilities on an island, according to whether flights were or were not said to reach it (to #1, "Can't tell, because Bean Island has an airport, but Bird Island might or might not have an airport"; to #2, "Yes, because there must be an airport on Bird Island, so the people from Fish Island can get there"; to #3, "No, Snail must be the one with no airport, so people from Fish Island can't get there"). This model-based approach, when correctly used, leads to correct answers to all three questions in the problem. It assumes information not given in the clues, however, and cannot be generalized to solve similar puzzles with different data.

Category IIIa (transition to abstract logic): logical explanation to question 2, that Bird Island can certainly be reached from Fish Island by way of a stop at Bean Island (to #2, "Yes, Fish to Bean to Bird"). Since the logical inference from the two positive statements (clues 1 and 3) needed for question 2 is easier, in our view, than the use of the negative statement (clue 2), question 2 does not make maximum demand on the subject's reasoning ability. We have therefore classified the logical answer here as being transitional to the abstract stage, rather than representing attainment of the abstract stage.

Category IIIb (abstract logic): logical explanations to questions 1 and 3 (to #1, "Can't tell because there is no information linking either Bean or Fish Island with Bird Island"; to

#3, "No, because a flight between Fish and Snail would make possible a route between Bird and Snail via Bean and Fish; this contradicts the second clue").

It is clear that these six categories must be viewed as tentative pending further study, in depth, of the reasoning of individual children and adults. The only difficulty we encountered was with a large number of explanations of the form "inadequate information" to question 1. The response rarely included a comment as to how or why the information was inadequate. We did not know whether the subject actually knew the logical reason, or did not and drew the conclusions on partially intuitive grounds. We therefore interpreted these answers in the light of the explanations offered to questions 2 and 3.

The descriptive terms "pre-logical", "concrete models", and "abstract logic" are used here to refer to sub-stages in the transition from concrete to formal thought, since our observations make this refinement necessary. Furthermore, we believe we have found here indications of a development that is analogous to the pre-operational—concrete operations—formal operations sequence,[1] in which reference to concrete models takes the place of actions on concrete objects.

RESULTS

A total of 449 subjects in six distinct groups participated in this investigation (Table 1). The groups range from fifth and sixth graders to NSTA members at the 1969 convention and college physics teachers at a regional meeting of the American Association of Physics Teachers. The percentage of subjects in each group placed in each category of explanation is reported in Table 2. For the purposes of this table, a subject was placed in the category of his explanations if all three fell within the same category. If they varied, and this was usually the case, an intermediate category was chosen. Thus, a logical answer to #2 (IIIa) and a repetition of clues to #3 (IIa) resulted in the assignment of the subject to category IIb.

It can be seen in the table that there is a gradual progress of the group median from category I for the "5–6" group to category IIa for the "7–9" group and finally category IIb for the "10–12" group. Curiously enough, there is little further progress even on the part of the adult groups. Perhaps a psychological set created by the introduction is limiting the logical thought.

More specifically, a large fraction of no explanations (N) by the "5–6" group disappears for the later groups. This is not surprising. Conversely, answers placed in the category

Table 1. The Subjects.

Subject group	Number	Designation of group
Suburban fifth and sixth graders	55	"5–6"
Suburban seventh, eighth, and ninth graders enrolled in a science class	78	"7–9"
Suburban tenth, eleventh, and twelfth graders in several college preparatory classes	98	"10–12"
Suburban twelfth graders in physics classes	66	"12P"
NSTA Convention participants at Piaget symposium	83	"NSTA"
American Association of Physics Teachers	69	"AAPT"
Total Subjects	449	

Table 2. Evaluation of Explanations (percent).

Category	Group					
	"5–6"	"7–9"	"10–12"	"12P"	"NSTA"	"AAPT"
N	24	3	1	9	10	0
I	40	39	9	15	4	3
IIa	11	26	13	6	18	6
IIb	18	18	62	47	54	51
IIIa	7	15	11	15	8	27
IIIb	0	0	3	8	6	13

of abstract thought (IIIb) are completely absent from the tested samples of fifth to ninth graders and appear only with high school groups. Since the abstract explanation is quite complicated to write down, it is likely that some of the younger children would display a better performance in an interview compared to a written examination. The combined levels IIIa and IIIb show surprisingly little variation from group to group, with the "5–6" being somewhat lower than the average and the high school physics group being somewhat higher but both at the borderline of statistical significance. Only the college physics teachers show a significantly increased percentage in the categories IIIa and IIIb.

A few interesting results are not revealed by Table 2. The high school groups, for instance, include about 30% model makers (within the scope of category IIb) while the "NSTA" group has less than 10% model makers. Almost all the adults answer logically (IIIa) to #2 but very many appeal to the clues directly (IIa) to items #1 and #3. Perhaps the cognitive style, rather than only the intellectual level of a subject, influences his response. Also, various groups use different techniques for record keeping. Diagrams are used by some, complete statements by others, abbreviated notes by most adults.

Because the total number of correct answers (that is, the answer pattern "Can't tell"— "yes"—"no"—which makes optimal use of the information given) is quite small—only 78 out of 449 subjects—and because the percentage varies only slightly from group to group (from 9% in the "5–6" group to 25% in the "AAPT" group), we have combined the six groups into one and the six categories into three. The numbers and percentages of subjects giving correct answers are reported in Table 3. As might be expected, subjects who explain their answers on a higher level of abstraction are much more successful in obtaining correct answers than the others. From zero correct answers in the N, I category the percentage rises to 60 in category III. On a random basis, one would expect about 3% correct answers (1 in 27) since there are three questions with three options each. Even the level II subjects perform somewhat better than random, but the excess is only at the borderline of statistical significance.

Table 3. Correct Answer Patterns.

Category	Number of subjects	Number of correct answers	Percentage of correct answers
N, I	106	0	0
IIa, IIb	254	25	10
IIIa, IIIb	89	53	60
Total	449	78	

DISCUSSION

It is clear from this study that intellectual development in abstract reasoning, as defined by the "Islands Puzzle," reached a plateau in the high school age group and did not progress much further. In addition, the plateau is at a disappointingly low level.

The following question of educational policy is thereby raised: is it a desirable educational objective that a larger fraction of the adult population (other than college physics teachers) should be able to answer the Islands Puzzle or an equivalent task on the level of abstract thought? We are pleased to be able to state this objective in such clear-cut behavioral terms, and we solicit the reader's opinion on the matter. If the answer is "yes", then the problem becomes one of designing a suitable educational program. Unfortunately, even the behavioral statement of the objective does not tell us how to proceed, and we again solicit the reader's suggestions.

Even though one can conceive other and better ways of posing the logical task than through the Islands Puzzle, we believe that this preliminary study calls attention to a neglected area of educational research which has significant bearing on large-scale improvements in scientific literacy. It is furthermore likely that tasks can be constructed to assess conservation reasoning, combinatorial logic, mathematical proportion, and other components of formal thought. What will these reveal about the intellectual level of our high school and adult populations?

We are indebted to Beverly R. Karplus and Margaret A. Karplus for assistance in the study. It was supported in part by a grant from the National Science Foundation.

APPENDIX

It is well known that even a brief teaching or "practice" exercise improves performance on many tests. Presumably the practice furnishes a pattern for the subject's subsequent performance. Even though more of the subject's ability is thereby brought to bear on the test, there is a question of just how applicable this ability is to situations where the subject is not cued by a practice item. We therefore consider it an open question as to whether a performance on the Islands Puzzle with or without a practice item is a more satisfactory indicator of intellectual development. Nevertheless, we have investigated the effect of a brief "practice" activity to explore this alternative.

The nature of the Islands Puzzle is such that many different forms of "practice" could improve performance. Since we were only concerned with identifying an effect, we chose to have the task administrator pose and answer a question analogous to puzzle question 3, which was the most difficult one for the subjects. The following statement was inserted after the introductory paragraph (Figure 1) and before the answer sheets with the pictured islands were distributed:

"I'll now give you one puzzle as an example, and I'll tell you my answer to this puzzle. Then, I'll give you another puzzle and a piece of paper on which you can write your answers.

"In my puzzle there are three islands, Island X, Islands Y, and Island Z." (Three islands are drawn in a triangular arrangement on the chalkboard. During the subsequent explanation, the islands mentioned are indicated by hand gestures.) "Now I'll give you two clues. The first clue is that people can go by plane between Islands X and Y. They can go in both directions, and it may be direct or with a stopover. The second clue is that people cannot go by plane between Islands X and Z. They cannot go from X to Z or Z to X either directly or by stopping somewhere.

"There is only one question left; what about plane trips between Islands Y and Z? Can people go by plane between Islands Y and Z? You could answer 'Yes' or 'No' or 'Can't

Table 4. Evaluation of Explanations with Practice (percent).

Category	Group		
	"5–6"	"7–9"	"10–12"
	(N = 44)	(N = 139)	(N = 114)
N	5	2	2
I	20	10	11
IIa	30	24	16
IIb	45	45	33
IIIa	0	13	24
IIIb	0	6	15

tell from just the two clues!' Now I'll tell you my answer, and then I'll explain the reason for my answer by using the two clues.

"My answer is 'No.' And the explanation goes like this: according to the first clue, people can go by plane from X to Y. Now, if there are planes to go from Y to Z, then people can go from X to Y and catch another plane from Y to Z. That way they could get from X to Z. But this contradicts the second clue, which said that people could *not* go by plane between X and Z. Therefore, people cannot go by plane between Y and Z.

Three groups of students, each in the same school as the corresponding group identified in Table 1, worked on the Islands Puzzle after the practice exercise. The results of their work are presented in Table 4 and may be compared with those for the first three groups in Table 2.

It is clear that all three groups show substantial gains. The number of students in category I has decreased dramatically. The median has shifted upward for each group, to category IIa for the "5–6" group, to category IIb for the "7–9" group, and above category IIb, for the "10–12" group. The single most noticeable change in the individual answers is the appearance of hypothetico-deductive statements (to #1, "Yes, if Bean can go to Fish, then I think it could go to Bird") even though the logic of these statements is frequently incorrect. Still, the steady advance with grade level from pre-logical to abstract logical thinking is very similar to the advance in the puzzle without practice. As might be expected, the number of students with a correct answer pattern has increased (Table 5, compare with Table 3), but is still far from complete.

One last point should be made. The students reacted to the practice puzzle with subdued but noticeable signs of dismay. This had not been the case at all when no practice was provided.

Table 5. Correct Answer Patterns with Practice.

Category	Number of subjects	Number of correct answers	Percentage of correct answers
N, I	42	2	5
IIa, IIb	185	55	30
IIIa, IIIb	70	51	79
Tatal	297	108	

Intellectual Development
Beyond Elementary School II*:
Ratio, A Survey

Robert Karplus and Rita W. Peterson

In the first report of this series of studies, we described the Islands Puzzle, a tool to assess abstract reasoning ability [1]. We have now completed a survey of students' ability to apply the concept of ratio (or proportion) in a problem requiring a change in the unit of length measurement. The concept of ratio is important because of its many scientific applications; as in the calculation of chemical equivalents, use of the ideal gas laws, definitions of density, speed, and acceleration, operation of simple machines, and analysis of electric circuits. In addition, the ability to apply ratios is an essential component of what Piaget has termed formal operational thought [2].

The problem is described in Figure 1. As in the Islands Puzzle, the students must give a specific answer and must also provide an explanation or rationale of their answer. In this report we shall only describe and categorize their explanations. The accuracy of the arithmetic, which left a great deal to be desired, was taken into account to help interpret explanations when necessary. Again, our procedure was to administer the problem to groups of subjects (727 individuals) and to collect written responses. Therefore it was impossible to investigate a particular subject's thinking beyond the level of the standard questions. In a pilot study with several subjects in the upper elementary grades we did conduct individual

*The second in a series; Part I appeared in the May 1970 issue.

Robert Karplus and Rita W. Peterson, Science Curriculum Improvement Study, Lawrence Hall of Science, University of California, Berkeley, California 94720.

"Intellectual Development Beyond Elementary School II: Ratio, A Survey" was taken from *School Science and Mathematics*, 1970, vol 70, pgs. 813–820.

interviews, however, and found no substantial difference from the types of written responses submitted in the group administration.

METHOD OF PRESENTATION

For all groups studied, the problem was administered by one of the authors. The subjects were given answer pages as described in Figure 1 and a chain of No. 1 "gem" paper clips containing seven to ten clips. The subjects were then shown a display chart with the same figure as on their pages on the front side and a scaled-up version of that figure on the back side. The small and large figures had been constructed so as to measure four and six jumbo paper clips, respectively. The experimenter (E) had a chain of eight jumbo-size gem paper clips.

With the subjects watching, the experimenter explained the problem and carried out the demonstrations as described in Figure 1. The display chart was then put away and the subjects proceeded to answer the questions. The measurements recorded on the chalkboard remained there for reference.

Subjects with questions or requests for more information were referred to the data on the chalkboard. The measurements were not repeated after the display chart had been put aside, nor were the subjects given access to the jumbo paper clips.

They were reassured that different people answered the questions differently, that there was not one correct way to solve the problem (a concern of many), that everyone could measure in his own way, and that great accuracy was not important. The completion of the problem with a normal-sized class group required about 15 minutes.

CATEGORIZATION OF THE RESPONSES

One partially subjective aspect of this study is our categorization of a subject's explanation. This task was approached independently by the two authors, and disagreements were reconciled until the categories resulted in complete agreement on classification. As we have pointed out, the fact that the task was presented to groups prevented detailed exploration of the reasoning of individuals. Some of the responses were very brief and sometimes difficult to interpret. In those cases, the arithmetic computation carried out by the subject usually gave additional clues about his reasoning and made it possible to assign the response to a category. A description of the categories follows:

Category N: no explanation or statement, "I can't explain."

Category I (intuition): An explanation referring to estimates, guesses, appearances, or extraneous factors without using the data. Examples of the predictions and related explanations are:

$9\frac{1}{2}$—The big man just looks that much bigger than the little man.
9—Because that's what I think it is.
$9\frac{1}{2}$—You use a ratio factor, but I don't remember how, so I guessed.
10—A guess.

Category IC (intuitive computation): The subject makes use of data haphazardly and in an illogical way. Examples are:

16—By multiplying.
10—I figured out that if he had bigger paper clips, you add 6 with the other 4 and you have 10.

$12\frac{1}{2}$—Half of 12 is 6, so you'd take the paper clips and measure him twice because he's longer than Mr. Short, so naturally it would be $12\frac{1}{2}$.

$10\frac{3}{4}$—Since it took 4 biggies for Mr. Short and $6\frac{1}{4}$ smallies, there is $2\frac{1}{4}$ difference, and it took 6 biggies for Mr. Tall, 2 more than for Mr. Short, so I added $2\frac{1}{4} + 2\frac{1}{4}$ together and got $4\frac{1}{2}$ and added that to $6\frac{1}{4}$ and got $10\frac{3}{4}$ for an answer.

Category A (addition): An explanation using all of the data, but applying the difference rather than the ratio of measurements. Examples are:

8—The little man was 4 of his and 6 of mine so I added 2.

$8\frac{1}{2}$—When you did it, the large man was 2 big paper clips bigger than your small man. So mine must be 8—if 6 smallies equal 4 biggies, then 6 biggies must equal 8 smallies.

Category S (scaling): The subject makes a change of scale when he predicts. He does not relate this operation to the scale inherent in the data, thereby failing to see the whole problem. He expresses a tentative attitude toward his estimate. Examples are:

12—The large man is two times bigger than the little man; the little man is 6, so I think it is 12.

12—I think it is twelve because in biggies it is six, and small ones are about half that size, and so I think it is about 12.

Category AS (addition and scaling): The subject focuses on the excess height of Mr. Tall, but scales up the excess number of jumbo paper clips by a factor of two to compensate for the size difference: An example is:

10—I think two smallies are as big as one biggie, so I added four smallies for the two extra biggies.

Category P (proportional reasoning): The subject uses proportionality and makes clear how the ratio is derived from the measurements on the two figures. He may or may not use the word "ratio." Examples are:

9—There is a mathematical problem in 4 big and 6 big, 4 is $\frac{2}{3}$ of 6, so it should be 9.

9—$6 \div 4 = 1\frac{1}{2}$, $6 \times 1\frac{1}{2} = 9$.

9—The ratio of the biggies is $2:3$, so you figure the small paper clips would also have the ratio $2:3$.

9.75—It was a direct ratio and proportion, small to large, small to large.

9—Set up a ratio, $6/4 = x/6$

9—$6/4 = x/6, x = 9$

There were, in addition sixteen papers on which the subjects used a ratio of $1\frac{1}{2}:1$ for the lengths of the paper clips, but did not explain further. Such a response would fall into Category S if the ratio were estimated, and into Category P if the ratio had been calculated from the data. Since we could not assign the explanations unambiguously, we eliminated these sixteen papers from the tabulation in the next section.

It is clear that a cross-sectional survey such as we have carried out cannot establish a genetic sequence as described by Piaget, [2] according to whose theory each category represents a developmental stage that emerges out of the previous one. While Category I evidently represents the most naive approach (preoperational) and Category P the most sophisticated (formal thought), the other groups cannot be serially ordered into a single sequence in an obvious fashion. In fact, Categories IC, A, and AS form one possible sequence; while Categories S, AS, and P form another. In the former, the subject first advances to handling the data systematically, according to the incorrect addition rather than to the correct ratio operation. In the latter, he first uses the correct mathematical operation but guesses or estimates the scale factor rather than deducing it from the data. It is conceivable that a student progresses through one *or* the other of these two sequences. If that is true, of course, the categories need not be combined into a single sequence. We have initiated a longitudinal study to resolve these questions. In the meantime, we present our findings but do not organize them into a developmental sequence.

RESULTS

A total of 727 subjects in six distinct groups participated in this investigation. They all lived in large metropolitan areas. About one-third were students of inner city schools, while the remainder attended suburban schools. They ranged in age from 9 to 18 years, and were enrolled in the fourth to the twelfth grades. The task was administered to most groups during a science or mathematics class. For the high school seniors, however, English and history classes were also included, because many of the students do not elect to take a science such as chemistry or physics.

Table 1 reports the percentage of subjects in each group placed in each category, after elimination of sixteen ambiguous responses. It is noteworthy that extremely few children gave no response, while on the Islands Puzzle [1] one-fourth of the elementary school students gave no explanation at all.

It can be seen from the table that the older urban and suburban groups were better able to solve the ratio problem than their younger colleagues. Yet the suburban subjects achieved substantial mastery of the task (Category P) by the end of high school, while the urban students showed relatively slight progress during the ages spanned by the sample populations. The difference between the two sixth grade groups is statistically insignificant, but the difference between the urban and suburban advanced high school students is substantial.

The table contains a clue about the developmental relation of Categories S and A. These two groups include closely equal fractions of subjects at all but the fourth and fifth grade levels, at which the scaling responses (Category S) significantly outnumber the

Table 1. Evaluation of Explanations (percent).

Category	Grades (Number)						
	4–5 Suburban (116)	6 Suburban (82)	6 Urban (95)	8–10 Suburban (75)	8–10 Urban (123)	11–12 Suburban (153)	11–32 Urban (67)
N	2	1	5	0	0	3	1
I	31	36	41	4	16	3	15
IC	9	5	12	11	21	1	12
A	16	26	20	25	30	10	31
S	33	26	16	24	25	3	20
AS	4	0	3	4	3	0	12
P	5	6	3	32	5	80	9

addition responses (Category A). From this result one may infer that scaling is a less advanced response than addition. The clue is a slight one, however, and the definitive resolution of the sequencing problem awaits a longitudinal study.

A few interesting observations are not revealed by Table 1. Most of the subjects in the Category S (scaling) estimated the length ratio of the paper clips, not the height ratio of the men. Furthermore, the estimated scale factor was equal to two in almost 90% of the scaling responses.

The table also hides the very large fluctuations in performance from class to class that we found at the elementary school level. In one class, more than two-thirds of the children responded in Category A; while in others this response accounted for a much smaller percentage. In one urban class, 22 out of 24 children placed in categories N, I, or IC; but in another, consisting of black students exclusively, Category AS and P responses were more numerous than in any one suburban class. These differences are suggestive of the learning patterns followed by the children and of the obstacles to learning that may be erected inadvertently by mathematics courses, by teachers, and by the children's cultural environment.

Finally, a substantial minority of the younger children were disturbed to find that Mr. Short measured between six and seven smallies. They asked for advice as to how they might cope with the fact that the measurement was not an integral number of clips. In response to such queries, the answer was, "Describe it the way it looks to you."

DISCUSSION

Few of the readers experienced in teaching mathematics and science will be surprised by the findings of this study. Nevertheless, it may be disappointing that successful proportional reasoning is not achieved earlier than the last years in high school, even though the subjects of ratio and proportion make their appearance in most mathematics programs in junior high school. An effort to improve the performance by cueing for the use of the data was not successful (see Appendix). It seems, therefore, that there is a serious gap between secondary school mathematics and science curricula and the students' reasoning ability.

This study was supported in part by a grant from the National Science Foundation.

REFERENCES

[1] Elizabeth F. Karplus and Robert Karplus. "Intellectual Development Beyond Elementary School. I. Deductive Logic." *School Science and Mathematics*, LXX (May 1970), 398–406.
[2] B. Inhelder and J. Piaget. *The Growth of Logical Thinking from Childhood to Adolescence.* Basic Books, New York, 1958.

APPENDIX

The educational literature suggests that subjects perform better on a task if its objectives is clearly stated in advance. The Piagetian view, however, is that a statement of objectives does not help raise the intellectual level at which a subject performs.

In view of their generally unsuccessful performance on the ratio problem, fifth and sixth grade pupils are an excellent population on which to test the effectiveness of defining the objective more clearly. We therefore administered the problem in a modified form to 91 additional fifth and sixth grade pupils, from a suburban school in the same community as

Table 2. Evaluation of Explanations With Cueing (percent).

Category	Group	
	Uncued (n = 127)	Cued (n = 90)
N	2	3
I	35	20
IC	6	28
A	23	16
S	27	24
SA	2	2
P	5	7

the original sample. This time the subjects were told that the measurements made by the experimenter using biggie paper clips were clues to help them solve their problem, and that they should use these clues to make their prediction and to explain it. The answer pages used by these students were identical to the pages used by the main sample; the entire difference lay in the verbal cueing of the subjects to use the clues in their reasoning.

The results of this modified approach are summarized in Table 2. It is clear that the clue did not substantially improve the quality of responses. There were fewer responses in Categories I and A and many more in Category IC, which had become the largest single group. The difference between the two groups is statistically significant at the 1% level. The many cued children in Category IC did make use of the data as they had been asked to do; their operations, however, were mathematically nonsensical. Examples are:

10—Because 6 + 4 biggies is 10.
 8—I added 6 + 4 − 2 and it just came out.
16—I got 6 for the small man and 4 + 6 = 10 + 6 = 16.
12—I made this prediction because 3 is half of six and 3 is small and six is big so I did 6 × 2.
12—Mr. Short is 6 smalls, he is also 4 biggies; Mr. Tall is 6 biggies, 2 is between 4 and 6, 2 × 6 is 12.

We therefore conclude that the cueing used was not genuinely helpful in raising the intellectual level at which the children performed on the ratio problem but merely led some to carry out meaningless arithmetical operations.

Figure 1. Ratio of length.

Oral Introduction: This problem is about Mr. Short, who is just like the man on your papers, and Mr. Tall (E. displays the two figures on the two sides of the display chart). I will measure how high the figures are with my chain of big paper clips—we'll call them "biggies" (E. displays his chain of eight jumbo clips) and write down the results (E. writes "Mr. Short:" and "Mr. Tall:" on the board). First we'll do Mr. Short—how high is he? (E. hangs paper clip chain from a pin at the top of Mr. Short's head. Class members count and chorus "four." E. writes "4 biggies" on the board.) Have all of you seen it? . . . Now we'll turn to Mr. Tall. (E. turns over chart to show Mr. Tall.) How high is he from the head to the ground? (E. hangs paper clip chain from a pin at the top of Mr. Tall's head. Class members count and chorus "six." E. writes "6 biggies" on the board.)

Have all of you seen it? (E. puts down the chart, and it is not used again. The jumbo paper clips are also put away.)

Now you do three things. First you measure Mr. Short on your papers, using your small paper clips—call them "smallies." Then you predict the height of Mr. Tall in smallies, and finally you explain how you made your prediction. Explain as best as you can what made you predict the number of smallies you chose.

Question and Answer Page:

(actual height: $7\frac{9}{16}''$)

How tall is this figure, measured with small paper clips?
Predict the height of the large figure, as measured with small paper clips.
Explain how you figured out your prediction.

Intellectual Development Beyond Elementary School III— Ratio: A Longitudinal Study*

Robert Karplus and Elizabeth F. Karplus

In an earlier report, elementary and secondary school students' ability to apply the ratio concept in a simple problem of measurement was described [1]. Subjects were asked to report a measurement, predict a result they could not measure, and explain their predictions. It was found that students' responses could be grouped into six categories, ranging from guessing to proportional reasoning. The authors speculated about a developmental sequence of these responses but were unable to come to any definite conclusion. We have now repeated the administration of the task to 155 sixth, eighth, and eleventh grade suburban students who had been tested two years ago as fourth, sixth and ninth graders. In addition to the participants in this longitudinal study, comparison data will be reported for 141 eighth grade students who had not participated in the earlier testing.

*This article is based on a paper presented at the IPN Symposium, Kiel, Germany, March 1972 under the title "Ratio, A Longitudinal Study." It is being published in the IPN Symposium Proceedings by Hans Huber (Bern/Stuttgart/Winn).

This study was supported in part by a grant from the National Science Foundation.

Robert Karplus and Elizabeth F. Karplus, Lawrence Hall of Science, University of California, Berkeley 94720.

"Intellectual Development Beyond Elementary School III—Ratio: A Longitudinal Study" was taken from School Science and Mathematics, 1972, vol 72, pgs. 735–742.

These fourth graders are using polar coordinates to describe the location of a flag on their playtground. One boy is determining the direction with the aid of a sighting device on the ground, while the other is about to pace off the distance to the flag from the starting point. Scaling their measurements when they draw a map of the flag position will involve them with ratios.

METHOD OF PRESENTATION

The method of presentation was the same as has been described in Reference 1. As before, each class group included between twenty and thirty students, some of whom had been tested in 1969 and others who had not. Completion of the problem required about 15 minutes. Very few students made reference to their familiarity with the task. In most cases the memory was faint, but one high school student did decline to make a prediction because she could not remember how to solve the problem and did not want to do it "incorrectly." Only one other student failed to provide an explanation of some kind.

A few days after the group tests, seventy students were interviewed individually, for one of four reasons: (1) they had been absent during the group administration, (2) their written responses appeared to be incomplete, (3) their responses appeared to regress compared to the original test, or (4) we wished to check the reliability of written responses of a few students selected at random from those who gave clear and complete explanations. The results of the interviews were compared with the written explanations and categorized as described in the next section.

CATEGORIZATION OF THE RESPONSES

The students' explanations were classified according to the six categories described previously [1]:

Category N: No explanation or statement, "I can't explain." Students who did this were interviewed when possible, and a more informative response was obtained in all but two cases. These were dropped from the analysis, leaving 153 subjects.

Category I (intuition): An explanation referring to estimates, guesses, appearances, or extraneous factors without using the data. All students who appeared to regress to Category I from a more advanced response two years ago were interviewed. Several of these students then gave a more informed explanation, but others repeated their written explanation. In all cases, the interview result rather than the written explanation was used for the later analysis.

Category IC (intuitive computation): An explanation using data haphazardly and in an illogical way. This group was quite small, and about half of them could be interviewed. Most of these gave a complete but somewhat complicated oral explanation that was classified in one of the other categories for the analysis.

Category S (scaling): An explanation based on a change of scale that does not relate to the scale inherent in the data. Most of the predictions changed the scale by a factor of two, either because one biggy was thought to be two smallies, or because Mr. Tall was thought to be two times as high as Mr. Short. About one-third of the total, however, used a scale factor of 1.5, thus raising the possibility that they had interpreted the data correctly without saying so. All of these students were interviewed, with the result that about half were reclassified into Category P (see below), one could not explain (Category I), and the others remained in Category S because they based their scale factor on their perceptions and not on the data.

Category A (addition): An explanation using all of the data but applying the difference rather than the ratio of measurements. This category was always very easy to identify; a few students were interviewed and all repeated the same explanation at that time.

Category AS (addition and scaling): An explanation focusing on the excess height of Mr. Tall, but scaling up the two excess biggies by a perceptually based factor (usually 2 or 1.5) to compensate for the difference in paper clip size. Several students in this group were interviewed and all repeated their explanations.

Category P (proportional reasoning): An explanation using proportionality and describing how the ratio is derived from the known measurements. The subject may or may not use the word "ratio"—very few students did in this sample, while many of the senior high school students had done so during the earlier survey. A few students in this category were interviewed, and all gave the same oral response. About one-fourth of the subjects used a geometrical rather than an arithmetical procedure to construct the ratio. They divided Mr. Short into fourths by visual estimation, then extended Mr. Short's height by two-fourths (equal to two biggies as they explained in the interview), and then measured the extended length with their chain of smallies.

It is clear that the two intuitive categories, where the written and oral responses frequently differed, are not as easy to define as the others. To what extent should written explanation be required? Should the encouragement given by a patient interviewer who asks repeated questions be discounted?

RESULTS OF THE LONGITUDINAL STUDY

To what extent do the data shed light on the pattern according to which students change from one category of response to another one? It is most informative to present the results of the longitudinal study in the form of a six-by-six matrix whose rows refer to the categories two years ago and whose columns refer to the present categories. This is done in Table 1, where the order of the rows and columns must not yet be given a precise significance.

Table 1. Matrix Comparing Students in 1969 and
1971 by Categories (Numbers of Students).

	(1971)						
	I	IC	S	A	AS	P	(Total)
(1969)							
I	12	4	11	8	3	5	(43)
IC	2	1	1	2	1	1	(8)
S	1	2	19	8	3	6	(39)
A	3	1	8	10	9	8	(39)
AS	0	0	0	0	1	4	(5)
P	0	0	0	1	0	18	(19)
(Total)	(18)	(8)	(39)	(29)	(17)	(42)	((153))

The first entry in the table (12) means that twelve children who answered intuitively in 1969 did the same in 1971. The next entry (4) means that four children who answered intuitively in 1969 did intuitive computation in 1971. The last entry in the top row (43) indicates the total number of children who answered intuitively in 1969, while the bottom entry in the first column (18) indicates the number of children who answered intuitively in 1971.

Remarkably, sixty-one of the 153 students—more than one-third of the total—showed no change in category over the two year period! The changes that did occur show clearly that Categories P and AS are more advanced than the others, since 36 students (28% of those in other categories) moved into P or AS, while only one student (4% of those in AS or P) moved out from P to A.

Regarding the two intuitive categories I and IC, which might be considered the most elementary, seven students (7% of those in other categories in 1969) moved into them, while thirty-two students (65% of those in I or IC in 1969) moved out. While this difference is substantial, it must be remembered that written intuitive responses alone—and that is the only data available from 1969—seemed to be not completely reliable. In spite of these reservations, we conclude that the present study supports the very natural conclusion that Categories I and IC represent the most naive approach to the ratio task.

The most interesting question raised in Reference 1 concerned the relationship of Categories S and A. Are they alternate or sequential, and if sequential, which precedes the other? The present study does not answer the question directly, since eight of the thirty-nine students originally in each of the two categories moved to the other one. It is true that more students remained in Category S than in A (19 compared to 10), and that fewer advanced from S to the more sophisticated Categories AS and P. Since neither of the two categories S and A deals with the ratio problem completely, it is possible that they represent a metastable operational equilibrium. In other words, the definite possibility exists that many students changed from S to A and then some changed back to S during the two years between testing, and that others changed from A to S and back to A. Our interviews of nineteen students in Categories S or A carried out a few days after the written tests, however, showed no evidence of such changes during the shorter period.

In analogy to the three developmental stages described by Piaget [2], we shall now group our categories into three levels. Level I includes Categories I and IC, Level II includes Categories S and A, and Level III includes Categories AS and P. Only for the latter are we able to claim that the two categories represent two substages, with AS being transitional to formal thought (IIIa) and P representing formal thought (IIIb) in so far as proportional

Table 2. Matrix Comparing Students in 1969 and 1971
by Levels (Numbers of Students).

	(1971)			
	Level I	Level II	Level III	(Total)
(1969)				
Level I = I + IC	19	22	10	(51)
Level II = S + A	7	45	26	(78)
Level III = AS + P	0	1	23	(24)
(Total)	(26)	(68)	(59)	((153))

reasoning is concerned. For the other two levels we are not prepared to identify substages with categories. The three-by-three intellectual development matrix referring to the Piagetian levels is presented in Table 2.

A total of 214 eighth graders participated in the investigation during 1971 because their school was interested in relating the mathematics and science programs to the students' proportional reasoning. The difference between the group of 73 retested subjects and the 141 subjects being tested for the first time, indicated that the first encounter two years ago did not provide any advantage. We shall here be concerned with 202 students who attended the same school for seventh grade and who provided a response to the ratio puzzle.

For instructional purposes, the seventh grade students had been divided into three groups of whom the teachers had different expectations. One group of forty-six "fast" students covered almost two years of instructional program to prepare them for the study of algebra in grade 8. A second group of 132 "average" students covered the usual seventh grade mathematics program. Finally, a group of 24 "slow" students concentrated on remedial topics concerning the number system and arithmetic. All of the groups spent a substantial effort on rational numbers (fractions), but none concentrated specifically on text chapters relating to the ratio and proportion concepts. Conversation with the teachers revealed that these concepts were treated in class, but were developed largely in conjunction with the solving of word problems.

The responses of all eighth graders and of the three spearate groups are classified in Table 3. It is quite clear that the three groups performed very differently, but it is not possible to conclude from these data whether the students' initial selection or the educational program during grade seven was more important in creating these differences. The fact that

Table 3. Performance of Eighth Grade Students, According to
Seventh Grade Group (Percentage of Responses).

	Fast $N = 46$	Average $N = 132$	Slow $N = 24$	Total $N = 202$
I	0	5	12	4
IC	2	1	8	3
S	7	31	72	30
A	15	24	4	20
AS	6	9	4	8
P	69	30	0	35

Table 4. Intellectual Development Matrix for Students in "Fast"
Seventh Grade Group (N = 16).

		(1971)				
		Level I	S	A	Level III	(Total)
(1969)						
Level I	= I + IC	0	0	1	1	2
	S	0	0	0	3	3
	A	0	1	1	4	6
Level III	= AS + P	0	0	0	5	5
(Total)		(0)	(1)	(2)	(13)	((16))

Table 5. Intellectual Development Matrix for Students in
"Average" Seventh Grade Group (N = 46).

		(1971)				
		Level I	S	A	Level III	(Total)
(1969)						
Level I	= I + IC	6	4	3	3	(16)
	S	0	3	4	1	(8)
	A	2	5	5	8	(20)
Level III	= AS + P	0	0	0	2	(2)
(Total)		(8)	(12)	(12)	(14)	((46))

the "slow" group was so heavily concentrated in Category S sheds interesting further light on the discussion of sequence. It suggests that Category S does indeed indicate a less advanced stage of reasoning than Category A. Unfortunately, the number of "fast" and "slow" eighth grade students in the longitudinal study is very small and makes separate intellectual development matrices difficult to interpret. Because of their possible interest, these are nevertheless presented in Tables 4, 5, and 6. Categories I and IC have been combined

Table 6. Intellectual Development Matrix for Students in
"Slow" Seventh Grade Group (N = 11).

		(1971)				
		Level I	S	A	Level III	(Total)
(1969)						
Level I	= I + IC	2	2	1	0	(5)
	S	1	5	0	0	(6)
	A	0	0	0	0	(0)
Level III	= AS + P	0	0	0	0	(0)
(Total)		(3)	(7)	(1)	(0)	((11))

into Level I, and Categories AS and P into Level III, but Categories S and A have been kept separate because they include more subjects and they hold a special interest.

The reader can see that the eleven students in the "slow" group made virtually no progress in the two-year period; they represent a serious educational problem. In the "fast" group, only three students failed to reach Level III, but their initial level was considerably more advanced than that of the "slow" group. It would seem that these students took good advantage of the intellectual stimulation they received. The students in the "average" group made some progress, but do not show the dramatic advance of their colleagues in the "fast" group. Of course, the "average" group was not so well prepared, with 35% students initially at Level I, compared to 12% Level I students in the "fast" group.

FURTHER QUESTIONS

The incomplete state of our conclusions concerning the scaling and addition categories suggests that this investigation be continued. There is a limit, however, to the number of times a student can be asked the same questions and still give a fairly spontaneous response. When working with younger children, Piaget appeared not to have any difficulty stemming from their remembering previous test experiences, but with adolescents in a school setting the matter is obviously more difficult. We have considered alternate but equivalent problems and have briefly investigated one form, but are not satisfied because it introduces additional factors that must be considered.

The process whereby students arrive at proportional reasoning also has not been clarified. The wide diversity of responses of the "average" group in Table 3 suggests that their common educational experience was not primarily responsible for the achievement of proportional reasoning by 30%. The relative uniformity of responses by the "fast" and "slow" groups, however, indicates that their syllabus may have included or excluded a key element, or that the advance selection process had been very effective.

A question of especially great interest derives from the substantial differences in the educational programs of various communities in the United States, and of various countries around the world. We have reported on a comparison of urban and suburban students in Reference 1, where the tragic shortcomings of urban education were illustrated. To our knowledge, however, the ratio task has not been used in other countries that have different student selection criteria and also different courses of study than the United States.

Finally, we turn to implications for science teaching. Many of the mathematical models in science involve proportionality. Yet to exhibit this proportionality experimentally, measurements have to be carried out quite accurately to reduce the errors and side effects that inevitably cause data to deviate from mathematical models. It would seem, nevertheless, that the investigation of levers or balances, shadows, electric circuits, solutions, density, and similar items could help students develop proportional reasoning.

REFERENCES

[1] "Intellectual Development Beyond Elementary School II: Ratio, A Survey," by Robert Karplus and Rita W. Peterson, *School Science and Mathematics*, December, 1970, pp. 813–820.

[2] B. Inhelder and J. Piaget, *The Growth of Logical Thinking from Childhood to Adolescence*, Basic Books, New York, 1958.

Intellectual Development Beyond Elementary School IV: Ratio, The Influence of Cognitive Style

Elizabeth F. Karplus, Robert Karplus,
and Warren Wollman

In order to resolve ambiguities remaining after our earlier investigations,[1,2] we have continued our investigation of students' ability to apply the ratio concept in a simple task of measurement and prediction. We can now report significant progress toward clarifying the relationships among students' responses to the task. It appears that many of the subjects who do not use proportional reasoning have access to several alternate procedures; they make use of one suggested by some particular aspect of the task presentation. This personal preference, we believe, reflects the individual's cognitive style[3] rather than a developmental level in the Piagetian sense.[4] These conclusions were derived from the use of a new form of the ratio task administered to 616 students in grades 4 through 9. We shall call the new procedure Form B, to distinguish it from the original version to be called Form A.

* AESOP (Advancement of Education in Sceince-Oriented Programs) is supported by a grant from the National Science Foundation.

Elizabeth F. Karplus, Robert Karplus, and Warren Wollman, AESOP,* Lawrence Hall of Science, University of California—Berkeley, California 94720.

"Intellectual Development Beyond Elementary School IV: Ratio, The Influence of Cognitive Style" was taken from *School Science and Mathematics*, 1974, vol 74, pgs. 476–482.

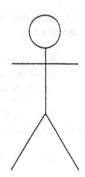

Figure 1. Question and answer page. "How tall is this figure, measured with paper clips? ___ Predict the height of the large figure, as measured with paper clips. ___ Explain how you figured out your prediction."

METHOD OF PRESENTATION

For all groups studied, Form B was administered by one of the authors. The subjects were given answer pages as described previously (Fig. 1) and a chain of No. 1 "gem" paper clips containing 8 to 15 clips.[1] The subjects were then told that there was a figure called Mr. Tall, similar to the Mr. Short on their papers but larger, at the experimenter's (E) office. E continued as follows: "I measured Mr. Short's height with large buttons, one on top of the other, starting with the floor between Mr. Short's feet and going to the top of his head. Four buttons reached to the top of his head." (At this point, E wrote 'Mr. Short: 4 buttons' on the board.) Then I measured Mr. Tall with the same buttons and found that he was six buttons high. (E wrote 'Mr. Tall: 6 buttons' on the board.)

"Now you do three things. First, you measure Mr. Short using your small paper clips—we call them smallies. Then you predict the height of Mr. Tall if you could measure him with smallies; this is the puzzle, because he is at my office and you can't really measure him. And then you explain how you figured out your prediction. Explain as best as you can how you figured out the number of smallies in your prediction."

Subjects with questions or requests for more information were referred to the data on the chalkboard and were encouraged to use their own ideas for answering the questions. E briefly scanned the papers when they were turned in and asked individuals with incomplete explanations to clarify their reasoning: about 20 per cent of the answers were extended in this fashion. The completion of the problem with a normal-sized class group required about 15 minutes.

CATEGORIZATION OF RESPONSES

The students' explanations were classified according to the categories similar to those described more fully earlier:[1]

Category N: No explanation or statement, "I can't explain."

Category I: An explanation referring to estimates, guesses, appearances, or extraneous factors without reference to the data. Examples of predictions and the related explanations are:

8—I figured he can't be too much taller.

9—I just thought Mr. Tall would be three more smallies bigger than Mr. Short.

Category IC: An explanation using the data haphazardly or in an illogical Way. Examples are:

$8\frac{1}{2}$—About one button is half a paper clip, so I added $2\frac{1}{2}$ to 6

12—I subtracted 4 from 6 and got 2. Then I multiplied 6×2 and got 12. I got the 6 from measuring Mr. Short.

Category S: An explanation based on a change of scale that the subject does not justify in terms of the data. An example is:

12—If Mr. Short is 6 smallies high, Mr. Tall must be twice as high.

Category A: An explanation focussing on a single difference (tall-short or paperclips-buttons) uncoordinated with other differences, and solving the problem by addition. An example is:

8—Mr. Tall was 2 buttons higher than Mr. Short, so I figured he was two paper clips higher.

Category AS: An explanation focussing on the excess height of Mr. Tall, and scaling up the two excess buttons by an unexplained factor to compensate for the size difference of buttons and small paper clips. An example is:

10—Because two paper clips are about the size of a button, I added four clips for the two buttons.

Category IP (incomplete proportion): An explanation making use of one ratio involved in the proportion, but not applying that ratio correctly. Examples are:

$6\frac{2}{3}$—I took 4 buttons and 6 buttons which is $\frac{2}{3}$. Then I added it to the paper clips, $6 + \frac{2}{3} = 6\frac{2}{3}$.

8—Because if it is 4 for short and 6 for tall, then Mr. Short is $\frac{1}{3}$ shorter and Mr. Tall must be $\frac{1}{3}$ or 2 paper clips bigger; so if Mr. Short is 6, Mr. Tall is 8.

$7\frac{1}{2}$—4 buttons equals six paper clips, so each button is $1\frac{1}{2}$ paper clips, and $6 + 1\frac{1}{2} = 7\frac{1}{2}$.

Category P: An explanation using proportional reasoning and relating the scale factor to the measurements. The answers falling into this category were divided into three sub-categories, as follows:

Subcategory PC (Proportion, concrete)—An explanation using the relation that one button is about 1.5 paper clips, where this relation is obtained by measuring one-fourth of Mr. Short's height with paper clips. On most of the papers there were pencils marks that indicated how Mr. Short had been divided into fourths, frequently after a trial-and-error procedure. Arithmetic errors were common. Examples are:

$9\frac{1}{2}$—I divided the $6\frac{1}{2}$ by 4 and then added to the $6\frac{1}{2}$ two of the fourths. (Pencil marks along Mr. Short.)

9—I worked out how long were the buttons with the paper clips, then I figured out 6 buttons; then I counted how many paper clips and got the answer. (Pencil marks along Mr. Short. E's question elicited the answer that each button was $1\frac{1}{2}$ paper clips).

$9\frac{1}{2}$—I used my fingers and made different sizes and found out what size button would fit 4 of them into Mr. Short. Then I measured 2 more of that same size.

Subcategory AP (addition and proportion)—An explanation focussing on the excess height of Mr. Tall, which is expressed in terms of paper clips by a factor based on the data. Examples are:

9—Divide 4 into 6 equals $1\frac{1}{2}$. There is two more buttons in Mr. Tall, so you add $1\frac{1}{2} + 1\frac{1}{2} + 6 = 9$.

$9\frac{3}{8}$—Since Mr. Tall is as big as Mr. Short plus $\frac{1}{2}$ more, I got half of Mr. Short ($3\frac{1}{8}$) and added it to all of Mr. Short and got Mr. Tall.

Subcategory R (application of ratio): An explanation using a proportion or deriving the scale ratio for the paper clips or heights directly from the data, and applying ratio in a proportion. Examples are:

9—6 is $1\frac{1}{2} \times 4$, so $6 \times 1\frac{1}{2} = 9$

9—Mr. Short is four buttons and six smallies; then every $1\frac{1}{2}$ smallies is a button; so Mr. Tall is 9 smallies tall.

9—Mr. Short to Mr. Tall is 4 to 6 buttons; it will be the same with paper clips, four is to six as six is to nine.

RESULTS

Five hundred of the subjects responding to Form B of the ratio task were from suburban schools in the San Francisco Bay Region, in the same community where the earlier investigations were carried out. The others came from urban schools in the region. Table 1 includes, in the first column of data, the percentages of subjects in the entire group placed in each category. Most striking is the exceedingly small fraction of responses in the S (scaling) category, which had included about one-third of the responses when Form A of the ratio task was administered.

Table 1. Evaluation of Explanations on Forms A and B (percent).

| | Grades (number) | | | | | | |
| | 4–9 | 4 | | 6 | | 8 | |
Category	Form B (610)	Form A (51)	Form B (63)	Form A (161)	Form B (73)	Form A (156)	Form B (173)
N	2	1	8	1	1	0	1
I	10	32	16	25	18	6	5
IC	6	5	4	6	11	2	6
IP	4	0	0	0	0	0	10
S	4	35	6	28	1	37	3
A	32	16	53	25	26	21	29
AS	5	6	2	6	7	8	4
P	37	5	12	9	36	26	42
PC	11	3	8	3	15	8	12
AP	10	0	4	1	11	4	14
E	16	2	0	6	10	14	16

For a more detailed description of the change in the pattern of explanations, please refer to the later columns of Table 1. There we list the percentages of student responses in each category for the two forms of the task, Form A with visual presentation of larger paper clips and Mr. Tall, Form B without. The subjects are the suburban fourth, sixth, and eighth graders in our sample, for whom we have data in comparable groups on both forms. The Category P responses to Form A have been rescored according to the three subcategories described in the previous section of this article.

It can be seen from the Table that the responses in Categories I and S, which depend partially on students' visual estimate of Mr. Tall and/or the large paper clips, decreased markedly on Form B compared to Form A; responses increased in Categories IC, IP, A, and P, which require conceptual processing of the data. Only very few students—eight per cent of the fourth graders—were so baffled by Form B of the task that they failed to respond. We had been concerned that Form B might be so abstract that many elementary school students would not respond at all, as happened on the Islands Puzzle,[5] but this was not the case.

DISCUSSION

What is the significance of these findings? Most importantly, it appears that on Form A of the task a substantial fraction of students—between 30 and 40 percent—preferred to provide an explanation derived principally from perceptions rather than relying on the measurements. When deprived of the opportunity to see Mr. Tall and the large paper clips, they did use the data with varying degrees of success. We therefore conclude that the intuitive and scaling responses to Form A (Categories I and S) may rather give evidence of a subject's attitude toward handling of the data in the task than of his cognitive level of competence. This explanation is consistent with our earlier difficulties in identifying a developmental sequence for Categories A and S. We believe that Form B of the ratio task is more satisfactory for studying cognitive development, because it compels the subjects to make use of the data.

We shall now discuss the developmental significance of the various responses. Only Category P responses indicate proportional reasoning on the ratio task, since they take all the data into account in a coordinated way. Responses in Subcategory PC are examples of a concrete operational approach, in that the subjects engage in manipulative or diagrammatic steps to represent the units of measurement and their size relation. Responses in Subcategories AP and R reflect proportional reasoning on the level of formal thought. One may speculate whether an explanation in Category R that makes explicit reference to the paper clip/button ratio is qualitatively somewhat less abstract than an explanation setting up the proportion and solving it without identifying the physical meaning of intermediate steps. We have, for example, found that more advanced students and scientifically trained adults prefer to use ratios of heights rather than the paper clip/button ratio.

Next we turn to Categories N and I, which include responses that cannot really be considered explanations. From our point of view, they must therefore be considered preoperational. Their combined number decreases from twenty-four percent in the fourth grade to six percent in the eighth.

Finally, we come to responses in Categories IC, S, A, IP, and AS. These make use of the data, though not always completely and not in a coordinated way. The present findings do not allow these categories, which include about half of all responses, to be ordered from the least to the most advanced, or to be placed in a developmental sequence. We consider the responses to be transitional and would expect the percentages to be affected by even minor differences in task presentation, such as the difference between Forms A and B. Most of these categories probably do not represent developmental stages. Rather, they may reflect recent school experience, partially assimilated. These transitional forms may therefore well provide suggestions about the nature of a teaching program that would help students progress toward the correct and consistent use of ratios. We should particularly like to point out the ten percent of eighth grade responses in Category IP. Some of these students are surely on the verge of proportional reasoning.

Category A is especially interesting, because a large fraction—about one-third—of all the responses were so classified. Subjects responding in Category A make use of all the data in a simple, well-defined way. However, their attention to differences rather than to total heights leads them not to raise the question, "If two added buttons correspond to two added paper clips, then why do the original four buttons correspond to six paper clips?" This hypothetico-deductive reasoning, characteristic of formal thought and implicitly included in proportional reasoning, is clearly absent from Category A responses.

From a careful examination of the responses to Form B and to similar tasks we have piloted, we suspect that the specific value of the ratio has an influence on students' explanations. Thus the ratio of 3/2, which lies between one and two, tends to increase the percentage of additive responses. For a ratio of 2/1, many students use proportional instead of additive reasoning. For a ratio of 5/2, additive responses are less numerous, and some students use approximate ratios of two or three, or become confused by the computation required. We conclude that Form B of the Paper Clips Task gives valuable indications of the subjects' proportional reasoning, but that it is not definitive.

It is also interesting to compare the present findings with the results of our earlier effort to cue subjects toward the use of the data by telling them to use the data while confronting Form A of the task.[1] The cueing had only slight effect on scaling responses, and resulted primarily in a shift from I to IC. In other words, the students' reaction to the verbal cue was not nearly as marked as their reaction to the absence of seeing Mr. Tall, even though many seem to have had the ability to conform to the cue's requirements. Why did they not do so? What aspects of a task determine the cognitive resources an individual will apply? Answers to these questions are important when an individual's functioning outside the school setting is considered, and when instructional objectives are formulated.

We are grateful to the Orinda Union School District, Orinda, California, for cooperating in this study.

REFERENCES

1. Robert Karplus and Rita W. Peterson, *School Science and Mathematics* Vol. 70, pp. 813–820 (December 1970).
2. Robert Karplus and Elizabeth F. Karplus, *School Science and Mathematics* Vol. 72, pp. 735–742 (November 1972).
3. H. F. Witkin, *Psychological Differentiation*, New York: John Wiley, 1962.
4. B. Inhelder and J. Piaget, *The Growth of Logical Thinking from Childhood to Adolescence*, New York: Basic Books, 1958.
5. Robert Karplus and Elizabeth F. Karplus, *School Science and Mathematics* Vol. 70, pp. 398–406 (May 1970).

Intellectual Development Beyond Elementary School V: Using Ratio in Differing Tasks

Warren Wollman and Robert Karplus

We have investigated the development of the concepts of ratio and proportion in several studies[1,2,3] for two principal reasons: first, these concepts are vital to an understanding of quantitative relations in science; second, they are an essential part of the integrated system of mental operations Piaget has called formal thought.[4] Our concentration on the Paper Clips Task[1,3] has enabled us to accumulate directly comparable information about many students, but it has also raised questions concerning the applicability of our observations to young people's use of ratio and proportionality when facing problems other than size relations of approximately 2:3. These uses may derive from different physical relationships, such as the turning rates of linked pulleys of differing diameters, or different numerical relationships, such as the simpler 1:2 ratio or the more complicated 5:9 ratio.

Other workers, especially Lunzer[5,6] and Lovell,[7] have investigated the development of proportional reasoning using matching lengths constructed of Cuisenaire rods,[8] a pantograph, balance beams, sets of number pairs with missing numbers, and similar items. They found that proportional reasoning unaccompanied by physical actions was rarely used by

*Supported by a grant from the National Science Foundation.

Warren Wollman and Robert Karplus, AESOP*, Lawrence Hall of Science, University of California, Berkeley, California, 94720.

"Intellectual Development Beyond Elementary School V: Using Ratio in Differing Tasks" was taken from *School Science and Mathematics*, 1974, vol 74, pgs. 593–613.

average subjects below age fifteen. Younger children solved some of the tasks by using successive addition, especially when they were shown concrete examples that illustrated the ratio inherent in the problem. The authors concluded that although their subjects had learned multiplication and division, they did not really appreciate the inverse relationship between them. Their subjects tended to avoid using division.

A question that may be raised concerning all these investigations is this: what accomplishments are evidence of proportional reasoning at the formal level? Certainly the application of a 2:1 or 3:1 ratio in an appropriate context is an instance of proportional reasoning. It is also clear, however, that many students who can deal with these numerically simple ratios use other techniques, such as addition, estimation, or even pure guesswork when faced with numerically more complex problems.

In our opinion, proportional reasoning, *as an element of formal thought*, is a highly adaptable technique that is used in appropriate mathematical and physical situations regardless of the numerical values of the data, the sequence in which they are presented, or the context. Furthermore, proportional reasoning can be used to test the applicability to a given physical situation of hypothesized mathematical models expressing direct or inverse proportionality. We believe, however, that the formulation of such mathematical models to describe the sizes of shadows, the operation of a balance, the turning rates of connected pulleys, and other phenomena requires elements of formal thought *in addition* to proportional reasoning. To focus on the latter, tasks must be devised that apply the ratio concept in familiar or easily imagined situations and do not require the understanding of light propagation, the law of moments, or other scientific principles.

In a recent publication, Lunzer, Harrison, and Davey[9] have reported that certain tasks requiring the application of propositional logic resulted in success rates that depended on details of the task presentation and material. The finding that the quality of reasoning depends substantially on content is inconsistent with the concept of formal reasoning as characterized in the preceding paragraph.

In this article, we report the results of our investigation of the responses of seventh- and eight-grade students to six problems that require proportional reasoning and represent differing degrees of concreteness. Our study of proportional reasoning therefore sheds further light on the generality of formal thought.

After a brief overview of the way students were assigned to tasks, we describe each task and the students' responses to that task. Then we turn to a correlation of performance on differing tasks, and finally discuss the significance of our data in the light of earlier work.

EXPERIMENTAL DESIGN

Our investigation of proportional reasoning employed four tasks on which subjects were asked to make inferences and to provide an explanation of their reasoning. These are the Paper Clips Task-Form B (described previously[3]), the Candy Task, the Ruler Task, and the Pulley Task to be described below. In addition, we used a short-answer paper-and-pencil test consisting of three geometrical items and four numerical items to extend the range of situations faced by the subjects.

We obtained responses from a total of 450 students, who included approximately equal numbers of seventh and eighth graders in two intermediate schools in Orinda, California. About 30 percent of the students were grouped in pre-algebra classes that were participating in an accelerated curriculum. Table 1 lists the numbers of students in each school responding to each task. The actual numbers to be reported when performance on tasks is compared in the results of this study will often be smaller than those given in Table 1 because some subjects responded to only one task.

Table 1. Numbers of Subjects Participating on the Tasks.

	Paper clips	Candy	Ruler	Pulley	Geom.	Numer.
School 1	191		92	65	199	198
School 2	212	238				

PAPER CLIPS TASK

The Paper Clips Task—Form B was administered to classroom groups in the way described previously, and the students' responses were classified according to the same categories.[3] This task requires identification and application of a 2:3 ratio. The results are given in the first column of data in Table 2. These percentages are very close to the findings we reported earlier for 173 eighth graders, who are included in the present sample. There were no instances of appreciable differences in replies between boys and girls on this or any other tasks.

CANDY TASK

Subjects (Ss) in classroom groups were given pages with information and questions (see Appendix) by one of the authors or a trained associate. The experimenter (E) read the entire paper to the students and then answered questions about the procedure individually while Ss were working. Administration of this task to a class required about fifteen minutes.

The Candy Task was intended to be similar to the Paper Clips Task in that it concerned familiar materials and a non-integer ratio on the principal item, question (a). It was more abstract than the Paper Clips Task in that it was entirely verbal and did not permit observation and manipulation of any materials. There were three other differences from the Paper Clips Task: (1) the ratio method was applicable because the candy mixture was uniform from bag to bag, rather than because the units of measurement were uniform; (2) bowlfuls of red and yellow candy were prepared arbitrarily by emptying bags and did not have the permanent identities of Mr. Tall and Mr. Short; (3) a second question (b)

Table 2. Evaluation of Explanations of Four Tasks (percent).

	Tasks			
	Paper clips N = 403	Candy (a) N = 238	Ruler N = 92	Pulley (b) N = 65
Category				
N, I	5	11	2	15
IC	5	13	8	11
IP	8	—	—	—
S	4	—	1	—
A	32	50	2	31
AS	6	—	—	—
P	40	26	87	43
PC	12	—	23	14
AP	13	3	—	11
R	15	23	64	18

involving an integer ratio ($20:10 = 2:1$) was also included, either before or after the principal question (a), to permit comparison of Ss' approaches to these two problems. The task was presented in two forms, with question (a) preceding (b) and with question (b) preceding (a), to identify any "learning" effect that might influence the responses.

With very minor modifications, the categories used in the Paper Clips Task could be adapted to the Candy Task:

Category N: No explanation or statement "I can't explain."

Category I: An explanation referring to estimates, guesses, or extraneous factors, or revealing a complete misunderstanding of the conditions of the task. Examples of answers to questions (a) or (b) and the related explanations are:

> a:blank—You can't tell because it depends on how many yellows were in a bag.
> a:12—Add it.
> a:12—It did not say there were any more yellows.
> a:29—If there are 32 candies in each bag, then the second bag had 15 red and 17 yellow. (Misunderstanding.)

Category IC: An explanation using the data in a haphazard or illogical way. Examples are:

> a:23—Each time the number in the yellows is subtracted from the number in the red bowl, that's how many there are in the yellow.
> a:17—$\frac{15}{3}$ = 5, and you would add that 5 on to the 12, which equals 17.
> b:8—I just subtract 12 from 20 and got 8.
> b:5—Assuming that she had 3 bags, she had 5 in each, 10 red, 5 yellow.

Category S: An explanation based on a red:yellow ratio the subject does not justify. An example is:

> a:19—There are about 2 yellows for every red. (Note the reversal!)

Category A: An explanation focusing on a single difference (initial red—initial yellow or final red—initial red) uncoordinated with other differences, and solving the problem by addition. Examples are:

> a:27—20 to 35 is 15, so you add 15 to 12 and get 27.
> a:27—There are 8 less yellows, so I subtracted 8 from 35.
> a:19—John had 8 more reds than yellows. So I subtracted 8 from 15 (additional reds) and added the answer to 12.

Category P: An explanation using proportional reasoning and relating the scale factor to the data. We established only two subcategories AP and R because the verbal explanations, in the absence of materials and diagrams, did not permit identification of the concrete approach in Subcategory PC.

Subcategoy AP (addition and proportion)—An explanation focusing on the excess of red candies, which is related to the additional yellows by a factor based on the data. An example is:

> a:21—Since 35 is the number of reds plus three-quarters more, I added the yellows (12) and three-quarters more (9) and got 21.

Subcategory R (application of ratio)—An explanation deriving the scale factor for the yellow candies from the data, and applying it by multiplication or division. Examples are:

Table 3. Responses on Question (b) of the Candy Task
(percent of students who did not use proportional
reasoning on question (a) of the task).

	Sequence of questions	
Category	(b), (a) (N = 72)	(a), (b) (N = 103)
N, I	11	18
IC	13	15
A	25	55
R	51	12

a:21—$\frac{12}{20} = \frac{3}{5}$, so 12 is $\frac{3}{5}$ of 20, $\frac{3}{5}$ 35 is 21.

a:21—4 goes into both 20 and 12; 20, 5 times, 12, 3 times. 5 goes into 35 7 times. $7 \times 3 = 21$.

a:21—I used the ratio $\frac{20}{35} = \frac{12}{?}$, $12 \times \frac{35}{20} = 21$.

a:21—If you guess he had 4 bags at first, $\frac{20}{4} = 5$, and $\frac{12}{4} = 3$, so then he had three more bags to put in to make 35, so $12 + 3 + 3 + 3 = 21$. (Note the iteration suggesting concrete operations.)

b: 6—$\frac{10}{5} = 2$, $2 \times 3 = 6$, same as above.

b: 6—$\frac{3}{5}$ of 10 is 6.

b: 6—If all the ratios are the same and it has $\frac{1}{2}$ as many reds it must also have $\frac{1}{2}$ as many yellows, which is 6.

The results for question (a) in the Candy Task are presented in the second column of data in Table 2. We have combined the responses for the two forms of the task because the percentages in the various categories depended only slightly on the order of presentation. Now, 63 Ss (26 percent of 238) used proportional reasoning on question (a), and all of them also used proportional reasoning on question (b). There was, however, a very substantial and statistically significant difference between the question (b) responses on the two forms by the 175 Ss (74 percent of 238) who did not use proportional reasoning on question (a). In Table 3, the data are compared for the 72 Ss who faced (b) before (a) and the 103 Ss who faced (a) before (b). Fifty-one percent, or slightly more than half of the former group, used proportional reasoning on (b), while only 12 percent, less than one-eighth of the latter group, did so. Additive responses on (b) were much more numerous when (b) followed (a) than when it came first. We conclude that application of a 2:1 ratio is certainly not indicative of proportional reasoning at the formal level for two reasons: (1) use of the ratio is not at all predictive of the use of proportional rather than additive reasoning on a more complicated task; (2) prior use of non-proportional reasoning interferes seriously with the application of the 2:1 ratio.

RULER TASK

The Ruler Task was administered as an individual interview. S and E worked with a two-foot long ruler marked in inches along one edge, in centimeters along the other. The two scales started at opposite ends of the ruler, so that a direct comparison of scales was not possible (Figure 1). The equipment was explained to the subjects, who were told that E was

Figure 1. Two-Foot ruler and wooden rod used in the Ruler Task.

interested in the method of reasoning, not in the exactness of the answers. Paper and pencil were made available to all Ss. With S watching closely, a twelve-inch-long rod was displaced two inches along the central groove of the ruler and E asked, "I moved the stick two inches; how many centimeters did that make?" S looked at the centimeter scale and replied "FIve." The ruler and stick were then taken away and S was asked, "If I moved the stick eight inches, how many centimeters would that be?" After answering, S was asked to explain his procedure and was reminded that two inches had been equivalent to five centimeters. Following this standard part of the interview, E brought up other displacements, reverse conversions from centimeters to inches, or related questions to explore the subject's reasoning further. The entire interview took from two to four minutes.

The Ruler Task was intended to be easy for Ss at the concrete level in spite of the non-integer relationship of the units. After observing the 2:5 ratio of the units, Ss could proceed by counting up how many two-inch lengths, each equal to five centimeters, were needed to make the eight-inch displacement. (In the Paper Clips Task, the given ratio of 4::6 had to be subdivided before it could be applied to the problem of 6:x.) The concrete nature of the ruler and familiarity of length measurement also helped.

With minor modifications, the categories used in the Paper Clips Task again served as basis for classifying the students' responses. Some Ss quickly switched their numerical answers just after they had uttered them, when asked to justify them, or when reminded of the two-to-five equivalence of inches and centimeters. In those cases, the second answers and associated explanations were classified. Examples of answers and the explanations offered follow:

Category IC:
40—Eight times five is forty.
30—Because two inches was five centimeters; there's six more, so six times five is thirty.
Category S:
16—Eight times two is sixteen.
Category A:
11—five is three more than two, eight and three is eleven.
Category P: We found it necessary to use only two subcategories.
Subcategory PC (an explanation iterating the observed relationship of inches and centimeters):
20—Two goes with five, four goes with ten, six goes with fifteen, eight goes with twenty (accompanied by the writing of the numerals).
Subcategory R (application of ratio):
20—One inch is two-and-a-half centimeters, so eight inches is twenty.
20—Eight divided by two is four, four times five is twenty.
20—Four two's equal eight, so four times five is twenty.

The results for the ninety-two students answering the Ruler Task are presented in the third column of data in Table 2. As expected, the great majority of the students applied proportional reasoning, either as ratio (Subcategory R) or by iteration of the given relationship

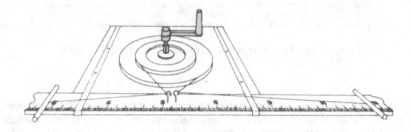

Figure 2. Pulleys, strings, and meter stick used in the Pulley Task.

(Subcategory PC). About one-fourth of these students had switched their answer after an initial incorrect reply, most often $8 \times 5 = 40$.

PULLEY TASK

The Pulley Task was administered as an individual interview immediately following the Ruler Task. Central to the task was a system of two pulleys (diameter ratio $3:2$) fixed to the same shaft that was free to turn in a bearing (Figure 2). A string was attached to each pulley in such a way that one string would around its pulley when the other was played out, and vice versa. The strings were guided along a meter stick so that displacements of the free ends could be measured and compared accurately. S and E each took the free end of one string and S pulled his until the end has moved ten centimeters. E told him that the other end has moved fifteen centimeters, and invited him to comment on this observation. Regardless of his reply, S was then asked to predict for question (a): "How far will my string move when yours moves six centimeters?" Then S explained his prediction. Following his reply, S was allowed to carry out the operation and observe the outcome (E's string moved nine centimeters). He was then asked to predict and explain for a displacement of his string of four centimeters; this constituted question (b). Further observations, predictions, and discussions of why the strings moved unequally concluded the interview after four to eight minutes.

This task was intended to be difficult and to challenge Ss who used proportional reasoning on the other tasks. At the same time, it was possible for a subject to compare his prediction with an experimental outcome, and then to tackle a new prediction problem where he might apply what he had learned—a significant difference from the Paper Clips Task that was confined to a single example and did not provide for feedback. The task was made difficult by the S's need to cope with these three complications: (1) the mechanical and size relationships of the two pulleys; (2) the correlation and conservation of lengths of the strings, as one string is shortened and the other lengthened by a different amount; and (3) the non-integer ratios of $10:15$ for the displacements within one experiment, and of $10:6$ when comparing the two experiments.

Once again, the same categories were used to classify the two responses each student gave. Here are some examples of their answers:

Category IC:
a:8—You move one-third more than I.
a:9—Nine and six if fifteen.

Table 4. Correlation of Responses on Questions (a)
and (b) of the Pulley Task (numbers of replies).

| | | Question (b) Categories | | | | | | Total |
		N, I	IC	A	PC	AP	R	on (a)
Question (a) Categories	N, I	1	0	5	0	1	0	7
	IC	0	2	0	0	1	0	3
	A	8	5	14	9	3	2	41
	AP	0	0	1	0	1	1	3
	R	1	0	0	0	1	9	11
Total on (b)		10	7	20	9	7	12	65

Category A:
    a:11—Your string went five more than mine, six and five is eleven.
    Subcategory PC:
        b:6—When I did ten, you got fifteen, and when I did six, you got nine; since six
            and four is ten, it is six, because nine and six is fifteen.
    Subcategory AP:
    a:9—Ten plus one-half more gives fifteen, so six plus one-half more gives nine.
    Subcategory R:
    a:9—Ten is two-thirds of fifteen, and six is two-thirds of nine.
    a:9—Ten over five is two, and fifteen over five is three; six over three is two and
        nine over three is three.
    a:9—Ten divided by two is five, times three is fifteen; six divided by two is three,
        times three is nine.

The results for the sixty-five subjects replying to question (b) of the Pulley Task are given in the last column of Table 2. With the additional information provided by their observational feedback to question (a) of the task, the distribution of responses is similar to the distribution on the Paper Clips Task. Only about half the students gave similar answers on (a) and (b), however. The others changed categories, as shown in Table 4. Most striking is the large number of Ss whose first explanation fell into Category A (almost two-thirds of the total), and who then observed the inadequacy of their procedure. Almost equal numbers of these either advanced to a form of proportional reasoning (fourteen Ss) or became confused by the contradictions and resorted to intuitive computation or guessing (thirteen Ss). The ones who advanced usually coordinated several concrete operations with the help of the added data, and did not establish a clear-cut ratio. Five of the Ss who had originally not given any explanation used the additive method on part (b), and one even used addition and proportion. Nevertheless, the qualitative impression gained during the interview was that most students were responding to a pattern of numbers and did not relate their approach to the physical properties and constraints of the pulleys.

    Their difficulty in grasping the physical relationships is also brought out by their comments in attempting to account for the unequal movement of the strings. Only one-fourth related their observation to the unequal size of the pulleys, and only 6 percent (four Ss) identified the pulleys' circumferences as being significant. The others mentioned unequal string lengths (29 percent), different starting points on the meter stick (15 percent), or did not provide and explanation. In other words, the mechanism was a mystery to most of the

Table 5. Number of Errors on Geometrical
and Numerical Items (percent).

Number of errors	Geometrical N = 199	Numerical N = 198
8 or more	13 (low)	7 (low)
7	3 (low)	4 (low)
6	4 (low)	15 (low)
5	7 (low)	13 (low)
4	9 (medium)	30 (medium)
3	7 (medium)	12 (medium)
2	10 (medium)	7 (high)
1	16 (medium)	6 (high)
0	32 (high)	6 (high)

Ss even though they could see it operate without any hidden parts. Their responses to question (b) were not correlated with these accounts.

GEOMETRICAL AND NUMERICAL ITEMS

Subjects in classroom groups were given pages with the instructions and questions as part of a booklet containing other materials also (see Appendix). They worked on these items without oral explanation, but were able to have questions answered individually. Completion of all the problems required about forty minutes, of which about twenty were devoted to the Geometrical and Numerical items. Even though many Ss did not answer all questions, the facts that they completed later parts of the booklet and that many handed it in early are evidence that their omissions were usually not caused by lack of time.

The Geometrical Items presented information in pictorial rather than numerical form, and were similar to exercises in mathematics texts. They were intended to identify Ss' ability to recognize a fraction of a whole under these conditions, and to describe this fraction pictorially and numerically. Note that some of the diagrams to be shaded were divided into equal parts, while others were divided into unequal parts.

The Numerical Items presented information in completely abstract form—number pairs out of context. They were intended as the most difficult tests of Ss' ability to apply proportional reasoning with no circumstantial clues. Like the Geometrical Items, they resemble textbook mathematics exercises. Similar items have been used by other investigators of proportional reasoning.[7]

Since the Ss did not explain their methods of arriving at their answers, the responses are less informative than those on the four tasks described earlier. Carelessness and haste on the students' part also introduced errors that cannot be distinquished from errors in reasoning. Frequency distributions of the number of errors are reported in Table 5. We have indicated "low," "medium," and "high" levels of performance that we shall use later to compare and interpret the results. The Geometrical Items were clearly much easier than the number pairs, with almost half the Ss making no more than one error on the former, while only one-eighth did as well on the latter. The fractions were stated more reliably than the figures were shaded, and the figures with unequal subdivisions provided more difficult than the others. Even these were answered correctly by more than half the students. About 17 percent of the Ss drew their own boundaries to subdivide some shapes, rather than following the boundaries in the diagrams.

Table 6. Nature of Errors on Numerical Items (percent).

	Part			
	(a)	(b)	(c)	(d)
additive	5	8	40	13
reversal	20	3	0	0
other error	15	9	33	40
blank	1	2	17	30
(correct)	(59)	(78)	(10)	(17)

On the Numerical Items, two-thirds of Ss made at least four mistakes; most of these had no correct answers at all on parts (c) and (d). By requiring two completions in each part, we were able to identify the erroneous reasoning of many students. This is summarized in Table 6, where we have listed specific forms of mistakes: (1) an incorrect additive procedure; (2) reversal of the rule, leading, for instance, to the answers 5 and 3 instead of 5 and 27 in (a); (3) "other" errors that could not be classified; and (4) blanks. In part (a), the major source of error was reversal of the rule in (iv), where the first rather than the second member of the pair had to be inferred. In part (c), the fractional answer in (iii) was found correctly only by 10 percent of Ss, but incorrect fractional results were given by an additional 6 percent. Additive procedures were especially numerous in (c), including a "method of declining differences" that led to answers of 9 and 15 in (iii) and (iv), respectively. The difficulty of the Numerical Items confirms results obtained by others.[7]

Success on the Geometrical and Numerical Items correlates significantly. Table 7 describes the correlation of "low," "medium," and "high" performance for the 196 students who answered both items. The asymmetry of the entries shows that even with our more lenient evaluation of the Numerical Items, they were a more difficult challenge for the seventh and eighth graders in our sample. In other words, a perfect score on the Geometrical Items was still not a very good predictor of success on the Numerical Items, but the reverse did hold.

SUMMARY OF TASKS

To provide an overview of the similarities and differences among the six tasks, we have used Table 8 to summarize the principal elements and relations that bear on proportional reasoning. The central columns contain the given data (A, B, and C), while the last column

Table 7. Correlation of Success on Geometrical and Numerical Items (number of subjects).

		Numerical items			
		Low	Medium	High	Total
Geometrical Items	Low	36	12	0	48
	Medium	30	38	8	76
	High	11	33	28	72
Total		77	83	36	196

Table 8. Summary of Proportional Reasoning Tasks, in the Form A:B = C:X.

Task	A	B	C	X
Paper Clips	Mr. Short's height in buttons	Mr. Short's height in paper clips	Mr. Tall's height in buttons	Mr. Tall's height in paper clips
	Mr. Short's height in buttons	Mr. Tall's height in buttons	Mr. Short's height in paper clips	Mr. Tall's height in paper clips
Candy	Initial number of red candies	Initial number of yellow candies	Final number of red candies	Final number of yellow candies
	Initial number of red candies	Final number of red candies	Initial number of yellow candies	Final number of yellow candies
Ruler	First displacement of rod in inches	Second displacement of rod in inches	First displacement of rod in centimeters	Second displacement of rod in centimeters
	First displacement of rod in inches	First displacement of rod in centimeters	Second displacement of rod in inches	Second displacement of rod in centimeters
Pulley	First motion of S's string	First motion of E's string	Second motion of S's string	Second motion of E's string
	First motion of S's string	Second motion of S's string	First motion of E's string	Second motion of E's string
Geometrical	Sample shape, whole	Sample shape, part	Test shape, whole	Test shape, part
Numerical (a, c, d)	A	B	C	X

indicates the prediction required (X). The Paper Clips, Candy, Ruler, and Pulley Tasks can be described by a proportion in two different ways, as indicated in the table. Though there is no logical difference between the two forms, there appears to be a psychological difference in that most Ss preferred one over the other, and we have always listed the preferred form first.

COMPARISON OF PAPER CLIPS AND CANDY TASKS

One group of 195 students completed both the Paper Clips and the Candy Tasks. In this comparison, we shall refer only to part (a) of the Candy Task, which is similar to the Paper Clips Task, though possibly somewhat more difficult because the numbers are larger. To

Table 9. Correlation of Responses on Paper Clips
and Candy (a) Tasks (percent, N = 195).

		Candy task part (a)			
		N + I + IC	A	AP + R	Total
Paper Clips Task	N + I + IC + S	4	7	1	12
	A + AS	8	21	5	34
	IP + PC	5	13	4	22
	AP + R	3	9	20	32
Total		20	50	30	(100)

simplify the correlation matrix and eliminate small groups, we have combined categories so as to leave only four: (1) N, I, IC, and S; (2) A and AS; (3) IP and PC; and (4) AP and R. Note that IP and PC have no analogue on the Candy Task. Table 9 presents the correlation matrix. It can be seen that Ss' performance on the two tasks correlates well, with two exceptions: (1) The intuitive categories on one task are not correlated with any category on the other task; (2) Category IP + PC on the Paper Clips Task, which we had identified as indicating a transition to proportional reasoning, correlates surprisingly well with Category A on the Candy Task. In other words the transition is only a very slight advance over additive reasoning: under the somewhat more abstract conditions of the Candy Task, most subjects regressed to additive reasoning rather than advancing to one of the subcategories of proportional reasoning.

COMPARISON OF PAPER CLIPS TASK WITH GEOMETRICAL AND NUMERICAL ITEMS

As we have indicated, the Geometrical and Numerical Items did not require an explanation, so that they had to be evaluated in terms of the accuracy of the answers. Even though certain patterns emerged from some of the incorrect answers, these do not correspond to the categories of explanations except for the additive rules used by many students on some of the Numerical Items. There is a significant ($p < 0.01$) but not dramatic correlation, in that 50 percent of 100 Ss using an additive procedure on the Numerical Items responded in Category A on the Paper Clips Task, while only 27 percent of 77 Ss not using an additive procedure responded in Category A.

Overall, the comparison of the Items with the Paper Clips Task does not show substantial correspondences and we shall not present correlation data in detail. As might be expected Ss who ranked high on numerical items (N = 36, see Table 6) tended to use proportional reasoning (70 percent), while those who ranked low on geometrical (N = 48) preferred additive reasoning (78 percent). Seven of the twelve students who responded in Subcategory PC to the Paper Clips Task were medium on both items, while the other five were ranked high on one or both items. Seven of the fifteen students responding in Category IP were in the high group, suggesting that this category includes Ss who may be able to reason well but had difficulty expressing themselves.

COMPARISON OF PAPER CLIPS, RULER, AND PULLEY TASKS

As can be seen from Table 2, the Ruler Task was the easiest among the three, while Pulley (b) and Paper Clips Tasks were somewhat more difficult and approximately equal. A total of fifty-two Ss responded to all three tasks, with somewhat larger groups participating in two of them. The numbers of replies in the various categories are therefore smaller than in the comparison discussed earlier, and correlations are less clear-cut. Only eight students consistently showed proportional reasoning in Subcategories AP or R on all three tasks. The others varied their approaches or failed to use proportional reasoning on any task.

Correlations of responses on the Paper Clips Task with each of the other two are shown in Tables 10 and 11. Five features deserve comment: (1) most of the students who responded in Subcategories AP and R on the Paper Clips Task did the same on the Ruler Task; (2) most of the students responding in Subcategory PC on the Ruler Task used additive reasoning on the Paper Clips Task; (3) on the Pulley and Paper Clips Tasks, there was little switching between PC and additive reasoning; (4) only 15 percent of the students responded in Subcategory AP or R on both the Paper Clips and Pulley Tasks; (5) a substantial fraction, about one-third of the students, responded in AP + R on either the Paper Clips or the Pulley Task, but not on both. These observations substantiate some of

Table 10. Correlation of Responses on Paper Clips
and Ruler Tasks (percent, N = 78).

		Ruler task				
		N + I + IC	A	PC	AP + R	Total
Paper Clips Task	N + I + IC + S	3	0	1	9	13
	A + AS	5	4	15	17	41
	IP + PC	0	0	3	12	15
	AP + R	1	0	4	26	31
Total		9	4	23	64	(100)

the conclusions reached earlier, such as the fact that concrete proportional reasoning (Subcategory PC) should be differentiated from Subcategories AP and R. At the same time, we have a clear indication that "formal" proportional reasoning (Subcategories AP and R) does not generalize simply from the Paper Clips to the Pulley Task in spite of the fact that the two tasks were of comparable difficulty.

COMPARISON OF PULLEY TASK AND NUMERICAL ITEMS

The Pulley Task question (b) resembles the Numerical items in that the subject has knowledge of two number pairs (associated displacements of the strings) when asked for the second prediction and explanation. Sixty-four Ss responded to both exercises, and only eight of these had two or more correct answers (out of a possible four correct) on parts (c) and (d) of the Numerical Items. All eight responded in Category P to the Pulley Task. In addition, there were nineteen students who responded in Category P on the Pulley Task, but had one or zero correct answers on parts (c) and (d) of the Numerical Items. Thus, once again, we arrive at the conclusion that the abstract construction of number pairs is very difficult for the seventh- and eighth-grade students with whom we worked and that proportional reasoning did not generalize from the Pulley Task to the Numerical Items.

DISCUSSION AND CONCLUSIONS

Four conclusions emerge quite clearly from the extensive collection of data reported in this article: (1) where concrete aspects existed, as in the Geometrical Items, the Ruler Task, and

Table 11. Correlation of Responses on Paper Clips
and Pulley Tasks (percent, N = 54).

		Pulley task part (b)				
		N + I + IC	A	PC	AP + R	Total
Paper Clips Task	N + I + IC + S	7	2	6	4	19
	A + A	7	22	4	4	37
	IP + PC	2	0	2	9	13
	AP + R	7	6	4	15	32
Total		23	30	16	32	(101)

the Paper Clips Task, more students answered the questions successfully; (2) a small number of students, about one-fifth of each group, applied proportional reasoning at the formal level consistently on several tasks; (3) a somewhat larger number of students, about one-fourth of each group, applied proportional reasoning at the formal level on some tasks but not on others, in an unpredictable way; (4) incorrect additive strategies were used by many students on several tasks. We shall not discuss these items further.

Our results with suburban students in California are generally, consistent with earlier findings reported in British studies.[5,6,7] Lunzer and Pumfrey[5] reported that even many nine-year-olds solved tasks similar to our Ruler Task by using matching lengths of Cuisenaire rods, often interating a basic relationship, such as two greens aligned with three reds. Their solution methods and also those of older subjects made use of additive and iterative procedures that were successful but only rarely involved them in what the authors termed "proportional reasoning"—similar to our Subcategories AP and R. Lunzer's and Pumfrey's use of the balance showed them, just as our use of the linked pulleys showed us, that the physical relationship behind the applicability of a proportion to even a simple system was grasped by few of the students who made a successful prediction.

The lack of generality of proportional reasoning may be interpreted in two ways. The first, in line with Piaget's theory, is that only a small number of students in the seventh and eighth grades had reached the level of formal thought—about 15 percent of the subjects. These used proportional reasoning consistently except on the Numerical Items. In this view, most of the students are transitional and apply proportional reasoning on some tasks and not on others, just as children who are transitional from preoperational to concrete thought apply conservation logic on some tasks but not yet on all. Another view recently expressed by Lunzer[10] is more open-ended, in that it does not provide for a definite stage of formal thought. Rather, there are formal operations, such as proportional reasoning and propositional logic, which are used by individuals when they are suitably motivated, have certain cues that suggest a formal rather than an intuitive approach, and/or are pressed to justify their conclusions. The matter of conceptual style enters here as well, as we have pointed out in another article.[3]

The widespread and frequently persistent use of additive reasoning can be seen especially clearly in Tables 4, 9, and 11, which deal with the Paper Clips, Pulley, and Candy Tasks. About 20 percent of the subjects responded in Categories A or AS to two of these. This observation suggests that additive reasoning on these tasks is a stable strategy and is not selected haphazardly by students who feel the need to respond but are not sure of what they should do. Several factors in the Ss school experience may contribute to this situation. First, addition is surely the arithmetic process with which students are most familiar and are most likely to use when relating data. Second, multiplication is frequently treated as repeated addition, and division as repeated subtraction; some students, therefore, may learn to identify the need to divide by verbal cues in an exercise ("How many paper clips are equal to one button?") rather than by the context as required in our tasks. Third, the remainder in division exercises is usually left as an additive item. The relevance of this custom was called to our attention when several students were asked to predict the conversion of twenty inches to centimeters after they had observed that six inches was equal to fifteen centimeters in a modified Ruler Task. They iterated the observed relationship of six to fifteen three times, reaching the equivalence of eighteen inches to forty-five centimeters, and then added the remainder of two (inches) to obtain forty-seven centimeters! In the Paper Clips and Pulley Tasks, too, the added units may have been perceived as remainder after a single "iteration." Additive reasoning, which appeared in the form of iterations on the original Ruler Task, did not lead to errors there (the remainder was zero).

Mathematics programs in the junior high school appear to contribute little toward proportional reasoning, even though ratio and proportion are topics usually taken up. One reason probably is that ratios are introduced as fractions and proportions as equivalent fractions, to be handled by certain standard procedures of multiplication and division. Yet the essence of ratio is that it may associate six paper clips with four buttons, twelve yellow candies with twenty reds, and fifteen centimeters of one length of string with ten centimeters of another. Curricula make little effort to interpret ratio, proportion, and the related division process in terms of such correspondences of measurements. In this use of division, the concept of remainder has no place.

It seems to us that the preoccupation with integers in both traditional and modern mathematics curricula leads to serious disadvantages when students have to use numbers to describe physical objects and have to carry out operations on these numbers. Furthermore, Cuisenaire rods, systems of coupled pulleys, and other mechanical devices that embody proportions would appear to be valuable teaching devices that could help students engage in proportional reasoning on the concrete level before being faced with similar problems on the abstract level. Combined math-science activities with such an emphasis should be considered seriously for the upper elementary or junior high school grades. At the same time, it may be that the foundations of the mathematics program in the primary grades will have to be rethought as well.

As final item, we should like to discuss a methodological matter. It is clear that our group administration of tasks, requiring Ss to write their responses, could lead to errors because individuals may not describe their reasoning accurately. For instance, if they are lazy, they may write "I don't know" instead of a complicated explanation. It is unlikely that the written reply will lead to an overestimate of their capacity. Conversely, in an interview there is the possibility that E may give information or cues and thereby help the subject. The correlations shown in Table 4 (two interviews), Table 9 (two written tasks), and Table 11 (an interview and a written task) are substantially similar and show relatively small entries for extreme differences in performance of the same subject on two tasks. Hence we conclude that the group approach does not reduce the reliability of the data seriously.

We are grateful to the Orinda Union School District, Orinda, California, for cooperating in this study.

REFERENCES

1. Robert Karplus and Rita W. Peterson, "Intellectual Development Beyond Elementary School II: Ratio, A Survey," *School Science and Mathematics*, 70 (December 1970): 813–820.
2. Robert Karplus and Elizabeth F. Karplus, "Intellectual Development Beyond Elementary School III—Ratio: A Longitudinal Study," *School Science and Mathematics*, 72 (November 1972): 735–742.
3. Elizabeth F. Karplus, Robert Karplus, and Warren Wollman, "Intellectual Development Beyond Elementary School IV: Ratio, The Influence of Cognitive Style," *School Science and Mathematics* (in press).
4. B. Inhelder and J. Piaget, *The Growth of Logical Thinking from Childhood to Adolescence* (New York: Basic Books, 1958).
5. E. A. Lunzer and P. D. Pumfrey, "Understanding Proportionality," *Mathematics Teaching*, 34 (1966): 7–12.
6. E. A. Lunzer, "Problems of Formal Reasoning in Test Situations," in *European Research in Cognitive Development*, edited by P. Mussen, Monograph of the Society for Research in Cognitive Development, Vol. 30, No. 2 (1965): 19–41.
7. K. Lovell and I. B. Butterworth, "Abilities Underlying the Understanding of Proportionality," *Mathematics Teaching*, 35 (1966): 5–9.
8. C. Gattegno, *A Teacher's Introduction to the Cuisenaire-Gattegno Method of Teaching Arithmetic* (New York: Cuisenaire Company of America, Inc., 1960).
9. E. A. Lunzer, C. Harrison, and M. Davey, "The Four-Card Problem and the Generality of Formal Reasoning," *Quarterly Journal of Psychology*, 24 (August 1972): 326–339.

APPENDIX

Candy Task

John has many bags of two kinds of candy. In each bag there are a certain number of red candies and a different certain number of yellow candies. All the bags are alike.

John opens the bags one at a time and puts the reds into a red candy bowl and the yellows into a yellow bowl. After doing this a while, there are 20 reds in the red bowl and 12 yellows in the yellow bowl.

(a) Then John empties some more bags and puts the candies into the two bowls until he has 35 reds in the red bowl. How many yellows does he now have in the yellow bowl?
Explain your answer.

(b) Mary takes some bages away from John, opens them, and empties them. She counts her reds and finds she has 10. Then she counts her yellows.. How many yellows does she have?
Explain your answer.

Geometrical Items

In these drawings a fraction of each figure is shaded in. Look at the figures and try to discover the fraction. Then shade in that fraction of the remaining figures. Write down the fraction in the space provided for it at the right of the drawings. Here is an example.

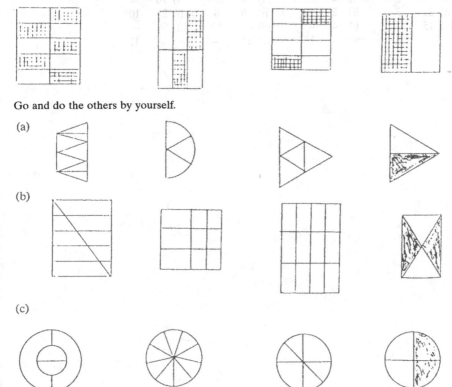

Go and do the others by yourself.

(a)

(b)

(c)

Numerical Items

Here are two pairs of numbers where the members of each pair go together according to a rule. Can you discover the rule?

i) 1   2
ii) 6   12

This the rule: The first number is one half of the second number. You could also say that the second number is two times the first number. Use this rule to complete these two pairs:

iii) 8   —
iv) —   9

The answer to iii) is 16. The answer to iv) is 4½ or 4.5. Here is another example

i) 6   9
ii) 12   15
iii) 4   —
iv) —   11

The answer to iii) is 7 because the second number is three more than the first number. You could also say that the first number is three less than the second. That makes the answer to iv) come out to be 8. Do the rest by yourself.

(a)		(b)		(c)		(d)	
i) 6	2	i) 14	21	i) 9	12	i) 10	4
ii) 21	7	ii) 2	9	ii) 6	8	ii) 25	10
iii) 15	—	iii) 4	—	iii) —	10	iii) 5	—
iv) —	9	iv) —	10	iv) 15	—	iv) —	12

A Survey of Proportional Reasoning and Control of Variables in Seven Countries

Robert Karplus, Elizabeth Karplus,
Marina Formisano, and Albert-Christian Paulsen

Educational researchers in many countries have been directing their attention at what Inhelder and Piaget (1958) have called *formal thought*. The development of formal thought, or at least the formation of formal reasoning patterns, takes place during the secondary school years and is, therefore, of great importance in the teaching of science and mathematics at this educational level.

For this investigation, two areas of formal thought, proportional reasoning and control of variables, were selected. Previous studies of these areas have helped to identify the gradual progress in adolescent reasoning for population samples in the United States, Great Britain, and Switzerland (Karplus & Peterson, 1970; Levine & Linn, 1977; Lovell & Butterworth, 1966; Lunzer & Pumfrey, 1966; Suarez, 1974; Wollman, 1975). Both areas are vital for the understanding of secondary science as shown by recent analyses of many science texts (Karplus et al., 1977). With new science teaching methods and course materials being introduced into schools continually, it seemed worthwhile to investigate student's reasoning in the areas selected.

Robert Karplus, Lawrence Hall of Science, University of California, Berkeley, California 94720. Elizabeth Karplus, Campolindo High School, Moraga, California.   Marina Formisano, University of Rome, Rome, Italy.   Albert-Christian Paulsen, Royal Danish College of Educational Studies, Copenhagen, Denmark.

A Survey of Proportional Reasoning and Control of Variables in Seven Countries, R. Karplus, E. Karplus, M. Formisano, and A.-C. Paulsen, *Journal of Research in Science Teaching*, **14**(5), 411–417, © 1977, Reprinted by permission of Wiley-Liss, Inc., a subsidiary of John Wiley & Sons, Inc.

As independent variables for the study, the students' gender, the country in which they were attending school, and socioeconomic status or achievement level (depending on the school organization in each country) were selected. The dependent variables were performance on a proportional reasoning task and on a control of variables task.

Only very limited information is available on the relation among these variables at the present time. A substantial socioeconomic effect on proportional reasoning has been reported by Karplus and Peterson (1970). In selective schools in Switzerland, better performance on proportional reasoning than in the United States has been observed (Suarez, 1974). Gender differences on several Piagetian tasks have been found by Lawson (1975).

These findings and other reports of student responses to Piagetian tasks led us to expect that the same response categories would be adequate for classifying the performance of students on our tasks in all technologically advanced countries. Therefore, we set out to determine how the independent variables mentioned above would affect the frequency distributions of student reasoning.

METHOD

Subjects

The subjects in our study were 13- to 15-year-old students in seven countries: Denmark (Copenhagen region, $N = 399$); Sweden (Gothenburg region, $N = 280$), Italy (Rome region, $N = 567$); United States (Northeast and North Central regions, $N = 1020$); Austria (Vienna region, $N = 595$); Germany (Göttingen region, $N = 319$); and Great Britain (London region, $N = 376$). Almost 1800 girls and an approximately equal number of boys participated. The age range was chosen to be younger than the school-leaving age in all the countries visited, so that the subjects would represent the entire population.

School organization in the seven countries varies considerably, and we tried to choose the populations for testing to obtain a fair survey appropriate to each school organization. We do not claim to have a representative sample of subjects for any country or even any town.

In Denmark, Sweden, Italy, and the United States, where the schools are comprehensive at the seventh- and eighth-grade levels, the population groups were classified as upper-middle class, middle class, working class, or urban low income, according to the socioeconomic level of the neighborhood served by the school (derived roughly from parental occupation, education, and/or quality of housing). In Austria, Germany, and Great Britain, the schools are selective, divided into types with differing entrance requirements, curricula, and school-leaving ages. These types served as the basis for our classification.

Task Presentation

The task for assessing proportional reasoning was similar to the Paper Clips Task—Form B described previously (Karplus, Karplus, & Wollman, 1974), but included two items rather than only one. The task for assessing control of variables was adapted from the procedure used by Wollman (1975) and included three items; it made reference to an inclined plane and spheres of two differing weights. The subjects had to separate the effects of sphere weight and release height on the dependent variable, how far a target sphere was pushed by a sphere rolling down the plane.

Since the tasks were used in five different languages, it was necessary to translate the pages with the questions and space for the students' answers. The supplementary oral explanation and demonstrations that accompanied the task presentations also were formulated in the five languages so they could be presented reproducibly. These preparations were carried out with the collaboration of science education research groups in each country.

For all groups studied, the tasks were administered during a 50-minute period by a team consisting of authors and collaborators. The tasks were presented by a person (E) fluent in the local language. After being introduced, E explained that the class would participate in a research project intended to show future teachers how students approach certain scientific and mathematical tasks. The two tasks were then presented in order, first proportional reasoning, then control of variables.

The description of problems and questions was read aloud from answer pages provided for the students. Individuals with questions were referred to the information on the pages and/or were invited to use their judgment to resolve their uncertainty. When papers were handed in, they were briefly scanned by a team member who asked students with incomplete answers to clarify their reasoning orally or in writing.

The tasks are described in greater detail by Karplus et al. (1975), who also give a complete characterization of the scoring.

Scoring of Responses

Scoring of the responses focused on explanations that Ss were requested to provide on the answer page. All papers were read by two or three of the authors, who assigned the students' responses to one of the categories listed below. In rare cases of disagreement, consensus on the interpretation of an unclear or incomplete statement was reached by the authors.

The following categories were used for the proportional reasoning task:

Category I (Intuitive): haphazard, illogical, or incomplete explanation.
Category A (Additive): focus on a single difference.
Category Tr (Transitional): partial or incomplete use of proportions.
Category R (Ratio): complete and correct use of proportions.

The control of variables task was scored numerically for convenience. Students earned one point for each item (up to three) on which they separated variables, and one additional point (up to two) for each item on which they stated the necessity of controlling variables. Thus, the scores ranged from zero to a maximum of five points.

RESULTS

Proportional Reasoning

The distributions of scores for the student samples from the 36 populations are too extensive to be reproduced here, but are included in another lengthier account (Karplus et al., 1975). About one-fourth of all subjects gave Ratio responses, about one-sixth used Additive reasoning, and the remainder were divided between Intuitive and Transitional categories.

To give an overview of the variety of distributions, six frequency patterns were established for the four response categories on this task. In group I were the seven male and four female samples whose responses were primarily Ratio and Transitional. This group was most advanced in proportional reasoning. Group IIA included the four male and one female

Table I. Proportional Reasoning and Control of Variables in Four Countries with Comprehensive School Systems.

Country population	N		Proportional Reasoning Pattern						Control of Variables Pattern					
	boys	girls	I	IIA	IIB	IIIA	IIIB	IV	I	II	III	IV	V	VI
Denmark														
Middle Class+	192	207		M		F						M, F		
Sweden														
Middle Class+	51	47		M, F						F	M			
Working Class+	90	92					M, F				M	F		
Italy														
Upper Middle°	92	69	M		F						M	F		
Middle Class	52	58		M *	F						M, F		M, F	
Working Class	99	97		M	**	F							M, F	
United States														
Upper Middle+	96	112								M, F	M, F			
Middle Class+	341	290									M, F			
Urban Low Income+	87	94						M, F					M **	F

+ Coeducational.
° Some classes coeducational.

* $p < 0.01$.
** $p < 0.001$.

samples with predominantly Ratio, Transitional, and Intuitive responses and including fewer than 11% Additive. Group IIB, with four male and six female samples, has an even distribution among all four categories with approximately the same "average" achievement as IIA, but more Additive and fewer Transitional and Intuitive answers.

The remaining three groups included very few Ratio responses. Group IIIA, consisting of one male and three female samples, had predominantly Transitional and Intuitive answers, with fewer than 11% Additive. The one male and three female samples in Group IIIB showed even distributions among the three categories, with more Additive and fewer Transitional or Intuitive responses than the samples in group IIIA. The one male and one female samples in group IV, finally, gave predominantly Intuitive answers.

The performance of the population samples according to these six patterns is shown in Table I for schools in comprehensive systems and in Table II for schools in selective systems. The statistical significance of a difference within a country, determined by a chi-square test, is reported only when it exceeds the 1% level ($p < 0.01$) because of the large number of potential comparisons.

Effects of gender can be observed in the Danish sample, all population groups in Italy, and some groups in Austria, Germany, and Great Britain. Curiously, these effects are not associated systematically with the absence or presence of coeducation. In all cases, the boys performed better than the girls. Furthermore, female samples show patterns characterized by large numbers of Additive responses (IIB, IIIB) more frequently than do boys' samples. These additive responses observed in substantial numbers in the United States (Karplus & Wollman, 1974; Kurtz, 1976) are given much less consistently by boys in the continental countries investigated.

Socioeconomic effects are most clearly observable in the United States, where they confirm the earlier reports of the very great difficulties experienced by Urban Low Income Students on proportional reasoning (Karplus & Peterson, 1970). In Sweden and Italy, the socioeconomic effects are statistically significant for the female samples.

Table II. Proportional Reasoning and Control of Variables in
Three Countries with Selective School Systems.

Country population	N		Proportional reasoning pattern						Control of variables pattern					
	boys	girls	I	IIA	IIB	IIIA	IIIB	IV	I	II	III	IV	V	VI
Austria														
Gymnasium	112	187	M, F							M, F				
Hauptsch.-A+	102	89	M	**	F							M	F	
Hauptsch.-B+					M		F							M, F
Germany														
Gymnasium+	60	61	M, F							M, F				
Realschule	60	59	M	**	F					M	**	F		
Hauptsch.+	53	54				M, F								M, F
Great Britain														
Direct Grant	48	39	M, F						M, F					
Grammar	61	56	M, F							M, F				
comprehensive△	85	87		M			F					M, F		

Coeducational.
Some classes coeducational.
*$p < 0.001$.

Possible socioeconomic differences in Austria, Germany, and Great Britain are masked by the selectivity of the schools. Girls' performance on proportional reasoning is more greatly dependent on school type than is boys'.

National comparisons must take into account the fraction of an age group attending various selective schools. In Austria, where almost 80% of Viennese students attend Gymnasium or Hauptschule Track A, this large fraction of male students displays the high degree of proportional reasoning of Pattern I. The high achieving populations in Germany and Great Britain are much smaller than this, with Swiss observations approximately matching the German (Suarez, 1974).

Control of Variables

As with proportional reasoning, six patterns were defined for performance on controlling variables by the 36 population samples. Group I included one male and one female population whose scores were almost exclusively in the range from 3 to 5 points on the task. Group II, marked by a fairly even distribution with a slight excess of scores 3–5, included four male and four female samples. The six male and three female samples in group III also had a fairly uniform distribution, but with a slight excess of scores 0–2.

The three male and five female samples in group IV scored heavily 0 or 1, with about 25% of scores 3 or higher. Group V included two male and two female samples who scored predominantly 0 or 1 points. Finally, the two male and three female samples in group VI scored predominantly 0 points.

The performance of the population samples according to the six patterns is shown in Tables I and II. It can be seen immediately that the effect of gender on control of variables reasoning in our task was much less than on proportional reasoning. For only two populations (Urban Low Income in the United States, Realschule in Germany) was the difference significant at better than the 0.01 level.

Socioeconomic effects among all three population samples in the United States were statistically significant at the 0.001 level, but in Italy the Upper-Middle and Middle-Class groups did not differ. In Sweden, the quite similar performance of the Middle and Working Class groups contrasts with the substantial difference between these on proportional reasoning.

The effect of the selective schools is much more noticeable on control of variables than on proportional reasoning in Austria and Great Britain. The very low performance of students in the Austrian Hauptschule-Track B (20% of the population) and the German Hauptschule (50% of the population) is similar to that of the Urban Low Income girls in the United States. It shows a surprising lack in these young people compared to their achievement on proportional reasoning.

The British Direct Grant school students were the outstanding population on this task. The remaining groups in all countries, except for those mentioned in the previous paragraph, achieved at approximately comparable levels, with the Upper-Middle and Middle-Class samples in the United States performing relatively better than they had on proportional reasoning.

CONCLUSIONS

The data in Tables I and II describe the effects of student gender, country, and socioeconomic status on achievement level. The effects of gender were not associated systematically with a particular form of school organization, nor with classes segregated by gender.

Significant differences that did occur always favored the boys. Girls tended to use additive reasoning more commonly than did boys.

Differences in achievement among countries were much smaller than differences among groups within a country. Austrian boys distinguished themselves on proportional reasoning, but Austrian students as a whole did not perform comparably well on control of variables. Otherwise, the overall level of accomplishment in all countries was rather similar. In two countries, Denmark and Germany, there was a much lower frequency of additive reasoning than in other countries.

Both socioeconomic status and selectivity of school affected student performance significantly, though the magnitude of the effect depended on the task and on the country. Since educational programs are adapted to the pupils in each school, any original differences among students at school entrance have been undoubtedly magnified during their years of attendance.

The findings on the two tasks, which have some resemblance to assignments that students face in their science and mathematics classes, have implications for teachers. First of all, it is clear that a substantial fraction of students between 13 and 15 years of age lack the ability to articulate proportional reasoning and/or control of variables. Science and mathematics programs in all but the top level of selective schools should take this diversity of student reasoning into account in so far as content selection, laboratory activities, and textbook choices are concerned.

Second, it appears from the small but significant country-to-country differences that teaching can have some influence on the development of reasoning by the students in the age range being investigated here. Applying proportional or control-of-variables reasoning is not the result of a process exclusively internal to the young people.

Third, the development of the reasoning patterns considered in this paper should be an important objective of teaching programs for 13- to 15-year-olds, an objective that is not achieved comprehensively by present school practice in any of the countries included in our study. Other investigations (Kurtz, 1976; Lawson & Wollman, 1976) have identified certain promising teaching approaches that have been presented elsewhere (Karplus et al., 1977). A key component of these approaches is to direct the attention of students at their own reasoning, the thought processes by which they have arrived at valid or invalid conclusions, correct or incorrect answers. Other aspects are based on (1) identification of the reasoning required for the understanding of science concepts, (2) arranging for active participation in learning by the students, and (3) providing concrete experiences at the beginning of new topics.

REFERENCES

Inhelder, B., & Piaget, J. *The growth of logical thinking from childhood to adolescence.* New York: Basic Books, 1958.

Karplus, R., Karplus, E. F., Formisano, M., & Paulsen, A. C. *Proportional reasoning and control of variables in seven countries.* Berkeley: Lawrence Hall of Science, 1975.

Karplus, R., Karplus, E. F., & Wollman, W. T. Intellectual development beyond elementary school IV: Ratio, the influence of cognitive style. *School Science and Mathematics,* 1974, **74,** 476–482.

Karplus, R., Lawson, A. E., Wollman, W. T., Appel, M., Bernoff, R., Howe, A., Rusch, J. J., & Sullivan, F. L. *Workshop on science teaching and the development of reasoning.* Berkeley: Lawrence Hall of Science, 1977.

Karplus, R., & Peterson, R. W. Intellectual development beyond elementary school II: Ratio, a survey. *School Science and Mathematics,* 1970, **70,** 813–820.

Karplus, R., & Wollman, W. T. Intellectual development beyond elementary school V: Using ratio in differing tasks. *School Science and Mathematics,* 1974, **74,** 593–711.

Kurtz, B. A study of teaching for proportional reasoning. Doctoral Dissertation, University of California, Berkeley, 1976.

Lawson, A. E. Sex differences in concrete and formal reasoning ability as measured by manipulative tasks and written tasks. *Science Education*, 1975, **59**, 387–396.

Lawson, A. E., & Wollman, W. T. Encouraging the transition from concrete to formal cognitive functioning—An experiment. *Journal of Research in Science Teaching*, 1976, **13**, 413–430.

Levine, D., & Linn, M. C. Scientific reasoning ability in adolescence: Theoretical viewpoints and educational implications. *Journal of Research in Science Teaching*, 1977, **14**, 371–384.

Lovell, K., & Butterworth, I. B. Abilities underlying the understanding of proportionality. *Mathematics Teaching*, 1966, **37**, 5–9.

Lunzer, A. E., & Pumfrey, P. D. Understanding proportionality. *Mathematics Teaching*, 1966, **34**, 7–12.

Suarez, A. Die Entwicklung der Denkoperation beim Verständnis Funktionaler Zusammenhänge. Doctoral Dissertation. Eidgenössische Technische Hochschule, Zürich, 1974.

Wollman, W. T. *Intellectual development beyond elementary school IV: Controlling variables, a survey.* Berkeley: Lawrence Hall of Science, 1975.

Robert, Beverly and Elizabeth Karplus
Summer, 1950

Intellectual Development Beyond Elementary School VI: Correlational Reasoning

Helen Adi, Robert Karplus, Anton Lawson, and Stephen Pulos

INTRODUCTION

The establishment of relationships among variables is basic to prediction and scientific explanation. Correlational reasoning—the reasoning processes one uses in determining the strength of mutual or reciprocal relationship between variables—is, therefore, a fundamental aspect of scientific reasoning.

Suppose, for instance, that a scientist is interested in finding out whether a correlation exists between the body weight of rats and the presence of a substance X in their blood. The establishment of a correlation requires an initial recognition of the four possible asso-

* AESOP (Advancing Education through Science-Oriented Programs).

This material is based upon research supported by the National Science Foundation under Grant No. SED74-18950.

Any opinions, findings, and conclusions or recommendations expressed in this publication are those of the authors and do not necessarily reflect the views of the National Science Foundation.

Helen Adi, Robert Karplus, Anton Lawson, and Stephen Pulos, AESOP* Lawrence Hall of Science University of California Berkeley, California 94720.

"Intellectual Development Beyond Elementary School VI: Correlational Reasoning" was taken from *School Science and Mathematics*, 1978, vol 78, pgs. 675–683.

Table 1. Organization of Data to Exhibit Possible
Correlation.

	Heavy weight	Light weight
Presence of Substance X	(a)	(c)
Absence of Substance X	(b)	(d)

ciations: (a) = heavy weight and presence of substance X; (b) = heavy weight and absence of substance X; (c) = light weight and presence of substance X; and (d) = light weight and absence of substance X. When variables can be dichotomized such as this, one may construct a 2 × 2 association table of the sort used to compute simple contingencies (Table 1).

In view of the fundamental role played by correlational reasoning in the investigative process, we asked ourselves the following question: How do high school science and mathematics students approach tasks that require correlational reasoning for successful solution? An answer to this question will indicate how students apply this important aspect of scientific reasoning and might suggest how this reasoning pattern could be enhanced through instruction.

METHOD

Procedure. Two tasks requiring correlational reasoning for successful solution were developed. One task involved a possible causal relationship between pill-taking and body size of rats while the other involved a possible coincidental relationship between tail color and body size of rats. Each task was presented to students in individual interviews.

Causal Task. The causal task was introduced as follows: "A scientist was studying the effects of some red pills on the weight of rats. One evening before he left his laboratory he forgot to lock the door of the rats' cage. That night the rats got loose. Some of them found and ate the red pills that were left on the laboratory shelf. The next morning the scientist returned to his laboratory and discovered what had happened.

Since eating the pills left a red color in the rats' mouths, the scientist was able to tell which rats had eaten the pills and which ones had not. He put the rats that had eaten the pills into one cage and the other rats into a second cage. Two weeks later he observed the rats to find out if there was a connection or relation between the size of the rats and whether or not they had eaten the red pills."

Following the introduction, the student was given the set of 3" × 5" cards. Each card represented one case [i.e., either a fat rat with a pill (a), a fat rat with no pill (b), a thin rat with a pill (c), or a thin rat with no pill (d)]. He or she was asked to examine the cards and discover if there was, in fact, a relation.

Students initially unable to proceed were given assistance in classifying the cards into the four categories. Further assistance was provided only to ensure that the problem was understood. Students were asked to explain how they had arrived at their conclusions.

Following the analysis of the first set of data cards, two additional sets of cards reflecting different distributions were presented successively. The student was told that rats in other laboratories had eaten some green or brown pills. Again the scientist wondered if there was a connection between eating the pills and the size of the rats after two weeks.

Table 2. Frequency Distributions of the Causal Correlational Reasoning Task.

Distribution	Cells[a]			
	(a)	(b)	(c)	(d)
(1) Positive Correlation	7	2	3	8
(2) Negative Correlation	2	8	7	3
(3) Zero Correlation	6	9	4	6

[a] The (a) and (d) cells represent confirming cases (e.g., (a) = fat rats with pills and (d) = thin rats with no pills), while the (b) and (c) cells represent the disconfirming cases (e.g., (b) = fat rats with no pills and (c) = the thin rats with pills). Confirmation and disconfirmation is considered with respect to the hypothesis of equivalence between the two variables.

The three distributions, shown in Table 2, represented (1) a positive correlation between size and pill taking, (2) a negative correlation, and (3) a zero correlation.

Coincidental Task. The task involving the coincidental relationship was administered in the same manner as the causal task. The following introductory paragraph was read: "A scientist recently conducted an investigation of the rats that inhabit a series of small islands in the South Pacific. On the first island that he visited he found that all of the rats were either quite fat or were quite thin. Also, all of the rats had either green tails or red tails. This made the scientist wonder if there might be a connection or relation between the size of the rats and the color of their tails. So he decided to capture some rats and observe them to see if there was a relation."

The student was then given the first of three sets of data cards. Again, each card consisted of one case [i.e., either a fat rat with a green tail (a), a fat rat with a red tail (b), a thin rat with a green tail (c), or a thin rat with a red tail (d)]. In all other respects the interview was conducted in the same manner as was the causal task. Following the analysis of the first set of the data, two additional sets, using other islands and different colors, were presented. The cell frequencies for these three sets of data were the same as for the causal task (Table 2).

Subjects. The subjects were 80 secondary school students (31 males and 49 females) enrolled in science and mathematics classes of a large high school located in a middle class community in the San Francisco Bay Area. Equal numbers of students were selected from ninth and twelfth grade classes. The ninth graders ranged in age from 13.7 years to 15.8 years (mean age = 14.7 years). The twelfth graders ranged in age from 16.3 years to 19.0 years (mean age = 17.5 years). Since the students were selected only from science and mathematics classes, the samples cannot be considered directly comparable or representative of the respective grades.

Each subject was administered either the causal or coincidental task, but not both. Interviews lasted about 15 minutes.

CATEGORIZATION OF RESPONSES

Notes on student responses were recorded during the interviews. Although a wide variety of responses were given by the students to justify their conclusions regarding correlations, it was possible to classify responses from both tasks into the following categories and subcategories:

Category NR: no relation among cell frequencies is considered.

NR-1—the subject does not accept the possibility of a relationship in view of the non-confirming cases, "No, there are rats who are fat who didn't have pills and rats who are skinny who did have pills."

NR-2—the subject describes various events qualitatively, "I don't know because there are some rats of each kind,"

NR-3—the subject describes various events quantitatively, "I think there is no connection, there are 7 fat rats with green tails and 3 with red tails, and there are two thin rats with green tails and 8 with red tails."

Category TC: the number of events in two cells is compared.

TC-1—the subject compares two cells with a common attribute, "The pill made them fat because there are 7 fat rats with pills and only two fat rats without pills."

Category FC: the numbers of events in all four cells are used to make two comparisons.

FC-1—the subject compares the events in two pairs of cells; "There are more red fat then green fat, and more green thin than red thin,"

FC-2—the subject combines two cells having one common attribute and compares one cell with the total, then does the same with the other pair of cells, "The majority of no-pill rats are skinny, the majority of pill rats are fat."

Category CO: correlation is described by a quantitative comparison using all four cells.

CO-1—the subject identifies and/or compares two ratios, "More small rats have green tails by 8/3, and more large rats have red tails by 7/2."

CO-2—the subject identifies and/or compares two percentages, "With black tails the rats are more likely to be small, 80%; with white tails the rats are most likely to be big, 70%."

CO-3—the subject compares the number of confirming cases $(a + d)$ to the number of disconfirming cases $(b + c)$, "The pill has a weight loss effect, because the eight thin rats with pills and the seven fat rats without pills make 15, and there are only 2 fat rats with pills and 3 thin rats without pills. These five rats are exceptions. So about 75% of the time it worked as diet pill."

The preceding categories and subcategories are differentiated by the complexity of the comparisons or operations used by the subjects in arriving at their conclusions. Category NR indicates the most naive approach to the task; although the concrete operations underlying the multiple classification of events is implied, no explicit comparisons between the numbers of events in the cells are made.

Category TC includes the multiple classification of events of Category NR, but it goes a step beyond by including a comparison of the number of events in two cells.

Category FC reflects still more complex and adequate reasoning in that two comparisons are made and the beginning of correlational reasoning appears. In Subcategory FC-1, the first comparison identifies, for example, the presence of more red-tailed fat reds than green-tailed fat rats, while the second comparison is between the number of thin rats with red or green tails. In Category FC-2, the typical subject uses the terms "majority" or "most of" to indicate the probability of occurrence of fat or thin rats relative to the combined number of all rats that did or did not eat the pill.

Category CO leads to a satisfactory solution of the problem with a clear identification of the mathematical relationships being used for the comparisons. This category includes the most advanced responses elicited by the tasks used in this study.

In summary then, we propose that there are at least four identifiable levels of reasoning in these tasks corresponding to the four categories: NR (no explicit quantitative comparisons between cells); TC (a single binary comparison or operation); FC (two binary comparisons without explicit comparison of the results); and CO (a comparison of the results of two binary comparisons or operations).

Further, we believe it is possible to equate these four levels of performance to Piagetian substages of intellectual development as detailed by Inhelder and Piaget (1958). Level NR performance indicates the use of concrete operational reasoning (the multiple classification of events); Level TC performance reflects the use of more advanced but still concrete or perhaps "post-concrete" reasoning; Level FC indicates the use of formal reasoning but only in an incomplete fashion, while Level CO reflects the complete and satisfactory use of formal reasoning.

RESULTS

Categories of Response

Responses of all 80 subjects were categorized according to the criteria given. Since each subject responded to three frequency distributions, 240 responses were classified. Table 3 presents the occurrence of responses in each category for both tasks. Inspection of the table shows only minor differences for the three distributions within each task. Also, both tasks appear to elicit about the same numbers of responses in Categories NR and CO. A substantial difference appears in regard to Categories TC and FC, however. The responses to the causal task fell more frequently into Category TC, while the responses to the coincidental task fell more frequently into Category FC—and about equally into the two subcategories FC-1 and FC-2.

Table 3. Number of Responses in Each Category for the Causal and
Coincidental Task Distributions.

Category	Causal task (n = 40)			Coincidental task (n = 40)		
	(1)	(2)	(3)	(1)	(2)	(3)
NR						
NR-1	1	1	1	2	1	0
NR-2	10	8	14	9	5	14
NR-3	0	2	0	1	1	0
TC						
TC-1	18	18	11	0	3	2
FC						
FC-1	3	3	3	9	15	10
FC-2	3	4	2	15	10	6
CO						
CO-1	1	1	6	2	2	6
CO-2	0	1	3	2	3	2
CO-3	4	2	0	0	0	0

Table 4. Number of Subjects at each Level of Performance.

Task	Level of performance			
	Concrete	Post-concrete	Incomplete formal	Complete formal
Causal (n = 40)	7	16	7	10
Coincidental (n = 40)	5	1	23	11

Overall Level of Subject Performance

Each subject's overall interview performance was placed in the four levels of reasoning, according to the correspondence of categories and developmental stages suggested earlier. In cases where the level was the same for all distributions, this level was simply considered to be his or her overall level. In cases where the level of performance varied from distribution to distribution, the highest level obtained was taken as an indication of overall level.

The number of subjects at each level of performance is shown in Table 4 for both tasks. The figures indicate that the coincidental task elicited somewhat better performance than the causal task. This difference in performance was found to be statistically significant ($X^2 = 22.2$, $p < .001$). It is a consequence of the more numerous responses in Category FC on the coincidental task compared to the more numerous responses in Category TC on the causal task.

Interestingly, all subjects responding in Category TC on the causal task compared two data cells differing only in one attribute (e.g., fat rats with and without pills), and never compared two cells that differed with respect to both attributes (fat rats with pills and thin rats without pills). This form of the answer suggests that the subjects were setting up a hypothesis (e.g., the pills make the rats fat) and then looked for confirming and disconfirming cases. The condition of the control population of rats that had not taken the pill may have been ignored by them as irrelevant to the hypothesis. Such limited consideration of evidence is characteristic of concrete reasoning.

The coincidental task apparently did not suggest the formulation of a hypothesis and thus did not distract the subjects from considering all the data. Most of the responses classified in Category FC did, however, emphasize the data cells with large frequencies, as in the example of an FC-2 response given earlier.

Age Differences in Performance

The twelfth graders performed at a higher level than the ninth graders on both tasks. These differences cannot, however, be taken as evidence of a general advance in reasoning from the ninth to the twelfth grade in this high school since the two samples were not comparable. The twelfth grade sample consisted of a substantial number of physics students, while the ninth grade sample came from science and mathematics classes that were required courses.

DISCUSSION

The finding that a significant number of students in this sample did not use correlational reasoning on our tasks is similar to the results of studies that have found large percentages of adolescents and adults using concrete reasoning patterns on tasks involving other formal operational schemes such as proportions, controlling variables, combinatorial reasoning and probability (e.g., Blasi and Hoeffel, 1974; Karplus and Peterson, 1970; Lawson and Renner, 1974; Niemark, 1975; Piaget, 1972). Since correlational reasoning plays a prominent role in the investigative process, our results should be a matter of substantial concern for science educators.

Indeed, the significance of our findings is not confined to the sciences. Problems requiring correlational reasoning are found in everyday experiences as well—is there a correlation between cigarette smoking and lung cancer, swine flu vaccination and paralysis, drunk driving and automobile accidents, vitamin C and resistance to the common cold?

What appear to be the elements of correlational reasoning? Although the nine subcategories of responses identified in this investigation probably do not form a unidimensional scale of increasingly complex and successful reasoning, there is reason to believe that they reflect the development of two parallel advances in reasoning. One dimension may center around the development of the concept of probability, while the other may center around the development the concept of proportionality. These parallel developments can best be conceived of as gradual progressions from very intuitive and limited awareness of the relationships involved to explicit and quantitative use of these concepts in problem solving.

Subcategories NR-2, TC-1 and CO-2 seem to reflect the development of the concept of probability from a complete lack of understanding of the probabilistic nature of phenomena ("There is no relation because there are some of each kind") to a very rudimentary understanding in Subcategory TC-1 ("There is some connection because there are more fat rats with pills than fat rats without pills"), to calculation and comparison of two percentages in Subcategory CO-2.

Subcategories FC-1 and CO-1 seem to reflect the development of the concept of proportionality. In Subcategory FC-1, the comparison is phrased in terms of "more" and "less", while in Subcategory CO-1 the ratios are explicitly stated.

Subcategory CO-3 may represent a joining of the concepts of probability and proportionality in such a way as to lead to successful problem solution. The comparison of the number of confirming cases to the total number of cases reflects the notion of probability (e.g., "Fifteen out of 20 rats fit the rule, the other 5 are exceptions"). Although proportions are not directly mentioned in this problem solution, they contribute to the conversion of fractions to percents at the conclusion (e.g., $15/20 = 75/100$, "So about 75% of the time it's a diet pill").

If the concept of correlations does involve these parallel sequences of development, then instruction designed to facilitate students' acquisition of correlation reasoning must take into consideration the students' ability to solve problems of probability and of proportionality.

In a very real sense, such instruction is likely to be effective only if it integrates topics that are studied separately in either science courses (cause-effect relationships, observational data gathering, hypothesis generation, experimental design, etc.) or in mathematics courses (probability, ratio, proportion, percent, correlation, etc.). If students were, in fact, given opportunities to raise questions of correlation, examine alternatives, gather data, and attempt to analyze their meanings, these science and mathematics concepts would immediately come into play in contexts meaningful to the students. The teaching approach proposed by Unified Science and Mathematics for Elementary Schools (USMES, 1970–76) has some of these characteristics, but is aimed at much younger students and does not deal explicitly with correlations. It would seem worthwhile to prepare similar programs for secondary school students and include direct attention to correlations.

REFERENCES

Blasi, A., and Hoeffel, E. C., "Adolescence and formal operations." *Human Development*, 17, 344–363, 1974.

Inhelder, B., and Piaget, J., *The Growth of Logical Thinking from Childhood to Adolescence.* New York: Basic Books, 1958.

Karplus, R., and Peterson, R., "Intellectual development beyond elementary school II: ratio, a survey." *School Science and Mathematics*, 70(9), 813–820, 1970.

Lawson, A., and Renner, J., "A quantitative analysis of responses to Piagetian tasks and its implications for curriculum." *Science Education*, 58(4), 545–559, 1974.

Neimark, E., "Intellectual development during adolescence." In F. Degan (Ed.) *Review of Child Development Research, Vol. 4.* Chicago: University of Chicago Press, 1975.

Piaget, J., "Intellectual evolution from adolescence to adulthood." *Human Development*, 15, 1–12, 1972.

USMES Guide, Newton, Massachusetts: Education Development Center, 1974.

Robert Karplus dancing at the Institute in Vienna, 1961.

Intellectual Development Beyond Elementary School VII: Teaching for Proportional Reasoning

Barry Kurtz* and Robert Karplus

Many status studies carried out during the past four years in the United States and abroad revealed that a substantial fraction of secondary school students did not use proportions successfully to solve simple constant ratio tasks (Karplus & Peterson, 1970; Wollman & Karplus, 1974; Suarez, 1974; Karplus et al., 1977a). Earlier research on proportional reasoning, as we shall call the operations involved in solving constant ratio tasks (Inhelder & Piaget, 1958), showed that problems of proportionality were relatively difficult for adolescents, especially if the ratios were not integers (Lovell & Butterworth, 1966; Lunzer & Pumfrey, 1966).

Yet proportional reasoning is vital for many scientific applications as well as for consumer mathematics, trigonometry, and an understanding of linear functions. We therefore undertook this study to test the following hypotheses:

* Present address: San Francisco State University, San Francisco, CA 94132.
   This material is based in part upon research supported by the National Science Foundation under Grant Nos. GZ2979 and SED74-18950.
   Any opinions, findings, and conclusions or recommendations expressed in this publication are those of the authors and do not necessarily reflect the views of the National Science Foundation.

Barry Kurtz and Robert Karplus, Group in Science and Mathematics Education and Lawrence Hall of Science, University of California, Berkeley, California 94720.

"Intellectual Development Beyond Elementary School VII: Teaching for Proportional Reasoning" was taken from *School Science and Mathematics*, 1979, vol 79, pgs. 387–398.

1. A brief teaching program, suitably designed, can enable ninth and tenth grade students in prealgebra classes to become proficient in proportional reasoning.
2. A form of this teaching program making use of manipulative materials will be more effective than a form using pencil and paper activities only.
3. Student attitudes toward this teaching program will favor the form with manipulatives over the pencil and paper form.

The target group of subjects was chosen because it was easily accessible in secondary schools and had been found to be particularly deficient in proportional reasoning. At the same time, many algebra students of the same age used proportional reasoning successfully. This suggested that the prealgebra students would be able to benefit from specially-prepared instruction even though they had obviously not gained an adequate understanding from the usual teaching on ratio and proportions in earlier grades.

The interest in manipulative materials derived from the possibility that experience with tangible examples of constant ratio relationships might be especially valuable. Such laboratory materials invite the active participation of all students, form a concrete basis for abstractions, and facilitate peer group interactions. Yet these materials also tend to be costly compared to other instructional media, require unusual storage and maintenance efforts, and are not traditionally part of secondary school mathematics instruction. Hence a second form of the teaching program without the manipulative materials was developed and tested as well.

The teaching program we developed was called *Numerical Relationships*. The next section of this report describes the program with special emphasis on how it differs from traditional teaching concerned with proportionality. We then turn to a presentation of the research design, the tests, and the results that were used to evaluate the effectiveness of the materials. We complete this report with conclusions that follow from our work. The entire instructional program and field test are described more fully elsewhere (Kurtz, 1976).

THE NUMERICAL RELATIONSHIPS PROGRAM

The *Numerical Relationships* (NR) Program was developed after a year of trials and pilot studies (Kurtz, 1976). It was based on research in proportional reasoning and on teaching strategies that had been found valuable in science and mathematics teaching (Biggs, 1973; Karplus & Lawson, 1974; Karplus et al., 1977b). Six features distinguish NR from the usual textbook approaches to proportions: (1) it makes use of laboratory activities; (2) teaching follows a learning cycle that proceeds from exploration to concept introduction to concept application; (3) NR directs the students' attention to the variables necessary to describe a relationship (4) the program gives students practice in distinguishing between the constant ratio, constant difference, and constant sum relationships that are frequently not discriminated adequately (Wollman & Karplus, 1974); (5) proportional reasoning is developed from a tendency observed in children to match certain subsets into a basic ratio and then to iterate this ratio (Lunzer & Pumfrey, 1966; Wollman & Lawson, in press); and (b) NR avoids algorithmic techniques such as the equality of the products of means and extremes.

The teaching materials for NR consisted of three parts: (1) Relationships Between Numbers, (2) Relationships in the Math Lab, (3) Relationships in the World. About sixty percent of the estimated student time was spent on situations in which a constant ratio governed the relationships between two variables, with the remainder divided equally between constant sum and constant difference problems. Substantial reliance was placed on giving the students practice with a larger variety of problems and thus building up their familiarity with the three distinct relationships encountered.

In Part 1, students were asked to identify the invariant relationship (constant ratio, constant difference, or constant sum) between two variables when given a table of data with some omissions. Then they had to write an equation for the relationship, infer table entries that had been omitted, graph the data, and note distinctive features of each family of graphs.

Part 2 was prepared in two versions, one with manipulative materials and one with pencil and paper lab activities. In both versions of Part 2, students were faced with ten practical situations. These consisted of mechanical devices in the manipulative version, diagrams and measuring instruments in the pencil and paper form.

One of the mechanical devices illustrating constant ratios was a spring. Students compared the length of the entire spring with the partial lengths to certain places marked along the spring. One of the pencil and paper activities required the students to draw sets of nested rectangles in a "picture frame" arrangement on paper with dots arranged in a square grid. They found that length and width showed a constant difference.

Each student was expected to complete at least six labs, four dealing with constant ratio and one each presenting constant difference and constant sum. By working at their own pace, faster students completed more labs and received more practice.

Part 3 led up to the presentation of proportion problems in the traditional format of three knowns and one unknown. In the early sections of this part, no numerical data were given; students were thereby led to identify the significant variables and their relationship and were not able to compute immediately. Next, one data pair was given in a specified situation (e.g. a woman buys fabric at a price of $10 for six yards) and students had to identify the two variables, give them letter names, generate additional possible data pairs compatible with the context, and finally write an equation to state the relationship between the variables. After this introduction to equations, students devised an equation as their first task in a problem (e.g. nine gallons of gasoline cost $6, or Cost = 6/9 Amount) and then solved it several times with slightly differing conditions to find correct data pairs.

Use of the NR Program required fourteen class periods of approximately 50 minutes duration.

RESEARCH DESIGN

The high schools participating in this research project were located in middle and upper middle class suburban communities. Four volunteer teachers taught the NR program in four high schools to a total of eight prealgebra classes. Six additional teachers provided access to prealgebra classes in the same schools with control subjects who studied simple one-variable equations. The arrangements and enrollments are summarized in Table 1.

In each of the schools, designated by the letters A to D, the same volunteer teacher taught both types of experimental classes. All were experienced secondary school mathematics teachers who received a special 16-hour training course to orient them to the NR materials and their pedagogical emphasis. The control teachers were informed about the research project but did not receive the special training.

Table 1. Assignment of Prealgebra Classes and Enrollments in the Experimental Design.

School	Number of classes	Control classes	(n)	Manipulative	(n)	Experimental classes pencil and paper	(n)
A	4	2	(56)	1	(26)	1	(26)
B	4	2	(52)	1	(30)	1	(27)
C	3	1	(33)	1	(34)	1	(13)
D	3	1	(36)	1	(31)	1	(34)
Total	14	6	(177)	4	(121)	4	(100)

Table 2. Overview of Test Batteries..

Item	Invariant	Pretest	Posttest	Delayed posttest
1	Ratio	Heights[a]	Volumes[b]	Pulleys[c]
2	Ratio	Pulleys	Gears	Wheels
3	Ratio	Enlargements	Enlargements	Enlargements
4	Difference	Ages	Ages	Ages

[a] Karplus, Karplus, & Wollman, 1974.
[b] Suarez, 1974.
[c] Wollman & Karplus, 1974.

The testing program made use of eight-item batteries of pre-, post-, and delayed posttests. The equivalence of these tests had been established during previous test-retest procedures with these items in schools that did not participate in the research study itself. The three series of tests were administered to the eight experimental classes. The posttest contributed to these students' math grade, but the pre- and delayed posttests did not. In the control classes only the pre- and posttests were administered, because student behavior led to the expectation that they would not cooperate with a third session of such a frustrating experience.

The numbers of students actually present for all test administrations was approximately twenty percent less than the total enrollments listed in Table 1. Further variable attrition for individual test items resulted from unclassifiable responses, which reduced the numbers available for statistical analysis.

THE TEST BATTERIES

The test batteries consisted of eight cognitive items and were administered by one of the authors during one class period. An attitude survey was administered separately by the classroom teachers in the authors' absence.

Six cognitive items were concerned with constant ratio relationships, one with constant sum, and one with a constant difference. In this presentation we shall use four of these items, three dealing with constant ratios and one with a constant difference (Table 2). Each of these items required an answer and an explanation. Two of the remaining four items were very similar to Items 1 and 4 (Table 2) and the others were concerned with secondary objectives (writing equations, drawing graphs). .

The three forms of Item 1 have been described elsewhere (Karplus, Karplus, & Wollman, 1974; Suarez, 1974; Wollman & Karplus, 1974). After the completion of this item, the students were given the correct answer as feedback during the test. The texts of sample forms of the other items, some of which were illustrated by diagrams, were as follows:

> Item 2 (Gears)—When a larger gear is turned 8 complete turns, a smaller gear connected to it turned 10 complete turns. In another experiment, the large gear is turned 20 complete turns. How many times did the smaller gear turn? Please explain carefully how you obtained this answer.
>
> Item 3 (Enlargement)—Here is a picture with a tall tree and a short tree. In this picture the short tree is 8 cm high and the tall tree is 10 cm high. The photographer made an enlargement of the same picture. In the larger picture, the short tree is 20 cm high. How high is the tall tree in the larger picture? Please explain carefully how you found this answer.

Item 4 (Ages)—Susan and Cathy are sisters. Susan was 6 years old when Cathy was twelve. Susan is now 9 years old. How old is Cathy now? Please explain carefully how you found this answer.

It should be noted that the gears and pulleys in the three versions of Item 2 were part of the apparatus used in the manipulative activities of one version of the teaching program. The enlargement problems were similar to measuring activities carried out by the students in the pencil and paper version of the course, and "age" problems were included in other sections. Item 1, however, was completely different from the content of the instructional materials.

Responses to the constant ratio items were classified into seven levels related to the categories established by Karplus and Peterson (1970). They are illustrated here by answers to a task with numerical values $4/6 = 6/x$.

1. Intuitive, illogical, or guess (10—1 added the 6 and the 4).
2. Unjustified scale factor (12—because 2 paper clips equals 1 button, and $6 \times 2 = 12$).
3. Additive (8—4 + 2 = 6, so 6 + 2 = 8).
4. Erroneous multiplicative (12—for every button there are $1\frac{1}{2}$ paper clips).
5. Correct response, combining additive and multiplicative (9—you added half of 4 to get 6, so add half of 6 to get 9).
6. Fully multiplicative (9—$4 \times 1.5 = 6, 6 \times 1.5 = 9$).
7. Using an algebraic equation (9—W = wide, N = narrow; W/N = 4/6; if W = 6, 6/N = 4/6, N = 9.)

Four levels were sufficient for classifying the student answers to the constant difference problem in Item 4. They are illustrated with answers to this item as given above:

1. Erroneous or no response (3—9 − 6 = 3.)
2. Constant ratio (18—Cathy is twice as old as Susan.)
3. Correct additive with explanation (15—Cathy was 6 years older, 9 + 6 = 15.)
4. Using an algebraic equation (15—S = Susan, C = Cathy, C = S + 6, so when S = 9, C = 9 + 6 = 15.)

All test papers were coded and separated so that the readers could not identify them as belonging to experimental or control classes. Two readers evaluated the papers independently and then compared results. The interrater agreements were near 95%.

The attitude survey asked students for their opinions regarding classroom activities in the NR program, their preference for various parts of the program, and their willingness to recommend the program to others. Classroom observations as well as conversations with students and teachers supplemented the written responses for an affective evaluation of the NR program.

RESULTS

In this section we shall report the results of the testing program in a way that communicates our finding succinctly. The complete results of the testing program have been described by Kurtz (1976). On all items that required use of proportions, the gains of the experimental group students from pre- to posttest were substantial and greatly exceeded small gains of the control subjects.

Table 3. Distributions of Subjects Among Seven Levels on Constant Ratio Item 1 (percent).

	n	1	2	3	4	5	6	7
Pretest								
Control	126	5	4	75	6	4	6	1
Manipulative	83	4	10	68	4	6	7	2
Pencil and Paper	72	3	8	72	0	11	6	0
Posttest								
Control	126	3	2	60	2	23	9	2
Manipulative	83	1	0	28	1	1	28	41
Pencil and Paper	72	0	0	36	1	7	8	47
Delayed Posttest								
Control	126	—	—	—	—	—	—	—
Manipulative	83	1	6	27	4	11	35	17
Pencil and Paper	72	3	1	28	1	13	28	26

First we shall describe the performance of the three groups of students (control, experimental manipulative, experimental pencil and paper) on the constant ratio Item 1, which was a transfer task for all subjects. Second, we shall give the data on the constant difference ages problem. Third, we shall report the distribution of a composite proportional reasoning score on all four items. Fourth, some of the attitude results will be described.

Table 3 presents the distributions of student performance on the three equivalent forms of Item 1. Several observations can be made: (1) the three groups exhibited no substantial differences on the pretest; (2) initially most students used an additive procedure on the item (Level 3); (3) the control group exhibited some gains on the posttest; (4) the experimental groups showed very substantial gains to algebraic procedures on the posttest; (5) the experimental groups showed substantial regression from using an algebraic equation to the multiplicative proportional category on the delayed posttest; (6) differences between the two experimental groups on both posttest and delayed posttest were slight.

In an analysis of control student performance on the two tests (Kurtz, 1976), it was found that almost sixty percent responded at the same level on both occasions. Thirty percent improved or regressed by one or two levels, and the remaining ten percent changed by more than two levels. The changes were not symmetrical, with somewhat more students moving up than moving down, but not to an extent that was statistically significant. This effect may be due to giving the answer to Item 1 as feedback during the test period and thereby "training" the control subjects slightly.

The statistical significance of the differences between the control and experimental groups on the posttest was evaluated by the Mann-Whitney U test (Siegel, 1956, pp. 116–127) to be $p < 0.001$. The statistical significance of the gains by the two experimental groups on the post- and delayed posttests as compared to the pretest was evaluated by the Wilcoxon T statistic (Siegel, 1956, pp. 75–83) for asymmetry. Again, the significance of the gains exceeds $p < 0.001$ in all cases. By comparison, the small gain of the control group from pre- to posttest mentioned above had a significance level of only $p > 0.1$ by the Wilcoxon test and is therefore unlikely to reflect a real effect.

The results on Items 2 and 3 were virtually the same as those reported above for Item 1. The experimental groups showed highly significant ($p < 0.001$) and closely equal gains from pre- to posttest, while the control group showed much smaller gains than on Item 1. As might be expected, small differences that were observed favored the experimental group that had encountered similar tasks during the teaching program, but these differences were not statistically significant.

We now turn to Item 4, the constant difference problem. The distributions of student responses are presented in Table 4. The following observations can be made: (1) at all times the vast majority of the subjects used a correct additive procedure; (2) the three groups performed equally on the pretest; (3) the control group showed a slight increase in constant ratio procedures reasoning on the posttest, possibly because of conditioning by the preponderance of constant ratio problems and the feedback on Item 1; (4) the two experimental groups showed indistinguishable distributions; (5) both experimental groups showed short-term gains in the use of algebraic equations on the posttest, but these gains disappeared almost completely on the delayed posttest.

We are now in a position to answer the question, "What is the gain in proportional reasoning by students who experienced the NR program?" If we consider only Item 1, then we can see from Table 3 that the percentage of students using proportional reasoning (levels 5, 6, and 7) changed from pretest to posttest as follows: in the control group, from 11% to 34%; in the group using manipulative instructional materials, from 15% to 70%; in the group using pencil and paper materials, from 17% to 62%. In other words, about half the students in the experimental groups gained during the learning experience. The other half already used proportional reasoning on the pretest (about one sixth of the total) or did not advance (about one third of the total).

To provide a more stringent measure of learning than this single item affords, we constructed three performance categories (high, medium, low) according to the following system applied to Items 1 through 4: for "high," the subject had to respond at level 6 or 7 on two of the proportional reasoning tasks, at level 5 or better on the third, and at level 3 or 4 on the additive Item 4; for "low," the subject had to score at level 5 or below on two proportional reasoning tasks, at level 4 or below on the third, and at any level on the additive item; the remaining combinations of scores earned he subject a "medium."

The distributions of test performance according to the three categories just defined are given in Table 5. All three groups showed closely equal performance on the pretest, with about 80 percent in the low category and fewer than 10 percent of the students in the high category. On the posttest, more than 50 percent of the students in the two experimental groups scored high, while fewer than ten percent of the control subjects did so. The difference between experimental and control groups on the posttest was significant at $p < 0.001$ according to a chi-square test. The superior and closely equal performance of the two experimental groups persisted on the delayed posttest.

Table 4. Distribution of Subjects Among Four Levels on Constant Difference Item 4 (percent).

	n	1	2	3	4
Pretest					
Control	133	2	2	96	0
Manipulative	99	1	7	93	0
Pencil and Paper	82	2	6	92	0
Posttest					
Control	133	2	13	86	0
Manipulative	99	1	12	53	34
Pencil and Paper	82	0	6	57	37
Delayed Posttest					
Control	133	—	—	—	—
Manipulative	99	1	11	78	10
Pencil and Paper	82	2	9	85	4

Table 5. Distributions of Subjects on Proportional Reasoning Categories (percent).

	n	Low	Medium	High
Pretest				
Control	138	82	11	7
Manipulative	99	78	18	4
Pencil and Paper	82	79	16	5
Posttest				
Control	138	75	18	7
Manipulative	99	11	32	57
Pencil and Paper	82	26	23	51
Delayed Posttest				
Control	138	—	—	—
Manipulative	99	24	22	54
Pencil and Paper	82	17	32	51

We now turn to the results of the classroom observations and attitude measures. The observations showed that the students were greatly interested in their tasks, more so during the manipulative lab activities than during any other part of instruction. In contrast to student enthusiasm and persistence shown in the manipulative math labs, the pencil and paper labs evoked only routine attention and occasionally boredom. Peer group interaction was high as teams of two worked together.

The teachers did not use all the recommended instructional techniques consistently. For instance, the exploration part of the learning cycle (Karplus et al., 1977b) was frequently shortened or omitted, teachers often showed students how to answer a question rather than letting them find out for themselves, and student assignments were not used systematically to give the students feedback.

In spite of these shortcomings, most students in both experimental groups gave highly favorable answers to the question, "Would you recommend this program for other classes?" The distribution of student responses to this question is given in Table 6. It favors the manipulative group by a small but statistically significant amount ($p < 0.01$ by a chi-square test). Other assessments of attitude support this finding that students preferred the manipulative version over the pencil and paper version (Kurtz, 1976).

CONCLUSIONS

This study has shown that it is possible to advance the use of proportional reasoning of many secondary school students by means of a well-designed teaching program requiring

Table 6. Subjects Recommending the NR Program to Classmates (percent).

Recommendation	Manipulative group	Pencil and paper group
Highly favorable	72	57
Mildly favorable	3	13
Undecided	1	2
Mildly unfavorable	13	22
Highly unfavorable	10	6

approximately three weeks of school time. Our first hypothesis is therefore supported. The "high" delayed test performance described in Table 5 provided evidence that the new problem-solving procedures had been internalized by about half the students in the participating suburban prealgebra classes, who had shown little success in applying proportional reasoning on the pretest.

At the same time, we observed that the formulation of algebraic equations, which were included in the teaching program, had not been similarly internalized: about forty percent of the students used this algebraic procedure on the posttest (Tables 3 and 4) but fewer than half of these continued to use them on the delayed posttest after an additional two months had elapsed. During this time, the students had continued their prealgebra course with no reference to ratio or proportion.

This difference in success on two aspects of the NR program needs comment, since the use of algebraic equations to represent the numerical relationships was an important part of the teaching program. Our interpretation is that the equations did not provide additional insight into the constant ratio and constant difference problems to which they were applied. After all, the relationships were relatively simple and could be handled mentally, so that few students found it necessary to write equations on the delayed posttest. A separate and more demanding application of equations—perhaps to problems that involve the simultaneous solution of two equations—may have more success. Yet our experience is relevant to much of secondary school mathematics instruction, in which students are required to use procedures, such as factoring polynomials, for which the context does not establish a clear need. We expect that long-term learning of such procedures will be limited to a small number of students.

We now redirect the reader's attention at the six distinguishing features of the NR program described earlier. Our research does not allow us to identify any one of these as being particularly responsible for the students' gains. In fact, we believe that all six are closely interrelated. Their common feature is to give the students more autonomy in their studies—more opportunities to look at relations from their own points of view, to raise questions that disturb them, and to seek answers to these questions in their own ways. In this way, the learning is more individualized, but the teaching is very far from being individually prescribed.

Further research might well be directed toward identifying the critical components of the instruction more precisely. Until this is done, teachers might well consider all the aspects we have described for the purpose of generalizing their use.

Were any differences between the manipulative and pencil and paper groups observed? As we have pointed out, the cognitive gains were equal for the two experimental groups. The attitude surveys and classroom observations made during the study, however, showed that the manipulative version was considerably more popular than the pencil and paper version. We therefore reject our second hypothesis but have found support for the third.

We conclude that the manipulative laboratory materials had a motivating value for many students and should therefore be considered in secondary school mathematics classes when student interest and attitude toward the subject are important issues. According to the evidence provided by this study, however, the additional cost of the manipulatives is not justified merely for attaining the learning objectives in the suburban prealgebra classes we used.

ACKNOWLEDGEMENTS

We are grateful to the Acalanes Union High School District and the Mount Diablo Unified School District for allowing us to conduct this research in district high schools. The cooperation of the volunteer teachers and the students in their classes was greatly appreciated. We thank Helen Adi, Leon Henkin, Elizabeth F. Karplus, Anton Lawson, Marcia Linn, Steven Pulos, and Warren Wollman for numerous helpful conversations and advice regarding this study.

REFERENCES

Biggs, E. "Investigations and Problem-Solving in Mathemetical Education." In A. G. Howson, Ed. *Proceedings of the Second International Congress on Mathematical Education*. London: Cambridge University Press, 1973.

Inhelder, B., & Piaget, J. *The Growth of Logical Thinking from Childhood to Adolescene*. New York: Basic Books, 1958.

Karplus, R., Karplus, E., Formisano, M., & Paulsen, A. C. "A Survey of Proportional Reasoning and Control of Variables in Seven Countries." *Journal of Research in Science Teaching*, 1977(a), *14*, 411–417.

Karplus, R., Lawson, A. E., Wollman, W., Appel, M., Bernoff, R., Howe, A., Rusch, J. J., & Sullivan, F. *Science Teaching and the Development of Reasoning*. Berkeley, CA: Lawrence Hall of Science, University of California, 1977(b).

Karplus, R., & Lawson, C. A. *SCIS Teachers Handbook*. Berkeley, CA: Lawrence Hall of Science, University of California, 1974.

Karplus, R., & Peterson, R. W. "Intellectual Development Beyond Elementary School II: Ratio, a Survey." *School Science and Mathematics*, 1970, *70*, 813–820.

Kurtz, G., "A Study of Teaching for Proportional Reasoning." Doctoral Dissertation, University of California, Berkeley, 1976.

Lovell, K., & Butterworth, I. B. "Abilities Underlying the Understanding of Proportionality." *Mathematics Teaching*, 1966, *37*, 5–9.

Lunzer, E. A., & Pumfrey, P. D. "Understanding Proportionality." *Mathematics Teaching*, 1966, *34*, 7–12.

Siegel, S. *Nonparametric Statistics for the Behavioral Sciences*. New York: McGraw Hill Book Co., Inc., 1956.

Suarez, A. "Die Entwicklung der Denkoperation Beim Verständnis Funktionaler Zusammenhänge." Doctoral Dissertation. Eidgenössische Technische Hochschule, Zürich, 1974.

Wollman, W., & Karplus, R. "Intellectual Development Beyond Elementary School V: Using Ratio in Differing Tasks," *School Science and Mathematics*, 1974, *74*, 593–613.

Wollman, W. T., & Lawson, A. E. "The Influence of Instruction on Proportional Reasoning in Seventh Graders," *Journal of Research in Science Teaching*, 1978, *15*, 227–232.

Intellectual Development Beyond Elementary School VIII: Proportional, Probabilistic, and Correlational Reasoning

Robert Karplus, Helen Adi*, and Anton E. Lawson**

Concern with the development of formal reasoning has been stimulated by the increasing recognition that formal reasoning is related to student success in secondary and college science and mathematics courses (Arons, 1976; Bauman, 1976; Griffiths, 1976; Herron, 1975; Karplus, 1978; Kolodyi, 1975; Lawson and Renner, 1975; Sayre and Ball, 1975; Shayer, 1972, 1973, 1974; Suarez, 1977). In this article we shall report on our study of six group-administered tasks that have not been previously used in the United States with student groups of widely differing ages. We seek to answer these questions:

* Present address: Department of Mathematical Sciences, Northern Illinois University, DeKalb, Illinois, 60115.
** Present address: Department of Physics, Arizona State University, Tempe, Arizona, 85281.
† (AESOP) Advancing Education through Science-Oriented Programs.
  This material is based in part upon research supported by the National Science Foundation under Grant No. SED74-18950.
  Any opinions, findings, and conclusions or recommendations expressed in this publication are those of the authors and do not necessarily reflect the views of the National Science Foundation.

Robert Karplus, Helen Adi, and Anton E. Lawson, AESOP† Lawrence Hall of Science, University of California, Berkeley, California 94720.

"Intellectual Development Beyond Elementary School VIII: Proportional, Probabilistic, and Correlational Reasoning" was taken from *School Science and Mathematics*, 1980, vol 80, pgs. 673–683.

Table 1. Number of Subjects at Each Educational Level.

	Grade 6	Grade 8	Grade 10	Grade 12	College	Total
Male	48	65	57	68	53	291
Female	53	39	45	44	33	214
Total	101	104	102	112	86	505

1. What categories are required for classifying the subjects' responses on tasks requiring proportional, probabilistic, or correlational reasoning?
2. How effective are these tasks for assessing these aspects of formal reasoning?
3. What implications for teaching are suggested by the observed distributions of student responses among the categories required?

SUBJECTS

The subjects were 505 students from sixth grade to college sophomores, ranging in age from 11.5 to 20.0 years. Somewhat over 400 of the subjects were enrolled in elementary and secondary schools located in a middle- to upper-middle-class suburban community in the San Francisco Bay area. The eighth graders were tested in English classes, the tenth graders in biology, and the twelfth graders in social studies to obtain representative samples not biased with respect to mathematics and science background. The college students were freshmen and sophomores enrolled in two physical science courses for non-majors at the University of California, Berkeley. The numbers and grade levels of the students are summarized in Table 1.

METHOD

The subjects were administered the tasks by one of the authors during a regularly scheduled class period. They were given a booklet with the written questions. The booklet had space for answers and justifications of the answers. The first four items were read aloud, with demonstrations (described below) accompanying the first three. The subjects read the remaining items, which included two tasks on propositional reasoning that are described in another publication (Lawson, Karplus, and Adi, 1978). The task administrator remained in the classroom to be available for answering individual questions and clarifying the tasks as requested by the subjects. Testing of each class required approximately 40 minutes.

A brief description of the six tasks follows:

Item 1: Proportions. (Cylinder task, Suarez and Rhonheimer, 1974). The subjects were shown two transparent hollow cylinders of differing diameters but equally-spaced graduations. A quantity of water occupying four units in the wider cylinder was poured into the narrow cylinder, which it filled up to six units. The task was to consider a quantity of water filling the wide container up to six units and to predict how high it would rise in the narrow cylinder.

Item 2: Probability. The subjects were shown two sacks labelled A and B, with A containing three white and three black cubes, B containing seven white and nine black. They were asked whether there was a greater chance of pulling a white block from sack A or sack B.

Item 3: Probability. (after Lawson, 1978). The students were shown a sack in which were placed twenty flat pieces of wood—three red squares, four yellow squares, five blue squares, four red diamonds, two yellow diamonds, and three blue diamonds. The subjects were asked to state the chance of drawing out a red or a blue diamond by reaching into the bag without feeling around.

Item 4: Correlations. The test booklet presented a picture of sixteen fat mice with black tails, six fat mice with white tails, two thin mice with black tails, and six thin mice with white tails. The subjects were informed that these mice were a sample captured by a farmer in a part of his fields. They were asked whether there was a relation between the size of the mice and the color of their tails.

Item 5: Probability. The subjects were informed that the farmer (Item 4) captured another mouse in a nearby part of the same field. They were asked, "What are the chances that this mouse is fat?"

Item 6: Correlations. The test booklet presented a picture of four large fish with narrow stripes, twelve small fish with narrow stripes, three large fish with wide stripes, and nine small fish with wide stripes. The subjects were asked whether there was a relation between the size of the fish and the width of their stripes.

CATEGORIZATION OF RESPONSES

Evaluation of responses was based primarily on the explanations by which the students justified their answers. Sometimes the answer itself also shed some light on student reasoning, but occasional arithmetic or counting errors were not taken into account. Response categories were established either in conformity with previous investigations or in accordance with the variety of student explanations encountered on the task in the present study. All three authors participated in formulating the categories. After this task had been accomplished, at least two of the authors read and evaluated all responses. Discussions resolved the disagreements that occurred in about 10 percent of the cases, especially when an answer was obscure or contained elements falling into two of the established categories.

Item 1: Proportions. Four categories used by Karplus, Karplus, Formisano, and Paulsen (1977) were adapted for this task, as follows:

Category I (Intuitive): no explanation, illogical computation, guess; e.g., "8, because the smaller cylinder won't hold as much water as the wide cylinder, so it has a tendency to rise."

Category A (Additive): procedure focussing on the difference of water levels; e.g., "8, in the narrow cylinder the water seems to rise two marks higher than in the wide cylinder, so I added 2."

Category Tr (Transitional): additive procedure focussing on a correspondence of amounts of water, e.g., "9, if the water rose 2 when the marking in the wide cylinder was 4, so for each two marks the water goes up 1. So from 6 it will go up 3 to 9."

Category R (Ratio): using constant ratio or converting units from wide to narrow cylinder; e.t., "9, for every mark in the large one it goes $1\frac{1}{2}$ in the small one, because 4 went up to 6."

Item 2: Probability. The following four categories were established for Item 2:

Category I (Intuitive); no or illogical explanation, guess, misunderstanding; e.g., "Sack A—there are more white than black."

Category AV (Absolute Value): comparing the absolute values of white or black blocks in the two bags; e.g., "Sack B—there are more blocks to take out of sack B."

Category 1C (One Comparison): comparing the numbers of white and black blocks in one sack; e.g., "Sack A—you have a fifty-fifty chance because there are three whites and three blacks."

Category 2C (Two Comparisons): comparing the numbers of white to black blocks in both sacks; e.g., "Sack A—there are the same of each color, in B there are more blacks than whites."

Items 3 and 5: Probability Reasoning. The following three categories were established for Items 3 and 5:

Category I (Intuitive): no or illogical explanation, misunderstanding; e.g., "Pretty good, because there are seven pieces you could try to get."

Category Ap (Approximate): approximate or qualitative description; e.g., "Very little—there aren't that many red and blue diamonds."

Category Q (Quantitative): quantitative description making reference to the number of each kind of block present; e.g., "7 out of 20—I added the red and blue diamonds, which were seven, and there are twenty objects in the bag."

Items 4 and 6: Correlations. The following categories including four established by Adi, Karplus, and Lawson (in press) were used for Items 4 and 6:

Category I (Intuitive): no or illogical explanation, misunderstanding; e.g., "No, because the mice don't have anything in common, they are two different breeds."

Category NR (No Relationship): observations are described but no relation among the cell frequencies is mentioned; e.g., "No, because there are some fish that are small with wide stripes and some that are big with wide stripes."

Category TC (Two Cells): the numbers of events in two cells are compared; e.g., "Yes, there are less dark-lined fish that there are thin-lined. I circled yes for that reason."

Category FC (Four Cells): the numbers of events in all four cells are used to make two qualitative or quantitative comparisons; e.g., "Yes, because most of the large mice have colored tails while most of the small mice have white tails."

Category Co (Correlation): the conclusion uses data in all four cells for two quantitative comparisons that are then compared; e.g., "No, although the group with thin stripes is larger, the ratio of large to small fish remains one to three in both groups of thin- and wide-striped fish."

RESULTS

Responses from all 505 subjects were categorized according to the criteria described above for the six tasks. In this section we shall present our findings in tabular form and briefly identify the most important aspects.

Our findings for Item 1 (proportions) in Table 2 indicate that most of the sixth graders used additive procedures (Category A), while the majority of the upper-middle-class high school students used proportional reasoning (Category R). In this respect, the present data are consistent with earlier results using the Paper Clips Task (Karplus, Karplus, and Wollman 1974; Karplus and Peterson, 1970; Karplus et al., 1975), which was found by Kurtz (1976) to be closely equivalent to the Cylinder Task. Though the college students also used primarily proportional reasoning, a substantial minority—20 percent—responded in Categories I or

Table 2. Response Frequencies on Item 1, Proportions (percent).

Category	Grade 6	Grade 8	Grade 10	Grade 12	College
I	17	22	6	6	7
A	73	59	28	20	13
Tr	7	10	13	12	6
R	3	10	55	63	74

I = intuitive, A = additive, Tr = transitional, R = ratio.

A. The very dramatic change from eighth to tenth grades in Table 2, however, has not been duplicated in other studies. We believe it represents a fluctuation in performance of the eighth graders rather than a sudden transition possibly resulting from high school experiences.

On the first probability task (Item 2), more than half the sixth graders used two comparisons to justify their conclusion that the chances of drawing a white block were larger in sack A (Table 3). Only 14 percent of these students compared the actual numbers of blocks and selected sack B. Another 24 percent expressed only one comparison to justify their answer (usually that the chances were "even" or "50-50" in sack A), perhaps implying but not making explicit their awareness of the unequal chances in sack B. For the older students, the success rate improved progressively, with almost all the college students describing both comparisons. In fact, almost half the college students actually computed the probabilities of drawing a white block for each sack to support their conclusion.

Because of the similarity of Items 3 and 5, we have combined the students' scores and report the frequencies for the composite categories identified in Table 4. Most interesting is the gradual increase in the percentage of students who responded in Category Q on both tasks; they displayed the required probabilistic reasoning consistently. The frequency increased from 23 percent in grade six to 79 percent in college. High school seniors, with 72 percent responses in this composite category, ranked close to the college students.

At the other end of the scale were subjects who responded illogically or with misunderstanding on both tasks. Their fraction decreased from 19 percent in grade six to ultimately 2 percent in college. In the next lowest category, consisting of students whose responses were rated I and/or Ap, there was a rather abrupt decrease from 25 percent to 5 percent between eighth and tenth grades.

The two correlations Items 4 and 6 were also very similar. Both concerned coincidental relationships (no cause and effect was suggested) between two characteristics of animal population samples. When creating a composite category on correlational reasoning, we responded especially to our observation of a substantial number of student responses in Category I (illogical, misunderstanding). This category had not been necessary when

Table 3. Response Frequencies on Item 2, Probability (percent).

Category	Grade 6	Grade 8	Grade 10	Grade 12	College
I	7	10	8	4	1
AV	14	10	0	2	2
1C	24	14	20	13	4
2C	55	66	63	81	93

I = intuitive, AV = absolute value, 1C = one comparison, 2C = two comparisons.

Table 4. Response Frequencies on Items 3 and 5, Probability (percent).

Category	Grade 6	Grade 8	Grade 10	Grade 12	College
(I, I)	19	14	9	7	2
(Ap, I), (Ap, Ap)	27	25	5	3	2
(Q, I), (Q, Ap)	31	25	29	18	16
(Q, Q)	23	36	57	72	79

I = intuitive, Ap = approximate, Q = quantitative.

correlations tasks were presented in individual interviews (Adi et al., in press), because repeated questioning and probing elicited specific information from each student. Since probing was not possible in the group administration, we sought to reduce the effects of accidental misunderstandings by placing each student in the higher of the two categories in which his or her response was classified on Items 4 and 6. The frequency distributions according to this scoring procedure are presented in Table 5 for the five educational levels.

The development of correlational reasoning from the sixth grade to college is most striking. None of the sixth graders gave evidence of correlational reasoning (category Co), but 50 percent of the college sample did. At the same time, the percentage of subjects who did not respond to the data of either task (category I) decreased from two-thirds of the sixth graders to only 14 percent of the college students. The 18 percent of tenth graders who used correlational reasoning on at least one of the two items was very close to the 20 percent of a ninth- and twelfth-grade sample who showed correlational reasoning when interviewed (Adi et al., 1978).

The younger students' explanations on the correlations items made it very clear that the term "relation" was difficult for them to understand as referring to an association between attributes. Many subjects interpreted it in the sense of a family relationship and identified the large mice or fish as parents of the small. Others concluded that the differences in appearance precluded a family relationship. Some tenth graders described in detail the genetic conditions under which the observed ratios of black- to white-tailed mice, or wide- to narrow-striped fish, would result from the mating of heterozygous parents.

DISCUSSION

We now turn to the question raised at the beginning of this article. The first of these, concerning the categories needed to classify the subjects' responses, was answered explicitly in an earlier section. Here we shall present an overview of these categories grouped

Table 5. Response Frequencies on Items 4 and 6, Correlations (percent).

Category	Grade 6	Grade 8	Grade 10	Grade 12	College
I	65	45	32	21	14
NR	26	12	27	17	15
TC	4	13	8	6	1
FC	5	23	14	19	20
Co	0	8	18	37	50

I = intuitive, NR = no relationship, TC = two cells, FC = four cells, Co = correlations.

Table 6. Assignment of Response Categories to Reasoning Levels.

Task	Formal	Transitional	Concrete	Inconclusive
Proportions (Item 1)	R	Tr	A	I
Probability (Item 2)	—	2C	AV	I, 1C
Probability (Items 3, 5)	(Q, Q)	(I, Q); (Ap, Q)	(I, Ap); (Ap, Ap)	(I, I)
Correlation (Items 4, 6)	Co	FC	NR, TC	I

according to the level of reasoning they reflect. If we distinguish between a formal level, a transitional level, and a concrete level of reasoning, then the categories can be grouped approximately as shown in Table 6. Because Category I on each task indicated an illogical response or a misunderstanding, it cannot be interpreted simply, as stated in the last column of Table 6. In our opinion, the 1C category on Item 2 is also inconclusive, because of the possibility that many students failed to express the second comparison even though they had it in mind.

The second question concerned the effectiveness of these group-administered tasks for assessing student reasoning. We have already referred to one limitation of the group approach, the impossibility of probing a subject's thoughts and overcoming an initial misunderstanding. A second limitation arises from possible interaction among students, students helping one another or copying. Other limitations derive from the needs to comprehend written instructions, express one's thoughts in writing, and concentrate on the tasks. Advantages of the group procedure are the absence of cueing due to body language of the interviewer, the reproducibility of task administration, and the relative ease with which large numbers of subjects can be studied.

Copying and students helping one another were kept in mind while we administered the tasks and read the responses. Since we did not rely on short answers but scored lengthy explanations, both of these unwelcome activities required substantial effort. We monitored the classrooms to discourage cooperation. We also looked for identical explanations on two or more papers and found exceedingly few. We estimate that student interaction affected fewer than 5 percent of the papers.

All the other problems of group tasks manifested themselves through responses in the "inconclusive" categories in Table 6. For the Cylinder Task we are concerned with I responses, which were 22 percent for the eighth grade, 17 percent for the sixth grade, and under 10 percent for the other samples. These percentages may be considered acceptable since the task was presented orally, unless one believes that probing would have led a substantial number of the students with I responses to display transitional or formal reasoning. We consider this outcome very doubtful.

On probability Item 2 the responses classified I occurred with a frequency of no more than 10 percent, a very acceptable small rate. In addition, however, about 20 percent of the secondary school students made only one comparison (Category 1C). It is very likely that many of these subjects, who selected sack A as their answers, would make the second comparison after a probing question (e.g., "Does that explain why you are more likely to pick a white block from sack A than from sack B?") On an interview, therefore, the students' responses might ultimately have been classified 2C. We admit that this outcome is a deficiency of the task, which could be remedied partially by urging the students to express their thought more fully. If this change were successful, then the frequencies of 2C responses would be increased above our observations.

Our composite scoring for probability Items 3 and 5 overcame some of the difficulties of student attention and expression by giving each subject two chances to succeed. Accordingly, the frequencies of I responses are all under 20 percent, small enough not to be a serious source of error.

The composite scoring for correlational Items 4 and 6 also gave the students two chances to succeed. In spite of this procedure, however, the frequencies of I responses for all grade levels but the college group were in excess of 20 percent—as high as 65 percent for the sixth graders. This is a serious matter and needs special consideration.

We have already explained that many of the younger students interpreted the questions as referring to a relationship between the animals rather than a relationship between characteristics of the animals. The former requires comparing the pictured animals with respect to any observable characteristic, a first-order operation at the level of concrete thought. The latter requires comparing the relative frequencies of two characteristics of the animals, a second-order operation at the level of formal thought (Lovell, 1971). The fact is that the younger students had difficulty even grasping the question well enough to recognize that the numbers of animals in the various categories were relevant to the analysis required of them.

We cannot say how well repeated questioning, that drew attention to the characteristics (fatness and tail color, size and stripes), would have succeeded in clarifying the task. Interviews with a few sixth graders revealed that some invented causal intermediaries to justify a relationship (e.g., "The white-tailed mice are weaker and cannot get food as well as the black-tailed mice, so they stay skinny.") Others did take into account the numbers, but none responded in Category Co. From these considerations and the nature of the responses we therefore conclude that Category Co frequencies in Table 5 are accurate, but that the other categories significantly underestimate student reasoning as it might be revealed in an interview.

Our third question in this article concerned the implications of the frequency distributions reported in Tables 2 to 5 for teaching. The relatively poor performance of many students on proportional reasoning even in high school suggests that the mathematics courses dealing with ratio and proportions do not provide sufficient instruction in applying these concepts. In view of the prevalence of the additive procedure, it seems to be particularly important that students have experiences which allow them to compare situations embodying a constant difference relationship (e.g., the ages of two children) and those obeying a constant ratio. A highly effective teaching program for proportional reasoning has recently been described (Kurtz, 1976; Kurtz and Karplus, 1979; Wollman and Lawson, 1978); it would seem to have promise to improve student competence in this area. Another approach might be better coordination between the teaching of ratio in mathematics classes and the application in science classes.

Probabilistic reasoning as defined by Tasks 2, 3, and 5 was accomplished much more successfully than correlational reasoning by subjects at all educational levels. Even sixth graders were not misled extensively into comparing the actual numbers of blocks in Item 2. Further, more than half applied a quantitative criterion to at least one of Items 3 and 5. It may be, therefore, that problem assignments dealing with probabilities could be very effective as applications of ratio. At the same time, a link between mathematics and natural or social science teaching could be established by using sampling activities in the latter courses to provide applications.

We have already pointed out that our subjects were least successful on the items requiring correlational reasoning. Since this reasoning pattern plays an important part in the investigative process, our findings suggest that science and social studies courses should pay more explicit attention to the analysis of data for correlations. Everyday news reports of the correlation between fluoride treatment and tooth decay, cigarette smoking and lung cancer, various dietary factors and heart disease, or drunk driving and automobile accidents can be a source of information whose implications deserve to be discussed.

Yet we would caution teachers not to emphasize activities heavily dependent on correlational reasoning in junior high school. Both proportional reasoning and probabilistic reasoning, which are necessary for an understanding of correlations, appear to be used quantitatively by only a small number of junior high school students. The tenth and eleventh grades appear to be a more appropriate time to introduce correlational reasoning assignments.

In concluding, we return briefly to the close connection of both mathematics and science teaching with the aspects of formal reasoning investigated in this article. The present separation of studies into mathematics, where numerical and algebraic relations are usually studied out of context, and science, where data are gathered but mathematical procedures are not usually stressed, may be a serious obstacle to the development of formal reasoning. We urge that new curricula tie these areas together. At the same time, the students must have opportunities to raise their own questions, formulate their own hypotheses, devise their own tests, and draw their own conclusions as they search for meaning in quantitative relationships. Further suggestions for the design of teaching programs incorporating these elements have been published elsewhere (Arons, 1976; Karplus, Lawson, Wollman, Appel, Bernoff, Howe, Rusch, and Sullivan, 1977; Lawson, 1975).

REFERENCES

Adi, H., Karplus, R. Lawson, A. E., and Pulos, S., Intellectual Development Beyond Elementary School VI: Correlational Reasoning. *School Science and Mathematics* 78(8), 675–683, 1978.

Arons, A. B. Cultivating the Capacity for Formal Reasoning: Objectives and Procedures in an Introductory Physical Science Course. *American Journal of Physics*, 44(9), 834–838, 1976.

Bauman, R. P. Applicability of Piagetian Theory to College Teaching. *Journal of College Science Teaching*, 6, 94–96, 1976.

Griffiths, D. Physics Teaching, Does It Hinder Intellectual Development? *American Journal of Physics*, 44(1), 81–86, 1976.

Herron, J. D. Piaget for Chemists. *Journal of Chemical Education*, 52, 146–150, 1975.

Karplus, E. F., Karplus, R., and Wollman, W. Intellectual Development Beyond Elementary School IV: The Influence of Cognitive Style. *School Science and Mathematics*, 74(6), 476–482, 1974.

Karplus, R. Opportunities for Concrete and Formal Thinking of Science Tasks. In Presseisen, B. Z., Goldstein, D., and Appel, M. H. (eds.) *Topics in Cognitive Development.* New York: Plenum Press, 1978.

Karplus, R., Karplus, E. F., Formisano, M., and Paulsen, A. C. A Survey of Proportional Reasoning and Control of Variables in Seven Countries. *Journal of Research in Science Teaching*, 14(5), 411–417, 1977.

Karplus, R., Karplus, E., Formisano, M., and Paulsen, A. C. Proportional Reasoning and Control of Variables in Seven Countries. Berkeley, CA: Lawrence Hall of Science, 1975.

Karplus, R., Lawson, A. E., Wollman, W. T., Appel, M., Bernoff, R., Howe, A., Rusch, J. J., and Sullivan, F. *Science Teaching and the Development of Reasoning.* Berkeley, CA: Lawrence Hall of Science, 1977.

Karplus, R. and Peterson, R. W. Intellectual Development Beyond Elementary School II: Ratio, A Survey. *School Science and Mathematics*, 70(9), 813–820, 1970.

Kolodyi, G. The Cognitive Development of High School and College Science Students. *Journal of College Science Teaching*, 5, 20–22, 1975.

Kurtz, B. A Study for Teaching for Proportional Reasoning. Doctoral Dissertation, University of California, Berkeley, 1976.

Kurtz, B. and Karplus, R. Intellectual Development Beyond Elementary School VII: Teaching for Proportional Reasoning. *School Science and Mathematics*, 79(5), 387–398, 1979.

Lawson, A. E. Developing Formal Thought through Biology Teaching. *The American Biology Teacher*, 37(7), 411–420, 1975.

Lawson, A. E. The Development and Validation of a Classroom Test of Formal Reasoning. *Journal of Research in Science Teaching*, 15(1), 11–24, 1978.

Lawson, A. E., Karplus, R., and Adi, H. *The Acquisition of Propositional Logic and Formal Operational Schemata During the Secondary School Years. Journal of Research in Science Teaching*, 15(6), 465–478, 1978.

Lawson, A. E. and Renner, J. W. Relationship of Science Subject Matter and Developmental Levels of Learning. *Journal of Research in Science Teaching*, 12(4), 347–358, 1975.

Lovell, K. Some Aspects of the Growth of the Concept of a Function. In M. Rosskopf, L. Steffe, and S. Tabac, Eds. *Piagetian Cognitive Development Research and Mathematical Education*, Reston, VA: National Council of Teachers of Mathematics, 1971, page 12.

Sayre, S. A. and Ball, D. W. Piagetian Development in Students. *Journal of College Science Teachings*, 5, 23, 1975.

Shayer, M. Conceptual Demands in the Nuffield O-Level Physics Course. *School Science Review*, 51, 186, 1972.

Shayer, M. Conceptual Demands of Nuffield O-Lovel Chemistry. *Education in Chemistry*, 8, 8, 1973.

Shayer, M. Conceptual Demands in the Nuffield O-Lovel Biology Course. *School Science Review*, 56, 381, 1974.

Suarez, A. *Formales Denken und Funktionsbegriff bei Jugendlichen.* Bern, Switzerland: Hans Huber, 1977.

Suarez, A. and Rhonheimer, M. *Lineare Funktion.* Zurich, Switzerland: Limmat Stiftung, 1974.

Wollman, W. and Lawson, A. E. The Influence of Instruction on Proportional Reasoning in Seventh Graders. *Journal of Research in Science Teaching*, 15(3), 227–232, 1978.

Conditional Logic Abilities on the Four-Card Problem: Assessment of Behavioral and Reasoning Performances[1]

Helen Adi, Robert Karplus and Anton E. Lawson

Abstract. The present study assessed the behavioral and the reasoning performances of 507 school and university students on the four logical principles of logical detachment, particular conversion, particular inversion, and particular contraposition. An adapted version of Wason's four-card problem was administered to all students in paper-and-pencil format and in group settings. Students were asked to respond to the logical questions and to justify their answers. Nine logical reasoning categories were identified. The results of behavioral and reasoning performances were compared across grade levels. Results indicated that many students who responded behaviorally correctly to the logical questions, provided incorrect reasoning justifications. The percentages of students who considered a conditional statement as hypothetical increased with age. No sex differences on the behavioral and the reasoning performances were reported.

The assessment of conditional reasoning abilities among adolescents and young adults has been the subject of various investigations (Carroll, 1975; Ennis and Paulus, 1965; Roberge, 1970; Roberge and Paulus, 1971; Wildman and Fletcher, 1977). However, with the intention of studying logical reasoning, the above investigators reported the mastery levels of their subjects on the related logical principles. The mastery of a logical principle was defined by the successful recognition of the validity of an argument which embedded that logical principle. Four principles of conditional logic were considered (after Ennis, 1976):

Educational Studies in Mathematics **11** (1980) 479–496. 0013-1954/80/0114-0479$01.80 *Copyright*
© 1980 *by D. Reidel Publishing Co., Dordrecht, Holland and Boston, U.S.A.*

(1) DETACHMENT   (2) Particular CONVERSION
     hypothesis: If p, then q hypothesis: If p, then q
     given: p given: q
     conclusion: q conclusion: p
(3) Particular INVERSION   (4) Particular CONTRAPOSITION
     hypothesis: If p, then q hypothesis: If p, then q
     given: not—p given: not—q
     conclusion: not—q conclusion: not—p

Of the above four principles, only logical detachment and particular contraposition are logically valid. Thus, the mastery of logical detachment for example, was assessed by the recognition of the validity of a given argument which embedded logical detachment. Such a
performance was directly associated with one's conditional reasoning ability.

The recognition of a valid logical conclusion does not necessarily imply that the
related logical principle was applied to arrive at one's conclusion. The principles of formal
logic by themselves do not reflect how people think (Copi, 1972), and the recognition of the
valid principles of conditional logic is not sufficient to describe the associated thought
processes. To investigate the conditional logic abilities of adolescents and young adults, the
present study assessed two performance outcomes: (1) the behavioral performances, which
referred to whether the deduced logical conclusions were correct or incorrect, and (2) the
reasoning performances, which referred to the subjects' justifications of how they had
arrived at their logical conclusions.

Previous studies investigating the development of behavioral performance on propositional paper-and-pencil logic tasks, among students of different age groups, indicated that
the percentages of mastery of the principle of logical detachment increased from about 50%
among fourth graders to about 90% among 12th graders; also, these percentages increased
from about 35% among fourth graders to about 60% among 12th graders for the mastery
of logical contraposition (Ennis, 1976; Roberge, 1970, 1972; Roberge and Paulus, 1971;
Wildman and Fletcher, 1977). Most of these studies reported that it was much easier for
students, across different age groups, to recognize the validity of logical detachment and
particular contraposition than the invalidity of particular conversion and particular inversion. Further, Ennis and Paulus (1965) reported that the greatest improvement in the
mastery of specific propositional logic principles, as students grew older, was that for the
invalid principles.

Jansson (1978) investigated the behavioral performances of a group of adolescents
on Wason's (1972) four-card problem, and he pointed out that the results were not sufficient
to describe the logical processing or the thought processes that the subjects were engaged
in while solving the four-card problem. In order to describe the thought processes subjects
were engaged in, both behavioral and reasoning performances could have been investigated.

The present study thus investigated the behavioral and the reasoning performances
of a group of adolescents and young adults on Wason's four-card problem, in an attempt to
assess their conditional logic abilities. In specific, the present study, attempted to answer the
following two questions:

(1) What are the behavioral performances of a group of adolescents and young
     adults on the four logical principles of detachment, particular conversion, par-
     ticular inversion, and particular contraposition?
(2) What are the reasoning performances of a group of adolescents and young
     adults on the four logical principles of detachment, particular conversion, par-
     ticular inversion, and particular contraposition?

Table I. Descriptive Statistics of the Subjects.

Grade	Frequency of males	Frequency of females	Total	Mean age (years)	SD
6	48	53	101	12.09	0.46
8	64	40	104	14.03	0.51
10	57	45	102	16.08	0.56
12	69	44	113	18.01	0.60
College	52	35	87	19.58	0.89

1. METHOD

1.2. Subjects

The subjects were 507 students. Over 400 of the students were enrolled in elementary and secondary schools in a middle to an upper middle SES suburban community in the San Francisco Bay area. The remaining students were enrolled in two non-major science courses at the University of California, Berkeley. The eighth graders were tested in English classes, the tenth graders required biology classes, and the twelfth graders in social studies classes to obtain representative samples not biased with respect to science and mathematics background. Care was taken to select schools within communities of the same socioeconomic level for each grade level. No randomization was used for student selection. The number and the mean age of students at each grade level is given in Table I.

1.2. Procedure

To investigate the associated thought processes in problem-solving situations, researchers have several procedural options to choose from. These options include:

(1) a "look ahead" procedural approach, prior to problem solving (Resnick, 1976).
(2) a "think aloud" procedural approach during problem solving, and
(3) a "justify your answer" or "how did you arrive at your conclusion?" procedural approach, after problem solving.

The "justify your answer" procedural approach was adopted for our investigative purposes to assess the associated thought processes of the subjects on an administered paper-and-pencil conditional logic task. Such a procedural selection was made primarily for pragmmatic purposes: we were not planning to interview subjects on a one-to-one basis.

A written adaptation of Wason's four-card problem (Wason and Johnson-Laird, 1972) was administered by one of the authors to all students during regular class hours. However, prior to the administration of this written version of the four-card problem, a formative evaluation of the instrument was conducted. This formative evaluation was conducted with a group of students, who were not participating in the study, in one-to-one interview sessions, and in group settings.

1.3. Task

The written adaptation of the four-card problem (Wason and Johnson-Laird, 1972), used in the present study was presented in Figure 1. The application of the four principles of

Here are pictures of four cards:

| E | K | 4 | 7 |

You know that each of these four cards has a letter on one side and a number on its other side.

Read the following rule:

> If a card has a vowel on one side, then it has an even number on the other side.

You want to find out whether the above rule is true or false.

You are asked to name those cards, and only those cards, that <u>need</u> to be turned over to test the rule.

Here are pictures of four cards:

| E | | K |

To test the rule, would you <u>need</u> to turn this card over?

☐ Yes   ☐ No

Why? _____
_____
_____

To test the rule, would you <u>need</u> to turn this card over?

☐ Yes   ☐ No

Why? _____
_____
_____

| 4 | | 7 |

To test the rule, would you <u>need</u> to turn this card over?

☐ Yes   ☐ No

Why? _____
_____
_____

To test the rule, would you <u>need</u> to turn this card over?

☐ Yes   ☐ No

Why? _____
_____
_____

Figure 1. The adapted version of Wason's four-card problem.

conditional logic: (1) detachment, (2) particular conversion, (3) particular inversion, and (4) particular contraposition was possible by responding to the cards: E, 4, K, and 7 respectively.

Given the rule: "if a card has a vowel on one side, then it has an even number on the other side", and given the four particular cards E, 4, K, and 7, students were asked (1) to respond by a "yes", or a "no" to whether they needed to know what was on the other side of each card in order to find out whether the rule was true or false and (2) to justify why they had made those particular respones. Both performance outcomes were expressed in writing. The "yes" or "no" responses were either correct or incorrect, and each constituted a behavioral performance. The justification given for the behavioral performance on each of the four cards defined a reasoning performance.

1.4. Scoring: Behavioral Performances

Each student was assigned a set of five behavioral performance scores: one score for each of the four cards, and one overall behavioral performance score. A score of one was assigned for a correct behavioral performance, and a score of zero was assigned for an incorrect behavioral performance. For the overall behavioral performance, a score of one was assigned when the behavioral performances on all four cards were correct; otherwise, a score of zero was assigned for the overall behavioral performance.

Correct behavioral performances consisted of (1) a "yes" response on card E, that is, wanting to know what was on the other side of card E, in order to verify the given conditional rule, (2) a "no" response on card 4, (3) a "no" response on card K, and (4) a "yes" response on card 7.

1.5. Scoring: Reasoning Performances

Each student was assigned a set of five reasoning performance scores: one score for each of the four cards, and one overall reasoning performance score. A scoring scheme was developed to classify the reasoning performances on the four-card problem. The development of the scoring scheme was as follows:

(1) A random sample of 50 students was selected from the group of 507 students; these 50 students were equally distributed among the five grade levels.
(2) Two raters separately analyzed the written justifications of the above 50 students on all four cards.
(3) Two sets of reasoning categories were independently identified by the two raters.
(4) The two raters compared their identified sets of reasoning categories; and accordingly, one classification scheme for the reasoning performances was formulated.
(5) A third rater validated the above classification scheme by scoring the reasoning performances of a sample of 30 students, who were not participating in the study, on the four-card problem.
(6) Recommendations of the third rater were considered in reformulating the operational definitions of the identified reasoning categories.
(7) Nine reasoning categories were identified, and accordingly, nine reasoning performance scores were assigned. For each student, one reasoning performance score was assigned for each of the four cards.

Two raters independently scored the reasoning performances of all 507 students. There was an 87% agreement between the two ratings. The differences in ratings were discussed and resolved in the presence of a third rater.

Only one of two scores was assigned for the overall reasoning performance of each student. A score of one was assigned for the overall reasoning performance when the reasoning responses on all four cards demonstrated correct conditional reasoning, and a score of zero was assigned for the overall reasoning performance when the reasoning response on at least one of the four cards was incorrect.

Table II. Percentages of Correct Behavioral Performance Scores on the Four-Card Problem.

Grade	Card E (detachment)	Card 4 (converse)	Card K (inverse)	Card 7 (contrapositive)	Cards E, 4, K, 7 (overall)
6 ($n = 101$)	64.2	51.5	51.6	35.7	2.0
8 ($n = 104$)	56.7	39.4	48.2	50.0	1.9
10 ($n = 102$)	82.3	44.1	45.0	50.0	9.8
12 ($n = 113$)	82.3	44.2	60.2	38.1	12.4
College ($n = 87$)	83.7	42.4	60.9	42.5	14.9
Total ($n = 507$)	73.8	42.0	53.1	43.2	8.1

2. RESULTS

At each grade level, the percentages of correct behavioral performances were highest on card E (logical detachment). The order of the magnitudes of these percentages on the other three cards varied from one grade level to the other (see Table II). The percentages of correct overall behavioral performances increased with grade level, but these percentages were generally low. The percentages of correct overall behavioral performances ranged from 2.0% for sixth graders to 14.9% for college students (see Table II). Such a finding was expected on the basis of earlier investigations using Wason's four-card problem (e.g., Jansson, 1978; O'Brien, 1975; Wason and Johnson-Laird, 1972). Chisquare analyses indicated no significant sex differences in the behavioral performances of students on any of the four cards.

2.2. Reasoning Performances

2.2.1. Reasoning Caterories. The identified reasoning categories consisted of the following:

Category 0: No written justification was given.

Category 1: The student gave redundant response. Example, "Yes, I want to turn this card to see what's on the other side", or "Yes, I want to turn this card to have more information", or "No, I don't need to turn this card, I have enough information".

Category 2: The student *described* the given information on the card and/or in the rule. Example, "No, I don't need to turn this card, seven is an odd number", or "No, I don't need to turn this card, the rule says nothing about odd numbers".

Category 3: The student *accepted* the rule as true, and applied it. Example, "No, I don't need to turn card E, because there will be an even number on the other side".

Category 4: The student *accepted* the converse, or the inverse of the rule as true, and applied it. Example, "No need to turn card 4, because there will be a vowel on the other side".

Category 5: The student attempted to *test* the converse, or the inverse of the rule. Example, "Yes, I want to turn card 4 to see if there is a vowel on the other side".

Category 6: The student attempted to *test* the implications of both possible outcomes of turning the card in relation to the rule, and/or expressed that the converse or the inverse of the rule need not be tested. Example, "No, I don't need to turn card 4 because no matter what is on the other side the rule may still be true", or "No, I don't need to turn card 4; the rule says nothing about its reverse".

Category 7: The student attempted to *test* the rule by looking for confirming evidence. Example, "Yes, I want to turn card E, to see if there is an even number on the other side".

Table III. Percentages of Correct* Reasoning Performance Scores on the Four-Card Problem.

Grade	Card E (detachment)	Card 4 (converse)	Card K (inverse)	Card 7 (contrapositive)	Cards E, K, 4, 7 (overall)
6 ($n = 101$)	24.6	0.0	2.0	5.0	0.0
8 ($n = 104$)	26.9	1.0	2.9	8.6	0.0
10 ($n = 102$)	48.0	10.8	12.7	18.6	4.9
12 ($n = 113$)	57.6	17.7	19.5	22.1	5.3
College ($n = 87$)	60.9	26.4	19.5	20.6	9.2
Total ($n = 507$)	43.0	10.7	11.2	15.0	3.7

*Reasoning categories 7 or 8 denoted correct reasoning performances on Cards E or 7, and reasoning category 6 denoted correct reasoning performances on Cards 4 or K.

Category 8: The student attempted to *test* the rule by looking for disconfirming evidence. Example, "Yes, I would like to turn card E; if I find an odd number on the other side, then the rule is false".

Of the above reasoning categories, the following defined correct reasoning performances: (1) category 7 or 8 for *card E*, (2) category 6 for *card 4*, (3) category 6 for *card K*, and (4) category 7 or 8 for *card 7*.

2.2.2. Correct Reasoning Performances

The percentages of correct reasoning performances on the four cards for the total group of subjects are given in Table III. At each grade level, the percentages of correct reasoning performances were highest for card E. However, these percentages of correct reasoning performances on card E were consistently lower than the corresponding percentages of correct behavioral performances. Thus, at each grade level there were students who demonstrated correct behavioral performances but did not express in writing correct reasoning performances. Percentages of correct reasoning performances were consistently lower than the corresponding percentages of correct behavioral performances on all four cards, and across all grade levels.

The percentages of correct overall reasoning performances ranged from 0.0% among the sixth graders to 9.2% among the college students. Chi-square analyses indicated no effects of sex differences on the correct reasoning performance scores on the four cards and across grade levels.

2.3. Behavioral and Reasoning Performances on Card E

The distribution of behavioral and reasoning performance scores on Card E (logical detachment) by grade levels and for the total group is given in Table IV. Of the total group of students 73.8% answered with a correct "yes" response on card E, but only 43.0% of the total group exhibited correct behavioral and reasoning performances ("yes" response with reasoning category 7 or 8).

Although 83.7% of the college students exhibited correct behavioral performances on card E, yet 24.0% of the college students provided incorrect reasoning justifications (categories 1, 2, and 3) for their correct behavioral performances. The percentages of students who exhibited incorrect reasoning for the correct behavioral response on card E increased to 39.6% among the sixth graders.

Table IV. Percentage of Behavioral and Reasoning Performance Scores on Card E (Lgical Detachment).

Behavioral performance	Grade	Reasoning performance score								
		8	7	6	5	4	3	2	1	0
Correct	6 (n = 101)	3.8	20.8	—*	—	—	3.0	11.9	24.7	—
	8 (n = 104)	2.9	24.0	—	—	—	8.6	6.7	13.5	1.0
	10 (n = 102)	8.8	39.2	—	—	—	2.0	3.9	27.4	1.0
	12 (n = 113)	8.0	49.6	—	—	—	1.8	10.6	11.5	0.9
	College (n = 87)	18.4	40.2	—	—	—	1.1	5.7	17.2	1.1
	Total (n = 507)	8.1	34.9	—	—	—	3.4	7.9	18.7	0.8
Incorrect	6 (n = 101)	—	—	—	—	—	19.8	5.9	7.9	2.0
	8 (n = 104)	—	1.0	—	—	—	20.2	5.7	9.6	6.7
	10 (n = 102)	—	—	—	—	—	9.8	1.0	3.9	2.9
	12 (n = 113)	—	—	—	—	—	12.4	0.9	2.6	1.8
	College (n = 87)	—	2.3	—	—	—	6.9	—	—	6.9
	Total (n = 507)	—	0.6	—	—	—	13.8	2.8	4.9	4.0

*Denoted 0.0%.

Table V. Percentages of Behavioral and Reasoning Performance Scores on Card 4 (Converse).

Behavioral performance	Grade	Reasoning performance score								
		8	7	6	5	4	3	2	1	0
Correct	6 (n = 101)	—*	—	—	—	22.8	—	8.9	8.9	1.0
	8 (n = 104)	—	—	1.0	—	16.3	—	6.7	12.5	2.9
	10 (n = 102)	—	—	10.8	1.0	10.8	—	1.0	19.6	1.0
	12 (n = 113)	—	—	17.7	1.8	8.8	—	4.4	9.7	—
	College (n = 87)	—	—	25.3	1.1	5.7	—	2.3	8.0	—
	Total (n = 507)	—	—	10.7	0.8	13.0	—	4.7	11.8	1.0
Incorrect	6 (n = 101)	—	—	—	26.7	2.0	—	8.9	18.8	2.0
	8 (n = 104)	—	—	—	26.0	6.8	—	7.7	11.5	8.6
	10 (n = 102)	—	—	—	26.5	2.0	—	3.9	20.6	2.9
	12 (n = 113)	—	—	—	34.5	2.7	—	5.3	10.6	4.4
	College (n = 87)	—	—	1.1	31.0	—	—	4.6	12.6	8.0
	Total (n = 507)	—	—	0.2	29.0	2.8	—	6.1	14.8	5.1

*Denoted 0.0%.

2.4. Behavioral and Reasoning Performances on Card 4

The distribution of behavioral and reasoning performance scores on card 4 (particular conversion) by grade levels and for the total group is given in Table V. Of the total group of students 42.0% answered with a correct "no" response on card 4, but only 10.7% of the total group exhibited both correct behavioral performances ("no" response) and correct reasoning performances (category 6).

Although 41.6% of the sixth graders gave correct behavioral responses on card 4, no one expressed in writing a correct reasoning performance (category 6). In fact, 22.5% of the sixth graders gave correct behavioral responses but justified their behaviors by reasoning category 4, which implied that they did not want to turn over card 4 because they had accepted the converse of the given conditional rule as true, and thus did not find it necessary to check what was on the other side of card 4. Table V also illustrates that the percentages of students who exhibited reasoning category 6 increased with grade level. At each grade level, more than 25% of the students responded with an incorrect behavioral response on card 4, and they justified their behavior by reasoning category 5—they were attempting to test the converse of the given conditional rule.

2.5. Behavioral and Reasoning Performances on Card K

The distribution of behavioral and reasoning performance scores on card K (particular inversion) for the total group and by grade level is given in Table VI. Of the total group of students 53.1% answered with a correct "no" response on card K, but only 11.2% of the total group exhibited both correct behavioral and reasoning performances ("no" response with reasoning category 6).

The percentages of students who exhibited both correct behavioral and reasoning performances on card K increased with grade level from 2.9% among the sixth graders to 19.5% among the twelfth graders and college students. Of the total group of students only

Table VI. Percentages of Behavioral and Reasoning Performance Scores on Card K (Inverse).

Behavioral performance	Grade	Reasoning performance score								
		8	7	6	5	4	3	2	1	0
Correct	6 (n = 101)	—*	—	2.0	—	14.9	—	19.8	11.9	3.0
	8 (n = 104)	—	—	2.9	1.0	9.6	—	21.2	10.6	2.9
	10 (n = 102)	—	—	12.7	—	3.9	—	20.6	7.8	—
	12 (n = 113)	—	—	19.5	0.9	4.4	—	28.3	5.3	1.8
	College (n = 87)	—	—	19.5	—	3.4	—	29.9	5.7	2.3
	Total (n = 507)	—	—	11.2	0.4	7.3	—	23.9	8.3	2.0
Incorrect	6 (n = 101)	—	—	—	9.9	2.0	—	7.9	24.8	4.0
	8 (n = 104)	—	—	—	11.5	3.9	—	14.5	14.5	7.7
	10 (n = 102)	—	—	—	20.6	—	—	2.9	26.5	4.9
	12 (n = 113)	—	—	—	15.9	3.6	—	5.3	11.5	3.6
	College (n = 87)	—	—	—	4.6	—	—	3.4	17.2	13.8
	Total (n = 507)	—	—	—	12.8	2.0	—	6.9	18.7	6.5

*Denoted 0.0%.

Table VII. Percentages of Behavioral and Reasoning Performance Scores on Card 7 (Contrapositive).

Behavioral performance	Grade	Reasoning performance score								
		8	7	6	5	4	3	2	1	0
Correct	6 (n = 101)	4.0	1.0	—*	5.0	1.0	—	6.9	17.8	—
	8 (n = 104)	4.8	3.8	—	3.8	1.0	—	15.4	19.2	1.9
	10 (n = 102)	17.6	1.0	—	6.9	—	—	3.9	18.6	2.0
	12 (n = 113)	15.0	7.1	—	2.7	—	—	1.8	10.6	0.9
	College (n = 87)	17.2	3.4	1.1	—	1.1	—	4.6	13.8	1.1
	Total (n = 507)	11.6	3.4	0.2	3.7	0.6	—	6.5	16.0	1.2
Incorrect	6 (n = 101)	—	—	—	—	18.8	—	18.8	17.8	8.9
	8 (n = 104)	1.0	—	1.0	—	4.8	1.0	22.2	15.4	4.8
	10 (n = 102)	—	—	2.0	—	6.9	—	21.6	15.7	3.9
	12 (n = 113)	—	—	7.1	—	0.9	9.7	24.8	15.0	4.5
	College (n = 87)	—	—	2.3	—	1.1	3.4	26.4	8.0	13.8
	Total (n = 507)	0.9	—	2.6	—	6.5	3.0	22.7	14.6	7.1

*Denoted 0.0%.

12.8% responded with an incorrect "yes" response on card K and gave reasoning category 5 justification. This percentage was higher on card 4; it was 29.0% for the total group of students.

2.6. Behavioral and Reasoning Performances on Card 7

The distribution of behavioral and reasoning performance scores on card 7 (particular contraposition) for the total group and by grade level is given in Table VII. Of the total group of students 43.2% answered with a correct "yes" response on card 7, but only 15.0% of the total group exhibited correct behavioral and reasoning performances ("yes" response with reasoning category 7 or 8). Of the total group of students 11.6% exhibited correct behavioral performances with the correct justification of having attempted to test the conditional rule by looking for disconfirming evidence (reasoning category 8), and 3.4% of the total group of students exhibited correct behavioral performances with the correct justification of having attempted to test the conditional rule by looking for confirming evidence (reasoning category 7).

Table VII illustrates that only 20.6% of the college students, 22.1% of the twelfth graders, 18.6% of the tenth graders, 8.6% of the eighth graders, and 5.0% of the sixth graders responded with correct behavioral and reasoning (categories 7 or 8) performances on card 7. Table VII also illustrates that 30.6% of the total group of students responded with reasoning category 1 on card 7. This is, 30.6% of the total group justified their behavioral performances by statements such as "Yes, I want to see what's on the other side of this card". In fact, 16.0% of the total group of students responded with a correct behavioral response on card 7, and justified their responses by reasoning category 1. A breakdown by grade level of those students who responded with a correct behavioral response on card 7 and who gave reasoning category 1 for their justifications showed that 13.8% of the college students, 10.6% of the twelfth graders, 18.6% of the tenth graders, 19.2% of the eighth graders, and 17.8% of the sixth graders belonged to this class of respondents.

3. DISCUSSION

Behavioral and reasoning performances of 507 students on the four principles of conditional logic: (1) logical detachment, (2) particular conversion, (3) particular inversion, and (4) particular contraposition, were assessed. One task with four related questions was administered to the students to assess their performance outcomes on conditional logic. The task used was an adapted, group-administered, paper-and-pencil version of the four-card problem (Wason and Johnson-Laird, 1972). The assessed behavioral and reasoning performances were based upon the student's written responses and justifications to their selection or non-selection of the four cards.

The investigative procedure of the present study had extensively been used in other studies for the purposes of assessing students' behavioral and reasoning performances in group settings (e.g., Bell, 1976; Karplus and Karplus, 1970; Lawson, 1978; and Lawson et al., 1978). However, to interpret the results of the present study, the limitations of having administered one paper-and-pencil task in a group setting, and of having analyzed the solutions, were considered. These limitations increased the likelihood of poorer performances by some students who could have performed better in a clinical setting where further probing into their reasoning was possible, and where no demands were put on their reading and writing skills. In spite of the above limitations, the present results of performance outcomes were not markedly different from results of other studies where the four-card problem was administered in clinical settings (e.g., Jansson, 1978; Wason and Johnson-Laird, 1972). However, in addition to the previously reported behavioral performance outcomes, reasoning performance outcomes were also reported in the present study.

Two different sets of tasks have been used in the literature to study logical reasoning: one set of tasks examines children's logic to the extent that logic mediates scientific inquiry, and these "scientific" tasks involve both the encoding of factual information and the application of logical rules of inference, while the second set of logic tasks consists of "propositional", strictly verbal, situations in which the child is studied as a logician where he/she is expected to deduce logical conclusions by relying on the mere verbal forms (Falmagne, in press). In studying the logical competency of the developing child, it is not sufficient to compare the child's logic against the logic of propositional calculus, but it is necessary to examine the heuristic mechanisms the child uses to arrive at his/her logical conclusions when faced with a "scientific" situation which entails the encoding of information, the generation of hypotheses, and the verification of these hypotheses. The four-card problem, used in the present study, examined both of the above aspects of logical competency by assessing behavioral and reasoning performances. However, when previous studies investigated behavioral performances on conditional logic tasks, it was reported that task differences resulted in differential effects on behavioral performances (e.g., Falmagne and Kenney, Note 2; Jansson, 1978; Johnson-Laird et al., 1972; Mast et al., Note 3). Some of the investigated task variables were differences in familiarity of content, representational modes, and representational formats. However, differential effects of task variables on reasoning performances have not been reported in the literature. Thus, in interpreting the reasoning performance outcomes of the present study, no claims were made as to their consistency in test-retest situations, or across a larger number or a wider variation of conditional logic tasks.

Different reasoning performances were identified when students were asked to justify their behavioral performances on the four-card problem. The identified reasoning performances represented: (a) hypothetical and correct reasoning (reasoning categories 6, 7, or 8), (b) hypothetical but incorrect reasoning (reasoning category 5), and (c) non-hypothetical and incorrect reasoning (reasoning categories 4, 3, 2, or 1). The non-hypothetical and incorrect reasoning varied from accepting the conditional rule and/or its converse as true statements (reasoning categories 3 and 4), to the description of the given cards or the rule

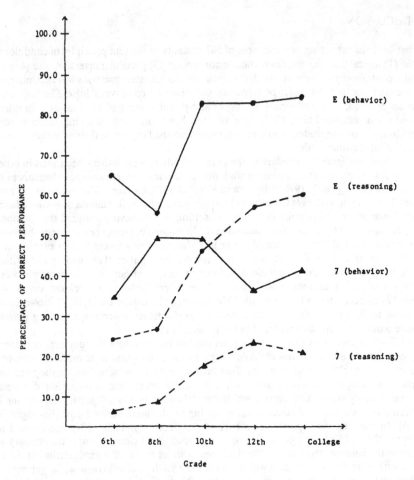

Figure 2. Percentages of correct behavioral and reasoning performance scores on cards E (logical detachment) and 7 (particular contraposition).

(reasoning category 2). Students who failed to provide written justifications for their behavioral performances were grouped together (reasoning category 0), and no further implication was made as to their reasoning beyond the fact that they did not justify their behavioral performances.

The principles of logical detachment and particular contraposition are logically equivalent, but when the related precentages of correct performance outcomes were compared, these two logical principles did not seem to be psychologically equivalent (see Figure 2). A similar conclusion was drawn when the performance outcomes on the two principles of particular conversion and particular inversion were compared (see Figure 3).

O'Brien (1975) reported that some students misinterpreted or rather deformed conditional statements by considering them as non-hypothetical and factual statements. This observation was confirmed in the present study by the identification of reasoning performance categories 3 and 4.

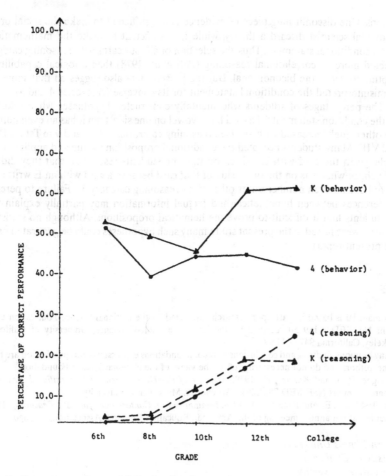

Figure 3. Percentages of correct behavioral and reasoning performance scores on cards 4 (particular conversion) and K (particular inversion).

Paris (1973), and Wason and Johnson-Laird (1972) reported that sometimes responses to conditional items were based upon a biconditional interpretation of the conditional operation. For example, Wason and Johnson-Laird (1972) reported that the selection of all four cards (E, K, 4, and 7) on the four-card problem occurred quite frequently among students of various ages. They explained such an occurrence as a misinterpretation of the conditional operation for the biconditional. Such a response mode was also frequent among our subjects, but the justifications indicated that many were *not* attempting to verify the biconditional. A frequent selection of all four cards occurred when the students wanted to collect as much information as possible (reasoning category 1). Many students attempted to collect both confirming and disconfirming evidences to the rule, but no indication was made that the provision of one disconfirming piece of evidence could fasify the conditional statement. In fact, such a performance outcome may very well be expected and accepted from a social scientist who may be attempting to establish the truth of a statement relating two conditions. One disconfirming piece of evidence is not sufficient to make the social or the experimental scientist discard a theory, while it is sufficient to make the mathematician discard a conditional statement. Thus, the selection of all four cards with reasoning category 1 reflected more of correlational reasoning (Adi *et al.*, 1978) than a logical capability of attempting to verify the biconditional. But the present data also suggested that some students misinterpreted the conditional statement for its converse (categories 4 and 5).

The percentages of students who mentally constructed hypotheses when asked to verify the condition statement "If a card has a vowel on one side, then it has an even number on the other side" increased with age (see reasoning caterories 5, 6, 7, and 9 in Tables IV, V, VI and VII). Many students accepted the conditional proposition as true and applied it. For example, given the card with E written on it, some students responded that they did not need to check what was on the other side of that card because a card with an E written on it had to have an even number on its other side (reasoning category 3). Failure to perceive the differences between hypothetical and factual information may partially explain why many students find it difficult to prove mathematical propositions. Although no statistical hypotheses were tested in the present study, many such hypotheses could be generated based on the present report.

NOTES

1. This material is based in part upon research supported by the National Science Foundation under Grant No. SED74-189 50 received by the Lawrence Hall of Science, University of California, Berkeley, California 94720, U.S.A.

   Any opinions, findings, and conclusions or recommendations expressed in this publication are those of the authors and do not necessarily reflect the views of the National Science Foundation.
2. Falmagne, R. J. and Kenney, H.: 1979, *Analysis of the Development of Propositional Reasoning* (Progress Report NSF: SED 78-22294), Worchester, Mass., Clark University.
3. Mast, T., Mason, E., and Bramble, W.: 1974, *Familiarity with Content and Syllogistic Reasoning.* Paper presented at the annual meeting of the American Education Research Association, Chicago.

Northern Illinois University
University of California
Arizona State University

REFERENCES

Adi, H., Karplus, R., Lawson, A., and Pulos, S.: 1978, "Intellectual development beyond elementary school VI: Correlational reasoning", *School Science and Mathematics* **78**, 675–638.

Bell, A. W.: 1976, "A study of pupils" proof-explanations in mathematical situations', *Educational Studies in Mathematics* **7**, 23–40.

Carroll, C. A.: 1975, "Low achievers" understanding of logical inference forms', in M. Rosskopf (ed.), *Children's Mathematical Concepts*, Teachers College Press, New York.

Copi, I. M.: 1978, *Introduction to Logic* (4th ed.), Macmillan, New York.

Ennis, R. H.: 1976, "An alternation to Piaget's conceptualization of logical competence", *Child Development* **47**, 903–919.

Ennis, R. H., and Paulus, D. H.: 1965, *Critical Thinking Readiness in Grades 1–12 (Phase 1: Deductive Reasoning in Adolescence)*, Cornell Critical Thinking Project, Ithaca, New York.

Falmagne, R. J.: "The development of logical competence: A psycholinguistic perspective", in R. Kluwe and H. Spada (eds.), *Development Models of Thinking*. Academic Press, New York (in press).

Jansson, L.: 1978, "A comparison of two approaches to the assessment of conditional reasoning abilities", *Journal for Research in Mathematics Education* **9**, 175–188.

Johnson-Laird, P. N., Legrenzi, P., and Legrenzi, M. S.: 1972, "Reasoning and a sense of reality, *British Journal of Psychology* **63**, 395–400.

Karplus, E., and Karplus, R.: 1970, "Intellectual development beyond elementary school I: Deductive reasoning", *School Science and Mathematics* **70**, 398–406.

Lawson, A.: 1978, "The development and validation of a classroom test of formal reasoning", *Journal of Research in Science Teaching* **15**, 11–24.

Lawson, A., Karplus, R., and Adi, H.: 1978, "The acquisition of propositional logic and formal operational schemata during the secondary school years", *Journal of Research in Science Teaching* **15**, 465–478.

O'Brien, T.: 1975, "Deformation and the four-card problem", *Educational Studies in Mathematics* **6**, 23–29.

Paris, S. G.: 1973, "Comprehension of language connectives and propositional logic relationships", *Journal of Experimental Psychology* **16**, 278–291.

Resnick, L.: 1976, *The Nature of Intelligence*, Lawrence Erlbaum Assoc., Hillsdale, New Jersey.

Roberge, J. J.: 1970, "A study of children's abilities to reason with basic principles of deductive reasoning", *American Education Research Journal* **7**, 583–596.

Roberge, J. J.: 1972, "Recent research on the development of children's comprehension of deductive reasoning schemes", *School Science and Mathematics* **72**, 197–200.

Roberge, J. J., and Paulus, D.: 1971, "Developmental patterns for children's class and conditional reasoning abilities", *Developmental Psychology* **4**, 191–200.

Wason, P. C., and Johnson-Laird, P. M.: 1972, *Psychology of Reasoning: Structure and Content*. Havard University Press, Cambridge, Mass.

Wildman, T., and Fletcher, H.: 1977, "Developmental increases and decreases in solutions of conditional syllogism problems", *Developmental Psychology* **13**, 63–636.

The Karplus children, circa 1964

Front row, from the left: Paul Andrew (b. 1957), Peter (b. 1962), David William (b. 1960). Back row, from the left: Margaret Alice (b. 1952), Beverly Ruth (b. 1950), Richard Samuel (b. 1953), Barbara Frazier (b. 1955).

CHAPTER 7

Teaching and the Conceptual Structure of Physics

Robert G. Fuller*

Almost from the beginning of his work with elementary school science education, Robert Karplus thought about how what he was learning in that venue applied to his work as a professor of physics. His work with Herb Thier on the SCIS project in 1963 lead him to offer his first thoughts on how the teaching of physics ought to be based on an understanding of the conceptual structure of physics. (See the reprint Karplus, 1966.)

Robert Karplus came back to this theme again, in one of the final papers that he published in the American Journal of Physics. (See the reprint Karplus, 1981). He wrote the paper in 1980. By that time, he had published an introductory textbook, *Introductory Physics: A Model Approach*, from which two selections are reprinted in Chapter 9. His textbook never made a second edition. He followed an unconventional sequence of topics, following his understanding of the conceptual structure of physics and trying to reduce the difficulties that students would have with the various concepts of physics. The three levels of scientific concepts that he proposes and discusses in his 1980 paper are reflected in his much earlier textbook. He was a strong advocate of operational definitions of physics terms. Even though his original work in

*University of Nebraska-Lincoln.

A Love of Discovery: Science Education—The Second Career of Robert Karplus,
Edited by Robert G. Fuller, Kluwer Academic / Plenum Publishers, New York, 2002.

physics was highly theoretical, he recognized the importance of offering concrete representations of physics concepts to beginning students.

In the 1970s, Bob and his co-workers, along with Arnold B. Arons, tried to direct the attention of the physics community to the growing body of information about the intellectual development of college students. (See the reprint Arons and Karplus, 1976.) The early papers by John Renner (McKinnon and Renner, 1971; Renner and Lawson, 1973) had provided substantial evidence that many college students in the USA did not systematically and reliably use the kinds of mature reasoning patterns necessary to master physics at the introductory level. By June, 1975, Bob had worked with a special committee of members of the American Association of Physics Teachers to develop and lead workshops on Physics Teaching and the Development of Reasoning (AAPT, 1975). (To read an essay from those workshop modules, see the reprint, Piaget in a Nutshell, in Chapter 8.). Another paper of his that was primarily intended for physicists and emphasized his interest in the model of intellectual development based upon the work of Jean Piaget was published in 1977. In that paper, the challenge to use physics teaching as a method to raise the intellectual development of students is presented. The Karplus learning cycle is suggested as an instructional technique to encourage students to develop more mature reasoning patterns. (See the reprint Fuller, Karplus and Lawson, 1977.)

I think it can properly be argued that the roots of the present day research in physics education carried out by physicists in physics departments is to be found in the early work of Robert Karplus and Arnold B. Arons.

REFERENCES

AAPT, 1975, Karplus, R., et al., *Physics Teaching and the Development of Reasoning, American Association of Physics Teachers.*

Arons, A.B., and Karplus, R., 1976, Implications of accumulating data on levels of intellectual development, *American Journal of Physics*, **44(4)**, 396.

Fuller, R.G., Karplus, R., and Lawson, A.E., 1977, Can Physics Develop Reasoning?, *Physics Today*, **30(2)**, 23–8.

Karplus, R., 1966, Conceptual Structure and Physics Teaching, *American Journal of Physics*, **34(8)**, 733.

Karplus, R., 1981, Educational Aspects of the Structure of Physics, *American Journal of Physics*, **49(3)**, 238–40.

Karplus, R. 1969, *Introductory Physics: A Model Approach.* W. A. Benjamin, Inc., New York.

McKinnon, J.W. and Renner, J.W., 1971, Are Colleges Concerned with Intellectual Development?, *American Journal of Physics*, **39(9)**, 1047–52.

Renner, J.W. and Lawson, A.E., 1973, Promoting Intellectual Development through Science Teaching, *The Physics Teacher*, **11**, 273–76.

Teaching and the Conceptual Structure of Physics

Proceedings of the Association

J5. Conceptual Structure and Physics Teaching. ROBERT KARPLUS, *University of California, Berkeley*. Jerome Bruner[1] has injected the idea of conceptual structure into educational considerations. Conceptual structure reflects the logical and phenomenological interrelations within an academic discipline. An example could be the relations of velocity, mass, force, impulse, and momentum to one another. In order to teach students, it is necessary not only to be aware of the internal relationships within a discipline, but one must also be concerned with the "surface layer," that is, the relationship of the discipline to the prior experience of the students. Considerations such as these are especially important when the students have had little prior experience in a field. The author has been interested in the teaching of physics to elementary school children, to their teachers, and to liberal arts students at the college level. He has found the traditional approach through Newtonian mechanics to be completely unsuitable. An alternate approach, in which the "surface layer" of the discipline is molded so as to have broad overlap with the students' prior experience, is being designed. One crucial technique is the concrete operational definition of the concepts in the "surface" of the conceptual structure, using operations that are familiar to the students. Only later are these concepts combined and used to generate more abstract ones. As one example, consider force as defined by mass times acceleration (abstract) vs force as defined by elastic deformation (concrete).

Reprinted with permission from Karplus, R., "Conceptual Structure and Physics Teaching," American Journal of Physics, 34, 733 © 1966.

[1] Jerome Bruner, *The Process of Education* (Harvard University Press, Cambridge, 1960).

Implications of Accumulating Data on Levels of Intellectual Development

Arnold B. Arons and Robert Karplus

For several years data have been accumulating regarding the fraction of various population groups at various age levels that have made the transition from "concrete operations" to "formal operations" in accordance with the increasingly familiar and empirically documented Piagetian Taxonomy.[1]

Using various Piagetian tasks (or modifications thereof) involving such aspects as conservation reasoning, control of variables, syllogistic reasoning, recognition of inadequacy of information, arithmetical proportional reasoning, etc., investigators have examined school, college, and adult populations ranging in age from about 13–45 yr.[2-8] Although the various

[1] J. Piaget and B. Inhelder, *Growth of Logical Thinking* (Basic Books, New York, 1958).
[2] J. W. McKinnon and J. W. Renner, Am. J. Phys. **39**, 1047 (1971).
[3] J. W. Renner and A. E. Lawson, Phys. Teach. **11**, 273 (1973).
[4] A. E. Lawson and A. J. D. Blake, AESOP report, Lawrence Hall of Science, University of California at Berkeley, 1974 (unpublished).
[5] R. Karplus, E. Karplus, M. Formisano, and A. C. Paulsen, AESOP Report No. ID-25, Lawrence Hall of Science, University of California at Berkeley, 1975 (unpublished).
[6] R. P. Bauman, in proceedings of the AAPT Summer Meeting at Boulder, CO, June 1975 (unpublished).
[7] D. Kuhn, J. Langer, L. Kohlberg, and N. S. Haan, J. Genet Psychol. (to be published).
[8] A. Arons and J. Smith, Sci. Educ. **58** (3), 391 (1974).

Arnold B. Arons, Department of Physics, University of Washington, Seattle, Washington 98195. Robert Karplus, Lawrence Hall of Science, University of California Berkeley, California 94720. (Received 14 October 1975).

Reprinted with permission from Arons, A. B. and Karplus, R., "Implications of Accumulating Data on Levels of Intellectual Development," *American Journal of Physics* 44, p 396. Copyright, American Institute of Physics [for *American Journal of Physics*], 1976.

investigations are beginning to reveal significant and interesting differences between social and economic groups, the grand averages have been emerging, with very little variation throughout the age and school level spectrum: about one-third have made the transition to formal operations, about one-third can be regarded as in the process of transition, and about one-third use primarily concrete patterns of reasoning.

The accumulating data can undoubtedly be interpreted in a variety of ways depending on the orientation and predilections of the interpreter, but we feel compelled to underline the following possible very grave implication for our society: If it is indeed true that one-third of the school population is formal operational by the age of about 14 while one-third is still concrete and that these proportions do not change substantially from then on in spite of further schooling (including at least some university levels), then we face the implication that our educational system is not contributing significantly to intellectual development (abstract logical reasoning). The one-third who become formal operational may well be *sui generis*, making the transition on their own regardless of the educational system, while the remainder are not being helped to make the progress that should be a major objective of formal education.

This is not to say that the educational system fails to develop certain basic skills such as reading, writing, and reckoning and various compendia of necessary facts and information (albeit there is much argument and criticism of the adequacy with which even these aspects are cultivated); our concern here is over the levels of intellectual development the educational system appears to generate.

If perpetuation and advancement of a demoncratic society do indeed demand the broadest participation of a thinking–reasoning citizenry, if intelligent participation does involve abstract reasoning on matters such as, for example, what constitutes *enlightened* self interest, if more people must be counted on to engage in decision making when confronted with incomplete, "on the one hand . . . and on the other hand" evidence shorn of reliance on a "pat" answer from an ultimate "expert" (as cogently argued recently by David),[9] then we *must* gear our educational system to greater effectiveness in enhancing intellectual development than the incoming data show it to exert.

If our suggested inference is correct, it seems to us that explicit awareness of the problem, and measures to attack it, must begin in the colleges and universities. These institutions educate the teachers for the educational system with which we are concerned. They must provide leadership in converting it from a passive one that merely allows the *sui generis* development of a small fraction to one that actively assists the intellectual development of the far larger proportion of the population we have every reason to believe is fully capable of abstract logical reasoning. We recognize that a number of institutions and scattered clusters of faculty members have initiated attacks on this problem. We emphasize, however, that significant progress can only result from far broader and more explicit awareness and from far more massive and wide spread efforts than have yet been activated in the realm of higher education.

[9] E. E. David, Science **189**, 679 (1975).

Can Physics Develop Reasoning?

Robert G. Fuller, Robert Karplus and Anton E. Lawson

The findings of Swiss scholar Jean Piaget suggest that it can—by helping people achieve a series of four distinct but overlapping stages of intellectual growth as they search for patterns and relationships.

The life of every physicist is punctuated by events that lead him to discover that the way physicists see natural phenomena is different from the way nonphysicists see them. Certain patterns of reasoning appear to be more common among physicists than in other groups. These include:

- focussing on the important variables (such as the force that accelerates the apple, rather than the lump it makes on your head);
- propositional logic ("if heat were a liquid it would occupy space and a cannon barrel could only contain a limited amount of heat, but this is contrary to my observations, so . . ."), and
- proportional reasoning (for example, the restoring force of a spring increases linearly with its displacement from equilibrium).

In recent studies of the reasoning used by students we have discovered among them qualitative differences similar to those between the reasoning patterns of physicists and nonphysicists.

Robert G. Fuller is a visiting professor of physics at the University of California, Berkeley and a research physicist at the Lawrence Hall of Science while on leave from the University of Nebraska–Lincoln, where he is a professor of physics: Robert Karplus is the acting director of the Lawrence Hall of Science and President of the American Association of Physics Teachers, and Anton E. Lawson is a research associate at the Lawrence Hall of Science, University of California, Berkeley.

Reprinted with permission from Fuller, et al., "Can Physics Develop Reasoning?", *Physics Today* 30(2), pp. 23–28. Copyright, American Institute of Physics [for Physics Today], 1977.

How can we understand these qualitative differences in reasoning? What role does physics play in the way reasoning develops in young people?

Along with a group of teachers in physics and other disciplines, we believe that some of the answers to these questions can be found in the work of developmental psychologists, especially that of the Swiss scholar Jean Piaget. We have helped start a modest movement, accordingly, to inform others of the relevant findings and theories of these social scientists.

To do so we have extended the psychologists' original investigations by dealing with their implications for the presentation of subject matter at the secondary-school and college levels. Textbooks, laboratory procedures, homework assignments, test questions and films may all be examined from the developmental point of view.[1]

In this article we shall describe those ideas in Piaget's work that we have found most useful; you may judge for yourself how valid they are. We shall conclude by suggesting ways in which you can use your expertise in physics and your personal contacts—whether you teach physics or not—to encourage others to develop their reasoning through their observations and analyses of physical systems.

STUDENT RESPONSES TO PUZZLES

To study the differences in reasoning used by students, we have devised a number of paper-and-pencil puzzles and given them to high-school and college students. Let us examine the following typical student responses to two of these, the Ticker-Tape Puzzle and the Islands Puzzle,[2] and discuss the differences in reasoning displayed in them by the students.

The responses to the Ticker-Tape Puzzle (see the Box on page 25) were collected from engineering and science students in an introductory physics course. Some of them had completed the term covering newtonian mechanics, others had not. Here are samples:

Fred (had used ticker tape)
1 B—Dots are spaced equally.
2 C—Dots are closing together, cart is going less distance in the same time.
3 A—Dots are getting farther apart, cart is moving farther in same time (accelerating).
4 D—Cart is falling through air; it has a rapid acceleration.

James (had not used ticker tape)
1 B—At constant speed, the same distance will be covered per unit time.
2 E—Deceleration means less velocity, so less distance per unit time.
3 D—Acceleration is exponential, ruling out A.
4 C—Assume a frictionless system, with brakes momentarily applied between dots five and six.

The responses to the Islands Puzzle (see the Box on page 26) were collected from a wide variety of adolescents and adults. These two are typical:

Deloris (College student, age 17)
1 "Yes, because the people can go north from Island D—because in the clue it could be made in both directions."
2 "No; I am presuming both directions doesn't include a 45° angle from B to C."
3 "Yes, because Island C is right below Island A."

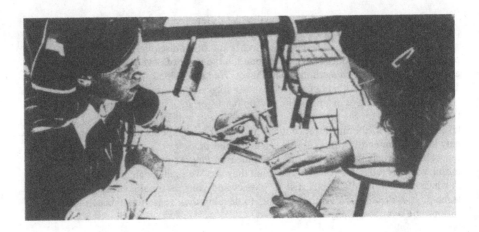

The wheels are turning as these two students compare the angles of rotation of three intermeshing gears. Their search for numerical relationships will help them develop proportional reasoning and understand when to apply this pattern of thought. The ability to handle functional relationships such as proportionality is a characteristic of formal reasoning, the fourth of Piaget's stages of intellectual development.

Myrna (College student, age 17)
1 "Can't tell from the clues given. The two clues don't relate the upper islands to the lower ones."
2 "Yes; they can go from B to D, and then to C, even if there are no direct flights."
3 "No, if they could go from C to A, then the people on B could go first to D, then to C, and then on to A. But this contradicts the second clue, that they don't go by plane between B and A."

You will notice some similarities between the responses of Fred (to the Ticker-Tape Puzzle) and Deloris (to the Island Puzzle). They both focus on the specific details of the puzzle. Fred makes direct correspondence between the arrangement of the dots and the physical examples given. Although he introduces the idea of "acceleration," he does not indicate that he has any more than a vague general idea of its meaning. In a similar way, Deloris concentrates on the spatial arrangement of the islands. Her explanations have more to do with her perception of the physical arrangement of the islands than with the clues given in the puzzle. Both Fred and Deloris appear limited in their reasoning to the specific details of a puzzle, and do not readily relate the facts of the puzzles to more general principles.

Consider, on the other hand, the responses of James and Myrna. Both of them have made conjectures to facilitate answering the questions. James, who had not previously used a ticker tape, begins his explanations with generalized concepts such as constant speed, deceleration, acceleration and a frictionless system. Even when his explanation is wrong ("acceleration is exponential") he demonstrates that he is reasoning within a system of deduction from hypotheses, in which a ticker tape can serve as one specific example representative of a more general principle.

Myrna, as she reasons about the Islands Puzzle, fits the clues into an overall scheme for explaining the air travel between the islands. She suggested a hypothetical trip, demonstrating the correctness of her answer by reasoning to a contradiction. James and Myrna display patterns of reasoning commonly used by physicists.

Even in the responses to these simple written puzzles, the qualitative differences in student reasoning are vividly displayed. For an understanding of these differences, let us turn to the work of Piaget.

THE DEVELOPMENT OF REASONING

Jean Piaget began his research on children in about 1920. The results of his work of primary concern to us are reported in the book, *The Growth of Logical Thinking from Childhood to Adolescence*.[3] In this book the responses of young people to various tasks concerning physical phenomena are described. These tasks included physics experiments such as those on the equality of the angles of incidence and reflection, the law of floating bodies, the flexibility of metal rods, the oscillation of a pendulum, the motion of bodies on an inclined plane, the conservation of momentum of a horizontal plane, the equilibrium of a balance and the projection of shadows.

On the basis of the responses, Piaget and his co-workers developed a theory for interpreting the development of what he considers to be universal patterns of reasoning. Pivotal to this theory is the concept of *stages of intellectual development*. The stages—there are four in the theory—are characterized by distinctive features in the patterns of a person's reasoning. It was hypothesized that each of Piaget's four stages serves as a precursor to all succeeding stages, so that reasoning develops sequentially, always from the less effective to the more effective stage, although not necessarily at the same rate for every individual.

Like a concept in any theory, a stage of intellectual development is a simplification that is helpful in analyzing and interpreting observations, somewhat like a point particle or a frictionless plane in mechanics. In this spirit, we should not expect that most people during their period of development will exhibit all the reasoning characteristics of, say, stage A for a certain period of time and then suddenly change to all the reasoning patterns appropriate to stage B. Rather, the development of a person's reasoning should be thought of as gradual, at a particular time showing the features of stage A on some problems while exhibiting certain features of stage B on others. The stage concept therefore may be more useful for classifying reasoning patterns than for describing the overall intellectual behavior of every particular person at a given time.

The first Piagetian stage is called *sensory–motor*. This stage is characteristic of children's thinking from birth to about two years of age. Piaget's work with infants provided an explanation for the humor of the "peek-a-boo" game:

The young infant appears to think that the only objects that exist are the objects that can be seen. The sudden "creation" of a large person by removing a blanket covering him does seem to be a funny event. Subsequent experiences provide the child with the opportunity to develop an awareness of the permanence of material objects.

The concept of permanence provides the basis for the child's need for language. If objects do exist when they are out of sight, then it is useful to have symbols (or words) to represent them. So the sensory–motor stage serves as the precursor for the next, *pre-operational*, stage.

During the pre-operational period the child is learning words and trying to fit his experiences of the world together. The pre-operational child lives in a very personal world with his own ego at the center ("The Sun is following me!"). He puts facts together to produce ad-hoc explanations, such as, "My dad mows the yard because he's a physicist."

The ticker-tape puzzle

The puzzle below is a task designed to display the variety of student reasoning patterns used in a typical physics classroom activity. It is taken from materials for the workshop on Physics Teaching and the Development of Reasoning offered at the 1975 AAPT–APS meeting in Anaheim, California (reference 1).

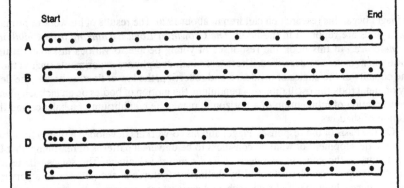

Start   End

Many physics labs allow you to study motion by making timer tapes like the five illustrated above. These are strips of paper attached to a moving object and passing through a timing mechanism that makes a row of small dots by striking regularly at equal time intervals, usually five to ten times per second.

• Have you ever used or watched such a device?
• Identify the tape that fits each of the examples below and justify your answers, taking special care to mention any tapes that a less experienced student might easily mistake for the correct one.

1. A student walking through the laboratory at constant speed   A B C D E
   Justification?
2. A cart gradually slowing down on a level plane   A B C D E
   Justification?
3. A cart rolling freely down an inclined plane   A B C D E
   Justification?
4. Explain how one of the two remaining tapes might have been made, and briefly justify your hypothesis.

The pre-operational child does not use causal reasoning. Some authors have used children's pre-causal explanations as the motif for humorous books. For Piaget, such explanations are clues as to how children think about the world in which they live.

The first two Piagetian stages are usually completed before a person is nine years old. The child's interaction with physical systems plays an essential role in his or her intellectual development during the first two stages. The role of physics in the development of reasoning in the elementary-school years was discussed in a special issue of PHYSICS TODAY.[4]

Sparks mark the position of the falling object on the ticker tape. The dot patterns can not be analysed readily by that third of US adolescents and adults who use only concrete reasoning.

CONCRETE REASONING

To explain the qualitative differences in the reasoning patterns of older students' responses to the two puzzles described earlier we must look to Piaget's third and fourth stages of intellectual development, *concrete reasoning* and *formal reasoning*. Certain characteristics help identify reasoning patterns associated with these two stages.

Here are some of the characteristics of concrete reasoning patterns; illustrative examples are added in parentheses:

Class Inclusion A person at this stage understands simple classifications and generalizations of familiar objects or events (can reason that all aluminum pieces can close an electric circuit, but not all objects that close a circuit are made of aluminum).

Conservation Such a person reasons that, if nothing is added or taken away, the amount or number remains the same even though the appearance differs (that when water is poured from a short wide container into a tall narrow container, the amount of water is not changed).

Serial ordering The person arranges a set of objects or data in serial order and may establish a one-to-one correspondence ("The heaviest block of copper stretches the spring the most.").

Reversibility A person using concrete reasoning mentally inverts a sequence of steps to return from the final to the initial conditions (reasoning that the removal of weight from a piston will enable the enclosed gas to expand back to its original volume).

Concrete reasoning enables a person to

The islands puzzle

The puzzle below is a written task designed to display the variety of deductive-logic strategies used by adolescents (reference 2).

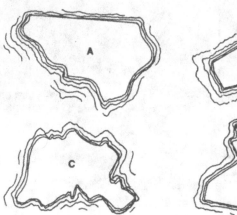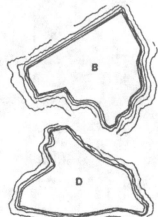

There are four islands in the ocean, islands A, B, C and D. People have been travelling these islands by boat for many years, but recently an airline started in business. Carefully read the clues about possible plane trips at present. The trips may be direct or include stops and plane changes on an island. When a trip is possible, it can be made in either direction between the islands. You may make notes or marks on the map to help use the clues.

First clue: People can go by plane between Islands C and D.

Second clue: People can not go by plane between Islands A and B.

- Use these clues to answer Question 1. Do not read the next clue yet.

1. Can people go by plane between Islands B and D?
   Yes _____ No _____ Can't tell from the two clues_____ Please explain your answer.

Third clue (do not change your answer to Question 1 now!): People can go by plane between Islands B and D.

- Use all three clues to answer Questions 2 and 3.

2. Can people go by plane between Islands B and C?
   Yes_____ No_____ Can't tell from the three clues_____
   Please explain your answer.

3. Can people go by plane between Islands A and C?
   Yes_____ No_____ Can't tell from the three clues_____
   Please explain your answer.

- understand concepts and simple hypotheses that make a direct reference to familiar actions and objects, and can be explained in terms of simple associations ("A larger force must be applied to move a larger mass.");
- follow step-by-step instructions as in a recipe, provided each step is specified (carry out a wide variety of physics experiments in a "cookbook" laboratory), and
- relate his own viewpoint to that of another in a simple situation (be aware an automobile approaching at 55 mph appears to be travelling much faster to a driver moving in the opposite direction at 55 mph).

However, persons whose reasoning has not developed beyond the concrete stage demonstrate certain *limitations* in their reasoning ability. These are evidenced as the person:

- searches for and identifies some variables influencing a phenomenon, but does so unsystematically (investigates the effects of one variable without holding all the others constant);
- makes observations and draws inferences from them but without considering all possibilities (fails to see all of the major sources of error in a laboratory experiment);
- responds to difficult problems by applying a related but not necessarily correct algorithm (uses the formula $s = at^2/2$ to calculate displacement, even when the acceleration is not a constant), and
- processes information, but is not spontaneously aware of his own reasoning (does not check his conclusions against the given data or other experience).

The puzzle responses given by Fred and Deloris are examples of concrete reasoning.

FORMAL REASONING

The following are characteristics of formal reasoning patterns and examples from the history of physics to illustrate them:

Combinatorial reasoning A person systematically considers all possible relations of experimental or theoretical conditions, even though some may not be realized in Nature (for example, using the spectral response of the eye to develop the three-element theory of color vision),

Control of variables In establishing the truth or falsity of hypotheses, a person recognizes the necessity of taking into consideration all the known variables and designing a test that controls all variables but the one being investigated (for example, changing only the direction of the light to detect the possible existence of the ether)

Concrete reasoning about constructs A person applies multiple classification, conservation, serial ordering and other reasoning patterns to concepts and abstract properties (for example, applying conservation of energy to propose the existence of the neutrino),

Functional relationships A person recognizes and interprets dependencies between variables in situations described by observable or abstract variables, and states the relationships in mathematical form (for example, stating that the rate of change of velocity is proportional to the net force),

Probabilistic correlations A person recognizes the fact that natural phenomena themselves are subject to random fluctuations and that any explanatory model must involve probabilistic considerations, including the comparison of the number of confirming and disconfirming cases of hypothesized relations (for example, arguing from the small number of alpha particles scattered through large angles from gold foil to suggest a nuclear model for the atom).

Formal reasoning patterns, taken in concert, enable individuals to use hypothesis and deduction in their reasoning. They can accept an unproven hypothesis, deduce its consequences in the light of other known information and then verify empirically whether, in fact, those consequences occur. Furthermore, they can reflect upon their own reasoning to look for inconsistencies. They can check their results in numerical calculations against order-of-magnitude estimates. James and Myrna, in their responses to the puzzles, gave evidence of using formal reasoning.

In the table on page 214 we summarize some differences between reasoning at the concrete and formal levels. It is quite clear that a successful physicist makes use of formal reasoning in his area of professional expertise. In fact, formal reasoning is prerequisite for producing quality work in physics.

Many theoretical and experimental issues relating to Piaget's work are still being investigated. Piaget's original notion was that all persons use formal reasoning reliably by their late teens. Yet recent studies strongly suggest that, although almost everyone becomes able to use concrete reasoning, many people do not come to use formal reasoning reliably. These persons often appear to be reasoning at the formal level and/or comprehending formal subject matter when they are actually only applying memorized formulas, words or phrases.

Concrete versus formal reasoning

In concrete reasoning, a person

- needs reference to familiar actions, objects and observable properties;
- uses classification, conservation, serial ordering and one-to-one correspondence in relation to concrete items above;
- needs step-by-step instructions in a lengthy procedure, and
- is not aware of his own reasoning, inconsistencies among various statements or contradictions with other known facts.

In formal reasoning, a person

- can reason with concepts, relationships, abstract properties, axioms and theories;
- uses symbols to express ideas;
- applies combinatorial, classification, conservation, serial ordering and proportional reasoning in these abstract modes of thought;
- can plan a lengthy procedure to attain given overall goals and resources, and
- is aware of and critical of, his own reasoning, and actively checks on the validity of his conclusions by appealing to other information.

From Module 9 of the *Science Teaching and the Development of Reasoning* workshop materials (see the Box on page 217).

The development of formal reasoning represents and extremely worthwhile educational aim. Formal reasoning is fundamental to developing a meaningful understanding of mathematics, the sciences and many other subjects of modern life. The finding, by a wide variety of studies,[5] that more than one third of the adolescents and adults in the United States do not employ formal reasoning patterns effectively presents a real educational challenge. What can be done about the significant fraction of the population that appears to be stuck at the stage of concrete reasoning?

SELF-REGULATION

As physicists, we can see the advantages to our profession of more wide-spread use of formal reasoning patterns. To see the role that physics would have to play in creating the necessary atmosphere for this, let us turn to another concept in Piaget's theory of intellectual development, that of self-regulation.

By comparing the extensions of a coil spring at various points, these students are gaining insight into proportionality; such formal-reasoning patterns are attained through self-regulation.

Self-regulation is the process whereby an individual's reasoning advances from one level to the next, an advance that is always in the direction toward more successful patterns of reasoning. Piaget considers this process of intellectual development as analogous to the differentiation and integration one sees in the biological development of an embryo, as well as analogous to the adaptation of evolving species.

A person develops formal reasoning only through the process of self-regulation. Concrete reasoning thus is a prerequisite for the development of formal reasoning.

The process of self-regulation is one in which a person actively searches for relationships and patterns to resolve contradictions and bring coherence to a new set of experiences. Implicit in this notion is the image of a relatively autonomous person, one who is neither under the constant guidance of a teacher nor strictly bound to a rigid set of precedents.

Self-regulation can be described as unfolding in alternating phases, beginning with *assimilation*. The individual's reasoning assimilates a problem situation and gives it a meaning determined by present reasoning patterns. This meaning may or may not, in fact, be appropriate. Inappropriateness produces what is called "disequilibrium," "cognitive conflict" or "contradiction," a state that, according to Piaget, is the prime mover in initiating the second phase—*accomodation*.

Accomodation entails

- an analysis of the situation to locate the source of difficulty and
- formation of new hypotheses and plans of attack.

Just how this is done varies from person to person and depends upon his analytical and problem-solving abilities. The results of these reflective and experimenting activities are new reasoning patterns that may include new understandings. In terms of assimilation and accommodation, self-correcting activites (accommodation) are constantly being tested (assimilation) until this alternation of phase produces successful behavior. The whole self-regulation process, directed at a stable rapport between patterns of reasoning and environment, is often called "equilibration" by Piaget.

Recall the self-regulation process that Count Rumford recounts in this essays on heat.[6] In Piaget's terms, Rumford experienced cognitive conflict by the extraordinary ability of apple pies to retain their heat, by the fact that heat had no effect upon the weight of objects and by the intense heat of the metallic chips separated from the cannons he bored. He could not assimilate these experiences with the caloric theory of heat, so he rejected that theory. He accommodated his reasoning to experience by developing the idea that heat was excited and communicated by motion.

The development of reasoning has two requirements: Exploratory experiences with the physical world, and discussion and reflection upon what has been done, what it means and how it fits, or does not fit, with previous patterns of thinking. This suggest that experiences gained through physics can play a key role in the development of reasoning and understanding.

ROLE OF THE PHYSICS COMMUNITY

Let us examine how physics could be used to foster self-regulation in a person. Two factors appear to be required:

- He must be faced with a physical situation that he can only partially understand in terms of old ideas and
- he must have sufficient time to grapple mentally with the new situation, possibly with appropriate hints, but without being told the answer—people must be allowed to put their ideas together for themselves.

The ideal situation would be one in which the problems experienced are felt to be solvable. The Piaget hypothesis is that a challenging but solvable problem will place persons into an initial state of disequilibrium. Then, through their own efforts at bringing together this challenge with their past experiences and what they learn from teachers or peers, they will gradually reorganize their thinking and solve the problem successfully. This success will establish a new and more stable equilibrium with increased understanding of the subject matter and increased problem-solving capability, that is, intellectual development.

One example of such a use of physics is an exhibit of a spring scale and an equal-arm balance mounted on the wall of an elevator in a public building.[7] The riders in the elevator noticed that the "weight" of the object on the scale varied while the balance remained stationary, a paradox that gave rise to some cognitive conflict. A small card beside the exhibit asked questions and offered hints to encourage the riders to accomodate to this experience.

Physics programs, done properly, can be effective means of promoting intellectual development. Such developmental-physics programs are not aimed at producing more physicists, but at enabling people to develop their potential for formal reasoning. This reasoning can serve them well in many aspects of our technological society.

If physics is an essential element in the growth of reasoning, why are persons so turned off by physics? It seems to us that the physics community has chosen to isolate itself from individuals using primarily concrete reasoning patterns. It has been suggested that *all* of the junior and senior high-school physics curricula that have been developed in the last 25 years have been intended for students who typically use formal reasoning.

True, modern secondary-school physics courses, such as PSSC Physics and the Project Physics course, have directed students toward laboratory experiments. Yet many of the experiments can only be understood within the hypothetical structure of the formal laws of physics. For example, the use of stroboscopic photographs to analyze the collisions of two objects appear to be at least as demanding as the Ticker-Tape Puzzle; yet we have seen that the solution to the Ticker-Tape Puzzle was inaccessible to students who used only concrete reasoning.

Workshops and programs based on Piaget's concepts

Workshops that focus on physics teaching and the development of reasoning have been offered at professional meetings and on individual college campuses. The workshop materials for examing instructional aids in various subject areas are available from several sources:

- Physics Teaching and the Development of Reasoning Workshop Materials, AAPT Executive Office, Graduate Physics Building, S.U.N.Y., Stony Brook, N.Y. 11794;
- Biology Teaching and the Development of Reasoning Workshop Materials, Lawrence Hall of Science, Berkeley, Cal. 94720;
- Science Teaching and the Development of Reasoning Workshop Materials (includes physics, chemistry, biology, general science and earth sciences), Lawrence Hall of Science, Berkeley, Cal. 94720, and
- College Teaching and the Development of Reasoning Workshop Materials (includes anthropology, economics, English, history, mathematics, philosophy and physics materials), ADAPT, 213 Ferguson Hall, University of Nebraska–Lincoln, Lincoln, Neb 68588.

Another such workshop is being sponsored by the American Association of Physics Teachers at the joint APS–AAPT meeting in Chicago this month.

College students are being encouraged to develop their reasoning in several programs, including:

- physical-science programs, such as those led by Arnold B. Arons, University of Washington (Amer, J. Phys. **44**, 834; 1976) and John W. Renner, University of Oklahoma (Amer, J. Phys. **44**, 218; 1976);
- the introductory physics laboratory course for engineering students developed by Robert Gerson, University of Missouri–Rolla, and
- two Piaget-based multidisciplinary programs for college freshmen, ADAPT at the University of Nebraska–Lincoln and DOORS at Illinois Central College, East Peoria.

In short, our fixation on the formal aspects of physics instead of its concrete experiences has made physics unnecessarily difficult and dry. We have removed the sense of exploration and discovery from the study of physics for the majority of students. Several generations of public-school students have been alienated from physics.[8,9]

What can you do to make the study of physics less a slave to the formal structure of the discipline and more of a servant to the development of reasoning? You can

- become more familiar with the applications of Piaget's ideas to learning from physics;

- learn about the present attempts to offer Piaget-based programs for large numbers of students;
- encourage your school or college to initiate some programs that focus on the development of reasoning rather than the mastery of content;
- assist service clubs and other groups to present physics to the citizens by means of displays, exhibits and media, and
- develop your skills as a facilitator of self-regulation in others.[10]

The Box on page 217 lists some sources of workshop materials, as well as current college programs based on the Piaget concepts.

THE HUMAN POTENTIAL

As a result of our professional experiences, we of the physics community may possess a valuable insight: that carefully planned interactions of persons with the experimental systems and concepts of physics can contribute vitally to the full human potential. Perhaps our efforts to increase the appropriate people–physics interactions are as important to the future of mankind as our continuing efforts to increase our fundamental understanding of physical systems.

*   *   *

This material is based upon work done as a part of AESOP (Advancing Education through Science-Oriented Programs), supported by the US National Science Foundation under Grant No. SED74-18950. The opinions are those of the authors and do not necessarily reflect the views of the Foundation.

REFERENCES

1. *Proceedings* of the Workshop of Physics Teaching and the Development of Reasoning (Anaheim, Calif. January 1975), American Association of Physics Teachers, Stony Brook, N.Y. (1975).
2. E. F. Karplus, R. Karplus, School Sci. and Math. **70**, 5 (1970).
3. B. Inhelder, J. Piaget, *The Growth of Logical Thinking from Childhood to Adolescence*, Basic Books, New York (1958).
4. PHYSICS TODAY, June 1972.
5. D. Griffiths, Amer. J. Phys. **14**, 81 (1976); G. Kolodiy, J. Coll. Sci. Teach. **5**, 20 (1975); A. E. Lawson, F. Nordland, A. DeVito, J. Res. Sci. Teach. **12**, 423 (1976); J. W. McKinnon, J. W. Renner, Amer. J. Phys. **39**, 1047 (1971); J. W. Renner, A. E. Lawson, Phys. Teach, **11**, 273 (1973); C. A. Tomlinson-Keasey, Dev. Psychol. **6**, 364 (1972).
6. *The Collected Works of Count Rumford* (S. C. Brown, ed.), Harvard U. P., Cambridge, Mass. (1968).
7. L. Eason, A. J. Friedman, Phys. Teach. **13**, 491 (1975).
8. P. de H. Hurd, School Sci. and Math. **53**, 439 (1953).
9. M. B. Rowe, The Science Teacher **42**, 21 (1975).
10. A. B. Arons, Amer. J. Phys. **44**, 834 (1974).

Educational Aspects of the Structure of Physics

Robert Karplus

The structure of physics, which consists of physical concepts and their relationships, can be treated as an instructional variable by teachers who make the following choices: defined versus derived concepts, operational versus paraphrase (formal) definitions, theoretical versus empirical relationships. Different forms of the structure of physics will emerge from treatments that differ in these respects. For beginning students, reliance on operational definitions and empirical relationships is recommended. For advanced students, paraphrase definitions and theoretical relationships lead to more efficiency and elegance.

When one listens to students' discussions of Newton's second law of motion, $\mathbf{F} = m\mathbf{a}$, one frequently gains the impression that they conceive it as an algebraic relation to be used for computing one of the three quantities (force, mass, acceleration) when the other two are known. For physicists, however, the law is a very general principle that relates the motion of a particle (or of the center of mass of a system) to the net force acting on the particle (or system). I believe that the students' view is undesirably limited and that the teaching of physics should have as one of its objectives the understanding of Newton's second law as a scientific principle rather than as an algebraic relation or even as a meaningless set of symbols.

In this paper I will describe some considerations regarding the content of instruction that seem to me to be important when one plans an educational program whose goals include the understanding of concepts and their relationships. The structure of physics, by which I have in mind physical concepts and their relationships as exemplified by Newton's second law, plays the role of an instructional design variable that imposes constraints but

Robert Karplus, Group in Science and Mathematics Education, Lawrence Hall of Science, University of California, Berkeley, California 94720. (Received 15 May 1980; accepted 1 August 1980).

Reprinted with permission from Karplus, R., "Educational Aspects of the Structure of Physics," *American Journal of Physics*, 49, 238–241, 1981.

also allows alternatives. These ideas originated in my work in elementary school science and the teaching of physics to nonscience majors.[1-4]

More recent research has indicated how organization of a piece of instruction can influence the way in which the subject matter is retained and applied.[5] In a structural analysis of sixteen college courses in various fields, Donald[6] has compared the differing approaches taken in physics, English, social psychology, and other subjects.

In this article, I shall begin with a quite general characterization of three levels of concepts that are relevant to physics teaching, even though only what I call the first of these is usually taken up explicitly in teaching. I then turn to the kinds of definitions that may be used to express one concept in terms of others or in terms of objects and operations. These ideas are then used to show how concepts on the first level may be related to one another in an instructional program. Basic to my suggestions for educational applications is a developmental approach to the learner, whose anticipated initial knowledge and preconceptions must be taken into account in teaching.

THREE LEVELS OF CONCEPTS

A scientific concept, as I use that term, is an idea such as length, system, vertical, speed, and variable that is used in thinking about natural phenomena. It appears useful to me to organize scientific concepts into three levels according to their generality (Table I). The specific physical quantities subject to quantitative or qualitative comparison of measurement are on the first level. They include force, mass, acceleration, energy, potential, electric charge, wavelength, temperature, and so on. These concepts can and should be incorporated into a logical structure that ultimately rests on definitions and undefined concepts. The concepts on this first level are usually studied in physics courses. Later in this paper I will give more details about such teaching.

At the third of the three levels I would place object, property, system, state of a system, interaction, evidence, variable, and other concepts that identify the current scientific philosophy with respect to the physical universe (object and system), its description (property and variable), causal explanation (interaction), and scientific proof (evidence). Intermediate between the two levels are more specialized descriptive concepts such as particle, motion, configuration, inertia, liquid, insulator, and medium which help to describe a system, and gravitation, electromagnetism, and elasticity, which identify the nature of an interaction.

These second and third levels are usually not included in the syllabus of physics courses even though they are fundamental to physics. Their absence, I believe, makes it very difficult for many students to classify the first level concepts in a way that illuminates their significance and relationships.

ATOMIC THEORY

One of the most powerful approaches for correlating large numbers of diverse observations is atomic theory, which is a central part of modern physics. According to this theory, inter-

[1] R. Karplus, *Introductory Physics: Preliminary Edition* (Benjamin, New York, 1966).
[2] R. Karplus, *Introductory Physics: A Model Approach* (Benjamin, New York, 1969).
[3] R. Karplus and H. D. Thier, *A New Look at Elementary School Science* (Rand McNally, Chicago, 1967).
[4] J. R. Eakin and R. Karplus, *SCIS Final Report* (University of California, Berkeley, CA, 1976).
[5] B. Eylon and R. Reif, Effects of Internal Knowledge Organization on Task Performance (unpublished).
[6] J. G. Donald, *Structures of Knowledge and Implications for Teaching*, Report #6 Center for the Improvement of Teaching (University of British Columbia, Vancouver, 1980).

Table I. Three levels of scientific concepts.

Third level (general)	object, property, system, reference frame, interaction, evidence, ...
Second level (intermediate)	particle, motion, configuration, inertia, liquid, insulator, ...
	gravitation, electromagnetism, elasticity, ...
First level (specific)	force, acceleration, energy, electric charge, potential, wavelength, temperature, wave interference, ...

preted in a general way, it is possible to discuss any physical process from the macroscopic and atomic points of view. Macroscopically, the process is described in terms of the actual objects and changes involved. On the atomic scale, it is discussed in terms of atomic configurations and interactions. A complete treatment of mechanical collisions, electric current, and other phenomena would involve both macroscopic and atomic aspects as well as a way of relating the one to the other. It seems to me that the concept of atom would be most appropriately placed at the intermediate level of the three introduced above, but the reader may have a different view.

DEFINITIONS

The definition by which a term like energy or pressure is given meaning may use a paraphrase, a description of characteristics, one or more illustrative exemplars, or a specification of the operations needed to measure or produce an exemplar. Most scientific concepts are usually defined by paraphrases, as in "energy is the ability to do work." In a paraphrase definition, the concept to be defined is expressed in terms of or in relation to other concepts for which a definition must be provided eventually. The advantage of paraphrase definitions is that, by their very nature, they relate concepts to one another, as energy and doing work in the example, or pressure, force, and area in the definition, "pressure is force per unit area." For many students, the disadvantage of paraphrase definitions is that they can be understood only if the terms and relations in them are understood or defined by using still other terms. For the expert in a field, who has broad knowledge that is being extended slightly by the definition, the paraphrase presents little difficulty.

Operational definitions, unlike paraphrases, make use of real objects and operations with these objects to produce, measure, or recognize an instance of the defined term. An operational definition of energy might be, "the energy of a system is measured by the mass of ice melted as the system comes to equilibrium with a mixture of ice and water." The informed reader will recognize that this definition actually identifies the thermal energy of the system with $0°C$ as the reference temperature, but with this limitation it is adequate though unconventional. Pressure might be defined operationally as follows: "the pressure of a gas is measured by the height of mercury in a barometer tube exposed to the gas." Note how the objects and actions—ice, water, melt, mercury, expose, barometer tube—replace the to-be-defined concepts appearing in the paraphrase definition. Since these items are tangible or observable, they can be comprehended by most individuals relatively easily. At the same time, the concept defined is not related to other concepts, and such relationships have to be established in other ways. These are the pedagogical advantages and disadvantages of operational definitions.

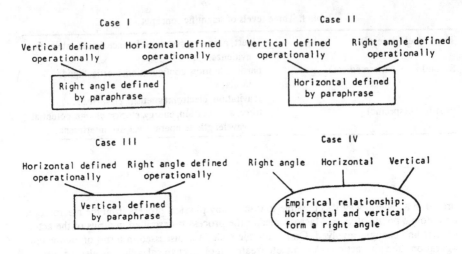

Figure 1. Concept cluster for vertical, horizontal, and right angle.

CONCEPT CLUSTERS

Consider the statement, "The vertical and horizontal directions form a right angle." Three concepts are related in this statement, and thereby form what I shall call a concept cluster. How can the three concepts and the statement be given meaning? Three possibilities are to use the statement to define any one of the three concepts in terms of the other two, which are then defined operationally. A fourth possibility is to define all three concepts operationally and then to recognize the statement as an empirical relationship among them (Figure 1).

Another concept cluster that would be encountered during the study of dynamics in a physics course is associated with Newton's second law. The key concepts are net force, mass, and acceleration. Unexpressed concepts that must also be considered are particle and force of interaction. Because particle acceleration is very difficult to define operationally, it may be wise to introduce a separate concept cluster associated with the definition of acceleration that is studied in the kinematics section of physics courses. This kinematic concept cluster includes time interval, reference frame, particle position, displacement, velocity, and particle acceleration. In this cluster, velocity and acceleration are defined by paraphrases in terms of displacements and time intervals in the usual way.

Coming back to Newton's second law $\mathbf{F} = m\mathbf{a}$, the reader will recognize that the situation is more complicated than it was in the horizontal–vertical example, but that one can proceed in the same ways: either use Newton's law to define one of the three concepts appearing in it, or define each of them separately and then treat Newton's law as an empirical relationship (Figure 2).

An additional feature, not yet illustrated, is the derivation of a theoretical relationship, which can be carried out in some concept clusters. Consider, for example, the conceptual relationships in Newtonian mechanics illustrated in Figure 3. At the top of the diagram are force, displacement, time interval, and mass, which are defined operationally. At the intermediate level are work, kinetic energy, velocity, and acceleration, which are defined by paraphrases in terms of the four operationally defined concepts, Newton's second law, $\mathbf{F} = m\mathbf{a}$, may be observed as an empirical relationship. Once the second law is established, the

3.5 The Debye-Hückel length, κ^{-1}, is (cf. Eq. 3.43 in the textbook):

$$\kappa^{-1} = \left(\frac{\varepsilon kT}{4\pi} \frac{1}{\sum\limits_i n_i^0 z_i^2 e_0^2} \right)^{1/2} \tag{3.12}$$

(a) Analyze the units of this equation. What are the units of κ^{-1}? (b) How should this equation be modified to have units in the *mksa* system? (GamboaAldeco)

Answer:

(a) The equation is written in the *cgs* system (cf. Exercise 2.3). The corresponding units are $[\varepsilon] \rightarrow$ dimensionless, $[k] \rightarrow$ erg K^{-1}, $[T] \rightarrow$ K, $[n_i^0] \rightarrow$ cm^{-3}, $[z_i] \rightarrow$ dimensionless, and $[e_0] \rightarrow$ esu (or electrostatic units). The corresponding analysis of units is,

$$\left[\kappa^{-1} \right] \rightarrow \left[\frac{\left(\text{erg K}^{-1} \right) \text{K}}{\left(\text{cm}^{-3} \right)(\text{esu})^2} \right]^{1/2} \times \left(\frac{1\,\text{esu}^2}{1\,\text{erg cm}} \right)^{1/2} \rightarrow \text{cm} \tag{3.13}$$

Therefore, κ^{-1} is given in cm.

(b) To write this equation in the *mksa* system, the numerator has to be multiplied by $4\pi\varepsilon_0$. Thus, the corresponding units are $[\varepsilon] \rightarrow$ dimensionless, $[\varepsilon_0] \rightarrow$ C^2 J^{-1}m^{-1}, $[k] \rightarrow$ J K^{-1}, $[T] \rightarrow$ K, $[n_i^0] \rightarrow$ m^{-3}, $[z_i] \rightarrow$ dimensionless, and $[e_0] \rightarrow$ C. The equation reads now,

$$\kappa^{-1} = \left(\frac{\varepsilon \varepsilon_0 kT}{\sum\limits_i n_i^0 z_i^2 e_0^2} \right)^{1/2} \tag{3.14}$$

The analysis of units of Eq. (3.14) is,

$$\left[\kappa^{-1}\right] \to \left[\frac{\left(C^2 J^{-1} m^{-1}\right)\left(J K^{-1}\right)K}{\left(m^{-3}\right)(C)^2}\right]^{1/2} \to m \qquad (3.15)$$

3.6 Calculate the thickness of the ionic atmosphere in 0.1 M solutions of an uni-valent electrolyte in the following solvents at 25 ^{0}C : nitrobenzene, with $\varepsilon = 34.8$, ethyl alcohol, with $\varepsilon = 24.3$, and ethylene dichloride, with $\varepsilon = 10.4$. (Cf. Exercise 3.24 in the textbook) (Constantinescu)

Data:

$c = 0.1 \, M$ $\varepsilon_{nitrobenzene} = 34.8$ $\varepsilon_{ethylene\ dichloride} = 10.4$

$|z_+| = |z_-| = 1$ $\varepsilon_{ethyl\ alcohol} = 24.3$ $T = 298 \, K$

Answer:

The thickness of the ionic atmosphere, κ^{-1}, in the *mksa* system is given by (cf. Eq. 3.14 in Exercise 3.5 and Eq. 3.43 in the textbook),

$$\kappa^{-1} = \left(\frac{\varepsilon \varepsilon_0 kT}{\sum_i n_i^0 z_i^2 e_0^2}\right)^{1/2} \qquad (3.16)$$

where n_i^0 is given in m^{-3}. If the concentration of the ions of the i-th kind, c_i, is given in mol liter^{-1} (= mol dm^{-3}) then,

$$c_i \left(\frac{mol}{dm^{-3}}\right) = \frac{n_i^0 \left(m^{-3}\right)}{N_A \left(mol^{-1}\right)} \times \frac{1 \, m^3}{1000 \, dm^3} \qquad (3.17)$$

Solving for n_i^0,

$$n_i^0 = 1000 \, c_i N_A \qquad (3.18)$$

Substituting n_i^0 from Eq. (3.18) into Eq. (3.16),

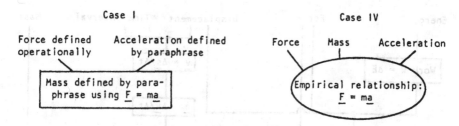

Figure 2. Concept cluster for Newton's second law of motion $F = ma$.

relation of kinetic energy to mass and velocity, $K = (1/2)mv^2$, can be derived theoretically in the usual way.

An alternate approach to Newtonian mechanics, illustrated in Figure 4, would make use of an operational definition of mechanical energy, such as, "The energy of a system is measured by the height to which the system can raise a standard weight in gravitational interaction with the earth."[7] In Figure 4, therefore, energy appears on the upper line, not derived from any other concepts. Work is now defined as change or transfer of energy, and the expression of work in terms of force and displacement is an empirical relationship. Observe how adding the operational definition of energy converted the paraphrase definition of work into an empirical relationship, a change similar to the one illustrated in Figure 1, Case IV as compared to Cases I–III.

The characteristics of other concept clusters at the introductory level are generally similar to these two examples. Central to a concept cluster is an empirical or theoretical relationship among several physical variables.

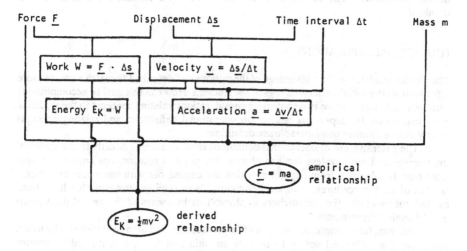

Figure 3. Concept cluster for kinetic energy E_k.

[7] See Ref. 2, p. 242.

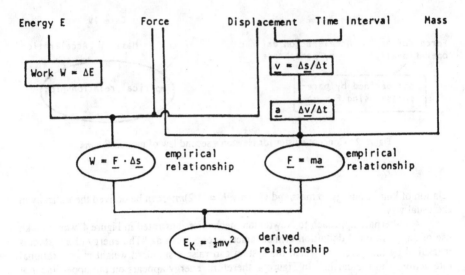

Figure 4. Alternate concept cluster for kinetic energy.

STRUCTURE OF PHYSICS

By linking concept clusters that have common physical variables, one can create larger structures that ultimately encompass all of physics. The approach to definitions, empirical relationships, and theoretical relationships has to be consistent among the clusters that are linked. As I have explained, there is considerable freedom in the choice of quantities to be defined and derived. The exact choices that are made will determine the structure that is obtained.

EDUCATIONAL APPLICATIONS

The physics teacher can take advantage of the options just described to match a course more effectively to the initial understandings of the students. How can this goal be accomplished? The approach must resolve two conflicting considerations: relating concepts to the students' prior experience by depending heavily on operational definitions, and relating concepts clearly to one another using paraphrase definitions.

The extensive use of operational definitions relates concepts directly to the students' experiences and more-or-less familiar objects. Yet physical relationships among concepts must then be obtained from experiments that are carried out with the errors and uncertainties of such procedures, from teacher claims about such experiments that have been carried out by others (i.e., researchers in physics), or by means of theoretical derivations and "thought" experiments.

One can, for example, define each of force, pressure, and area operationally by using, respectively, a calibrated spring to balance an unknown force, a calibrated barometer exposed to a gas of unknown pressure, and small unit-area squares to cover a surface of unknown area. The relationship that pressure is the force per unit area must then be found empirically by comparing force and pressure measurements or theoretically by reasoning about the forces that act on the mercury at equilibrium in a barometer tube. "Deriving" the kinetic energy equation from two empirical relations as in Figure 4 also is mathematically weak.

The opposite approach, extensive use of physical relationships to generate paraphrase definitions, connects the concepts clearly to one another. Yet poor familiarity with the concepts in the definition is likely to make the students' grasp of the newly defined concept tenuous. Applications to real phenomena will be hard to visualize.

Returning to the pressure example, pressure can be defined as force per unit area, a paraphrase, after force and area are defined operationally. The relationship of force to pressure is then clear, but the process by which the barometer measures pressure through the forces acting on the mercury takes some effort to understand—the same effort as the theoretical reasoning referred to above.

Regardless of the teaching approach taken, it would appear that necessary linking of the concepts in a cluster requires teaching that ultimately deals with the entire cluster as an entity. Any sequential procedure starting from one concept and leading to others suffers from the shortcomings illustrated.

An application of the so-called spiral approach to teaching might lead to use of both of the procedures I have described. During a student's first encounter with the subject matter, most concepts are defined operationally. Relationships are taken from experiments that are actually carried out or that are reported to the class—preferably the former. Then, during a student's second encounter with the same topics, a more limited number of concepts is defined operationally and an increasing number of concepts is defined by paraphrases making use of physical relationships. Thus the two approaches to pressure could both be used: the first in elementary school or junior high school teaching; the second in senior high school and college. Hereby the learners' developmental needs are recognized and the demands of compactness and elegance in the subject matter given second priority.

In conclusion, the reader may wonder to what extent the structure of physics itself might beneficially be part of the course content. My experience with a few trials suggests that teachers be very cautious in making introductory level students face directly the alternative structural relationships I have described. These could, however, be very useful at a more advanced level, say during a third turn of the "spiral approach," after students have encountered two or more different organizations of the structure.

American Association of Physics Teachers 1980 Oersted Medalist: Robert Karplus

The Oersted Medal is the oldest and highest honor awarded by the American Association of Physics Teachers. It commemorates Hans Christian Oersted, an exemplary physics teacher, who made his great discovery of electromagnetism in the course of a college physics lecture. Today we award the 46th Oersted Medal to one of America's greatest leaders in physics teaching and in research in physics education, to Robert Karplus, professor of physics at the University of California at Berkeley.

For more than 20 years Robert Karplus has focussed his professional work on the problem of preparing elementary school children to understand the physical sciences. He was among the first to see that the success of older students and adults in physics and other physical sciences is largely determined by their contacts with fundamental physical concepts during their first school years. In the School Science Improvement Study (SCIS), which he directed for some 16 years, he has created an elementary science curriculum, which is sound in its conceptual basis, which emphasizes the acquisition of skills essential for comprehending science, and which provides an excellent foundation for later study of science in school and university.

There is a remarkable parallel between Robert Karplus, the 46th Oersted Medalist, and Gerald Holton, the 45th Oersted Medalist. Both came to the United States from Vienna. Both received their graduate degrees at Harvard University. Both earned their physics Ph.D.'s in 1948 and both became assistant professors of physics at Harvard. But in 1958 Karplus broke this lock step by transferring to Berkeley where he has fluorished ever since, except for years as visiting professor at MIT and at the University of Maryland.

Robert Karplus began his academic career in chemistry. In fact, he received two degrees in chemistry. But then he became an expert in quantum electrodynamics. For a while, around 1960, it looked like magnetohydrodynamics might capture his affections, but in the turmoil over science education in those years, he turned to science teaching as his major work.

Reprinted with permission from the *American Journal of Physics*, 49(9), 810, 1981.

Unlike most scientists who bemoaned the state of science teaching in the post-Sputnik era, unlike even most who worked to improve the nation's scientific literacy, Robert Karplus began his work by going himself into the elementary classroom and by working directly with the children he sought to teach. He acquired first-hand knowledge of how science is taught and how children learn or fail to learn. The SCIS elementary science curriculum was the result. In the course of this work—or, perhaps, even earlier—he recognized the significance of Jean Piaget's studies in developmental psychology for science teaching. It is largely through his efforts that American science teachers have become aware of the applicability of Piaget's concepts to their teaching at all levels.

Having become involved in the problems of how to teach physical science to young children, Karplus also became involved in research on the development of reasoning. He was among the first to understand—and I quote his own words, written in 1961—"that the childish systems of knowledge are not used by children only; many of their components

[1] R. Karplus, Am. J. Phys. **30**, 1–9 (1962).

remain with the individual throughout his life."[1] In a long series of papers beginning in 1970, mostly written in collaboration with Elizabeth F. Karplus, his wife, he has concentrated on intellectual development beyond the elementary school, with particular emphasis on deductive logic and ratio reasoning. His work has become well known in the physics community through articles in the American Journal of Physics and in *Physics Today*. Even more influential, perhaps, have been the "Workshop on Science Teaching and the Development of Reasoning," which Karplus and several collaborators prepared in the mid-70s. Through them many hundreds of science teachers at both school and college level have been introduced to the profound implications of research in the development of reasoning for their own teaching.

Robert Karplus has been Associate Director of the Lawrence Hall of Science at Berkeley since 1969. He is Director of Intellectual Development Project. From 1978 to 1980 he chaired the Graduate Group in Science and Mathematics Education. He has been a Fulbright Fellow and twice a Guggenheim Fellow, and is a Fellow of the American Physical Society. He received the AAPT Distinguished Service Citation in 1972 and the Distinguished Service Award of the National Association of Science Teachers in 1978. He served as President of AAPT in 1977. Last October he was awarded an honorary Ph.D. by the University of Gothenburg.

For his many contributions to physics teaching at all levels and especially for his work in revealing the implications for physics teaching of research in the development of reasoning, the American Association of Physics Teachers is honored to award the Oersted Medal for 1980 to Robert Karplus.

Presented by James B. Gerhart, University of Washington, Seattle. Past President and Chair of the AAPT Awards Committee.

The response of Robert Karplus to this award is reprinted in Chapter 5, pages 90–95.

CHAPTER 8

Scientific Literacy and the Teacher Development Gap: Karplus' Challenge

Jane Bowyer*

OVERVIEW

The late 1950's witnessed the beginning of two revolutions, one in science with the launch of the first space satellite, and the other in education with the launch of a major paradigm shift in the teaching of elementary science. The two were intimately connected. The Soviet Union's dramatic launch of a satellite prompted politicians, educators, and scientists to focus on the lack of a scientifically literate public in the United States and its implications for national security and scientific prowess. The challenge was how to create a scientifically literate population in the shortest possible span of time.

Karplus, at that time an internationally recognized physicist, was quickly drawn to the center of the educational revolution. As the father of seven children he had become extremely interested in education. Ultimately it was Karplus who led the way in transforming the elementary school science

* Abbie Valley Professor of Education and Department Chair, Mills College, Oakland, CA 94613.

A Love of Discovery: Science Education—The Second Career of Robert Karplus, Edited by Robert G. Fuller, Kluwer Academic / Plenum Publishers, New York, 2002.

curriculum in terms of content, process and pedagogy. For over 20 years he passionately pursued the question of what scientifically literate people in the 20th century need to know, and how educators can help them to know it.

A PERSONAL NOTE

But before I begin my story, a note about how Karplus and I met. I was a sixth grade teacher in our neighborhood elementary school where Karplus and his colleagues were trying out some of the elementary science curriculum ideas they were developing. Fortunately, my students and I got to be a trial classroom. This began a 12 year relationship which profoundly influenced my professional development and life's work. At Karplus' invitation, I left classroom teaching and joined his SCIS Curriculum Development Project. Since I had never been a science teacher, I was in many ways the perfect SCIS in-house novice. This was especially important given the teacher development challenges Karplus was to face. For the next five years I "carried the clay" for Karplus and his colleagues as they went from classroom to classroom and workshop to workshop in their efforts to create and disseminate their uniquely original elementary science curriculum. Ultimately Karplus became my Ph.D. thesis advisor. My dissertation evaluated the cumulative effects of the SCIS curriculum on the development of scientific literacy in children. He then became my mentor as a professor involved directly in teacher preparation and development. The lessons I learned from Karplus remain at the heart of all I do whether it's teaching a class of perspective science teachers, co-authoring a science education textbook for teachers, leading an NSF local systemic change project designed to get more and better science into 1400 elementary teachers' classrooms in Oakland, California, or developing a Mid Career Mathematics and Science Teachers Preparation Program at Mills College to get more qualified science teachers into our high school classrooms. A sample of some of the work Karplus and I did with high school science teachers in northern California using his materials can be found at the end of this chapter.

TEACHER DEVELOPMENT CHALLENGES

In my opinion, of all the challenges Karplus faced in his combined roles as science curriculum researcher/developer, the creation of a major paradigm shift in the teacher's role was the most important and the most difficult. And, the stakes were high! If Karplus was not successful in identifying and addressing the major issues involved in science teacher development, the goal of achieving scientific literacy for all students would fail. In retrospect, it proved to be easier for students to learn the major concepts and processes of science using the SCIS curriculum (when it was optimally taught) than it was for teachers, particularly elementary teachers, to make the transitions necessary to teach this material. The reasons for this are numerous and important to understand if we are to fully appreciate the scope and enormity of Karplus' work.

The major barrier facing Karplus in developing and implementing a scientific literacy curriculum related to values and attitudes. His decision to begin with the very youngest school children presented a major problem. Teaching serious science systematically to first and second graders was not only a radical idea but for the most part one that was unwelcomed by teachers, administrators and parents. From their perspective, science was a curricular frill like art or music. Reading, writing and arithmetic were the staples of the elementary curricula; teachers were prepared to teach these subjects and students were expected to learn them.

The first problem Karplus had to address concerned teachers' knowledge of science. There were few teacher preparation programs at our universities, or in-service staff development programs in school districts, that prepared elementary teachers to teach scientific literacy. It doesn't take an educational scholar to know that successful teachers must have expert knowledge of the subject matter they are teaching. For many junior and senior high school science teachers (usually males) this was not a problem since they had studied science in high school and college. However, for most elementary teachers (predominately female) who were teaching in the 60's and 70's when Karplus' science curriculum was being developed, science was foreign. They simply had not studied it in school; their elementary school experiences were for the most part limited to their teacher's favorite nature-study activity (commonly bird watching or stargazing) and their high school science coursework was minimal. Those classes were traditionally populated by boys except for the occasional course in general science and/or biology. The majority of female teachers were commonly excluded from basic science college courses because they did not have the appropriate high school prerequisites. Thus a major challenge for Karplus was to address the issue of developing teacher access to scientific understanding and knowledge.

A second challenge for Karplus related to teachers' knowledge of the nature of the scientific enterprise. Teachers with little knowledge of science were understandably unaware of the processes involved in discovering scientific knowledge. These processes were as important for scientific literacy as knowing the major concepts of science. If teachers could teach their students the processes of science by providing opportunities for them to become scientist-discoverers in their own right, scientific literacy would be enhanced. However, for this to occur Karplus concluded that the teachers themselves would have to understand the processes of science. They could use this new knowledge to drastically change how they taught. No longer would they *reveal* the facts of science in the traditional didactic fashion; students would have opportunities to *discover* major science concepts through a guided discovery process. This would mean that teachers would transform their classrooms into temporary laboratories that used equipment and materials rather than text-books as their major information sources. The children's thinking and work would be shifted to center-stage and would replace the teacher at the front of the class. Teaching would no longer be dominated by teacher-talk and student-listening. Right answers would become less important than observing, hypoth-esizing, experimenting, and reasoning about puzzling phenomena. Student collaboration, like scientific collaboration, would be valued in the learning process, not condemned as copying. To support teachers in making these radical changes Karplus' challenge was to create opportunities for them to per-sonally discover the nature of the scientific enterprise.

A third issue for Karplus was precipitated by new research on the devel-opment of logical thinking in students. These new insights greatly influenced his curriculum development work. Unfortunately, most teachers were unaware of these studies. This model was generally not part of the curriculum taught during the teacher preparation in this era. Typically teachers were taught a behaviorist learning model influenced to a large extent by rat maze and pigeon training experiments. This model was particularly useful in informing teach-ers about ways to teach material involving rote learning. In contrast, the primary research-based model that influenced Karplus' work was developed by the Swiss epistomologist Jean Piaget and his collaborators. Piaget's star-tling discoveries about children and adolescents' reasoning patterns were based on studies of infants and young people in the process of thinking and problem solving. This research had enormous implications for the conceptual understanding of science.

The challenges facing Karplus were only equaled by those facing the teachers. From their perspective, they would have to simultaneously learn new science content, processes, and pedagogy, while working an eight to ten hour day teaching. The adage, "trying to change a tire while the truck is still moving" applied.

Friends and colleagues who knew Karplus also knew that the hardest challenge was always the one that he chose to tackle with the most vigor. And friends also knew that in his passionate drive to understand and learn, he would bring a zeal and energy to the task that was almost endless. In my work with Karplus I soon learned that his impatience with himself as a learner was only equaled by his deep respect for teachers and their knowledge and skills.

How Did He Do It?

In the early sixties Karplus was a prominent physicist with an international reputation, and the scientist rumored to be the University of California's best hope for an in-house Nobel Prize in physics. At the same time, it is important to know what Karplus was not. Very clearly, Karplus was *not* an elementary school teacher! He had never taken a course in teacher preparation, never done a day of K-12 student teaching, and except for occasionally volunteering in his own children's classrooms, had spent no time in K-12 classrooms. He was a complete outsider in terms of any experiential or intellectual knowledge of teacher development.

This knowledge deficit was serious. Of all the factors involved in designing a curriculum for the creation of a scientifically literate population, the area of teacher development proved to be Karplus' biggest challenge. He needed to quickly gain a knowledge base in the discipline of teaching, learn the conditions and context in which teachers operate, and understand specifically the barriers teachers might encounter as they attempted to teach a very different sort of science curriculum.

It is significant that the first person Karplus selected for his curriculum development team was a former science teacher turned school principal, Herb Thier. Thier became an in-house invention partner in the area of teacher development. Thier's contributions provided Karplus day-to-day input from a perspective that was informed, experienced, and expert! As a result of the Karplus/Thier relationship bold, and visionary solutions to the enormous challenges the teacher development variable posed were conceived and implemented.

I believe that Karplus learned about teacher development using the same approach he and his colleagues developed for teachers to use in teaching his science curriculum. Known as the learning cycle it included three phases: *exploration, invention* and *discovery*: (1) *explorations* to scope out the major variables in the field, (2) *inventions* from reading the works and interacting with educational researchers, and (3) *discoveries* from his own classroom experiments.

The first or *exploration* phase of Karplus' learning utilized a technique that was familiar to him as a research scientist. He went directly to the classroom and immersed himself in the culture. He tirelessly pursued answers to endless questions concerning all aspects of classroom phenomena. When we walked into a classroom to observe the teaching of an SCIS lesson that had been newly developed, or perhaps revised and was being retaught, Karplus did not quietly take a seat at the back of the room. It never occurred to him that this might be the polite thing to do in terms of the school culture. Instead he immediately walked over to a group of students, and proceeded to ask thoughtful questions concerning answers they had on their data sheets, or designs they had developed with their experimental apparatus. And Karplus did listen. He listened with the same intensity that you could imagine him listening to a colleague describing unexpected results from a physics experiment. He was visibly pained when teachers described frustrations. He was elated, beaming in fact, when a teacher joyfully discovered a student who understood a concept because of a particular experiment he/she had conducted. Karplus was open to the disequilibrated feelings associated with not knowing, being totally puzzled, and desperately wanting to understand.

The second, or *invention* phase of Karplus' learning consisted of finding out what expert practitioners and researchers in the field had to say. Questions generated from the exploration phase of his learning process focused his study. Among the giants whose work was most influential in Karplus' thinking were Pestalozzi, Dewey, Brunner, Vygotsky and Piaget.

The third *discovery* phase of Karplus' learning came from applying knowledge he gained from the exploration and invention phases. His methodology consisted of teaching, revising, reflecting, and then reteaching. This ultimately resulted in the optimum placement of concepts, processes, and pedagogical practices that maximized the opportunities for student understanding.

KARPLUS' CONTRIBUTIONS TO TEACHER-DEVELOPMENT

Karplus brought a research-development model to curriculum development that was totally new. Prior to his approach curriculum developers were typically subject matter experts or teachers who spent time, during usually the summer, writing for textbook publishers. Their new revisions were primarily the old texts with the addition of the latest scientific discoveries. The approach that Karplus used in determining the content to be included was based on teacher and student data collected in the field.

This methodology yielded a set of products that was to ultimately transform the landscape of science education and classroom teaching. Only few major concepts were identified to be taught each semester; they were the most fundamental for understanding the discipline. These were then located within the curriculum in terms of the developmental levels of the children. For example, formal concepts were reserved for the older students.

In an effort to address teacher development issues, teacher guides were developed that provided science background information relating to the science concept to be taught. Examples of open-ended questions were provided to simulate student's thinking. A lesson plan design prototype, the learning cycle, was employed in all SCIS materials to model a constructivist approach to learning. Evaluation supplements were designed to give teachers information regarding their student's understanding of the concepts. Materials the students could use to create optimum events for learning key concepts were provided as a major staple of the curriculum. Instructions on the acquisition, distribution and organization of the laboratory materials were provided. Finally, the psycho-social and pedagogical underpinnings of the curriculum were provided for the teacher.

In addition to these efforts, Karplus and his colleagues created films that showed model lessons being taught in the classroom and Piagetian style interviews which dramatically demonstrated about the learning stages of the students. Finally, dissemination workshops involving teacher leaders, principals, and science educators were developed. These workshops were an ongoing, integral part of the Karplus curriculum development work. Additional workshops for teachers were (and still are) provided at professional meetings for elementary science teachers, physics teachers, chemistry teachers, and high school teachers. Special materials for these were developed using the same careful methodologies Karplus brought to all his work.

CONCLUSIONS

Karplus was a giant. He began his work in schools as a complete novice but his persistence, curiosity, and passionate drive soon placed him at the forefront of the science education reform. He was brilliant, stubborn, and worried. The stakes were high. He cared passionately that teachers be able to guide students in discovering the methods of science and excitement of discovery.

Karplus worked harder, prepared more, and was more attentive to details than anyone I have ever known. It was a compulsion . . . there was so much to learn, so much to do, and so little time. It was not unusual to get a call from him at 6:30 Sunday morning asking if the materials we were to take into the classroom Monday included the latest revisions developed the previous week.

Karplus never stopped thinking about children and teachers. He continually tried to figure out how they might come to a better understanding of the basic scientific constructs. He truly hoped they would retain an appreciation of scientific explorations throughout their lives.

Karplus' journey into the realm of teacher development continued for nearly twenty years and although it's been nearly a half century since he began his work, the contributions he made to the field of science teacher development continue to form the cornerstone of our work today.

Inservice Staff Development: A Piagetian Model

Jane Bowyer and Robert Karplus

A substantial shift in emphasis from preservice to inservice teacher education is currently taking place in the United States. The National Teacher Corps, is redirecting its yearly budget of $40,000,000 from the college student to the classroom teacher. The newest and largest federally funded education project, the Teacher Center legislation, is spending its 1979 budget of $80,000,000 on staff development for professionally employed teachers.

In California, the recently passed bills AB 551 and AB 65 provide funds for inservice staff development. The State Department of Education, August, 1977, organized a Department of Staff Development to provide leadership in coordinating inservice education for nearly 200,000 certificated personnel in California public schools.

A primary reason for the reallocation of educational resources from preservice to inservice training is the decline in school enrollments, resulting in less teacher turnover. To ensure continued vitality in public schools, teachers need to interact with current research and development ideas that have practical classroom application. In the past, a major source of this information has been the newly hired teacher coming into a school system from a college or university. The shrinking teacher population means that innovative ideas are no longer available at a fast enough rate to insure quality teacher change.

The expanded role of inservice education in providing growth opportunities for teachers suggests that a critical examination of existing staff development practices is essential. This paper describes an inservice workshop model currently being used with high school

Jane Bowyer, Department of Education Mills College.   Robert Karplus, Department of Physics University of California at Berkeley.

Reprinted with permission of Caddo Gap Press for *Teacher Education Quarterly* (formerly *California Journal of Teacher Education*). 5(3), 48–54, 1978.

and college teachers to increase their understanding of ideas suggested by the research and theory of the Swiss psychologist, Jean Piaget.

THE WORKSHOP MODEL: SCIENCE TEACHING AND THE DEVELOPMENT OF REASONING

The most significant ideas to affect education in this decade relate to the work of Jean Piaget and others who have studied the development of logical thinking. Their research has shed light on difficulties experienced by many secondary school students in understanding and using complex concepts. According to the theory, problems result when students fail to use thinking strategies which include proportional reasoning, propositional logic, separation and control of variables, and other elements of what Piaget has termed formal thought.

Piaget has focused attention on the patterns of reasoning (mental operations) an individual uses when solving a problem, making a prediction, or drawing a conclusion. Concrete operations deal with physical objects (i.e., mental grouping of similar objects and/or mentally reversing an observed rearrangement of blocks to accurately imagine the original configuration). Formal operations deal with relationships or hypothesized objects (i.e., proving a proposition by establishing that its negative leads to a contradiction; forming a hypothesis and deriving from it a testable prediction; or extrapolating from data on the basis of a consistent hypothesized functional relationship). The essence of Piaget's developmental theory is that an individual uses formal operations to organize experiences only after coping with the same subject matter successfully using concrete reasoning patterns.

Recent investigations indicate that few high school students use the strategies of formal thought confidently and reliably. (Karplus and Karplus, 1970; Karplus and Peterson, 1970; Shayer, 1972; Renner, 1973; Kolodyi, 1975; Lawson and Wollman, 1976) Unfortunately, most of the curricula and teaching approaches used in high schools expect students to use formal reasoning patterns. Studies indicate that even at the college level, fewer than half of the students consistently use formal thinking. (Renner and Lawson, 1973; Griffiths, 1976)

It appears that secondary school courses and teaching approaches must be modified if they are to provide for the diversity of individual reasoning patterns. Students functioning at the level of formal thought need to be challenged to use and further develop their intellectual resources. Students not functioning at the level of formal thought must be helped to develop the necessary thinking strategies in order to effectively understand concepts.

DEVELOPMENT AND DESCRIPTION OF THE WORKSHOP

The spark for development of this inservice model came from the American Association of Physics Teachers (AAPT), whose members had been alerted to the significance of reasoning in high school and college physics courses by lectures and articles prepared by John Renner of the University of Oklahoma. Robert Karplus, who had extensive experience in curriculum development and the application of Piagetian research, was asked to lead an effort to produce a workshop that could be presented at a national meeting of the Association.

The staff development workshop materials were designed to familiarize teachers with Piagetian ideas regarding intellectual development as a first step toward remedying many of the difficulties students have in learning science. If applied to teaching, this familiarity will lead to the following improvements:

1. Teachers will become more aware of and better understand the diverse cognitive backgrounds and abilities of the students in their science classes.

2. Teachers will be able to reorient their courses so as to reach and improve the reasoning strategies of students.
3. Teachers will have a new framework for assessing students and diagnosing their learning problems.
4. In evaluating students, teachers will place greater stress on their abilities to justify conclusions in terms of physical theories and experimental data, and less on an accumulation of facts and the ability to substitute numbers into formulas.
5. Teachers will be in a better position to evaluate science texts, course syllabi, and the need for laboratory work.
6. Teacher development of new courses, textbooks and instructional materials (films and modules) will lead to products that facilitate the students' thinking strategies.

The materials in the workshops included a paperback study guide for each participant, a film, and some simple experimental apparatus. The guide for the workshop on *Science Teaching and the Development of Reasoning* is organized into three parts consisting of eleven modules. The titles of the modules follow:

Part I: REASONING PATTERNS
      1. How students think?
      2. Concrete and formal reasoning patterns.
      3. "Formal Reasoning Patterns"   (film)
      4. Science texts and reasoning patterns.
Part II: TEACHING FOR SELF-REGULATION
      5. Self-regulation and the learning cycle.
      6. The laboratory and self-regulation.
      7. Concrete and formal concepts.
      8. Tests and self-regulation.
Part III: WHAT CAN YOU DO?
      9. Teaching strategies and goals.
    10. Suggested readings.
    11. Workshop planning and management.

The workshop takes an individualized approach to learning. Modules are self-paced and include group activities as well as individual work opportunities. During the workshop, teachers engage in reading, writing responses, analyzing students' thinking and textbook materials, viewing a film, experimenting, and discussing ideas.

The role of a workshop leader is non-directive except during discussions. Generally, staff act as organizers of materials, managers of traffic flow, and sources of information when requested. The length of a workshop and the number of participants vary. If only one part of the study guide is used, a minimum of three hours is adequate; if all modules are used, eight to twelve hours are recommended. A ratio of one staff person for seven teachers is suggested.

In Part I of the inservice guide four modules are provided to familiarize the teacher with concrete and formal reasoning patterns. Examples of high school students' approaches to certain standard science tasks are given to illustrate various reasoning strategies. In all cases a student's explanation or justification of his answer is considered in addition to the answer itself. Two major groups of reasoning patterns are identified and designated as concrete and formal, in consonance with two developmental stages described by Piaget. A film entitled "Formal Reasoning Patterns" allows the teacher to observe interviewers working with students between the ages of twelve and seventeen on four Piagetian tasks. Successful completion of the tasks requires the student to use the following patterns of formal thinking: proportional reasoning, separation and control of variables, combinatorial techniques, and the application of proportional reasoning to the physical phenomenon of the balance beam. The last module in Part I gives teachers the opportunity to examine selections from modern science texts and laboratory activties in order to analyze the reasoning patterns required of the students.

In Part II of this model, the teacher becomes familiar with self-regulation, the process whereby an individual advances from one stage of intellectual development to the next. Laboratory experiences with mirrors and "backward" writing, followed by a formal essay which functionally defines self-regulation in terms of the mirror experience, are the means whereby the teacher is introduced to this concept. Opportunity is provided for the teacher to learn how to classify individual science concepts as requiring concrete or formal thinking. Application of the self-regulation idea to student tests is also provided. Tests are designed to evaluate the students' reasoning and also help them engage in self-regulation by reflecting on their own thinking.

DISSEMINATION STRATEGIES

The dissemination strategy built into the workshop materials (Part III) is intended to maximize the multiplier effect obtainable from leadership exercised by workshop participants. Workshop presentations are organized and staffed by persons who have been workshop participants as part of an inservice or staff development effort. For that reason, the study guides are self-instructional with minimal dependence on expert workshop leaders. An effective presentation depends on a staff that is skillful in conducting workshops but need not have professional training in developmental psychology.

Although information concerning Piaget's work is available in educational and psychological research journals, these journals are usually read by other researchers, not practicing teachers. The materials and guides presented in this paper are goal directed and help teachers understand and apply Piagetian concepts to their teaching. Clearly this model identifies and fulfills a need.

IMPLICATIONS OF THE MODEL

Efforts need to be made to transform this unique inservice model to fit the specific discipline requirements of elementary and non-science high school teachers. Obviously, the science content examples in this model will not meet the needs of a teacher of European history. Piaget's theory suggests that if an idea is to be understood, the learner's prior experience with the concept and his/her cognitive thinking patterns must be adequate and appropriate. This applies whether the learner is a child learning science or a teacher learning about Piaget. Extension of the current Piagetian model and/or application of its design to other staff development efforts offers an alternative for meeting the increased demand for teacher inservicing.

REFERENCES

Collea, Francis. *Physics Teaching and the Development of Reasoning*. Distributed by the Lawrence Hall of Science, University of California, Berkeley, California, 1977.
Fuller, Robert G. *Multidisciplinary Piagetian-Based Programs for College Freshmen*, University of Nebraska-Lincoln, 1977.

Fuller, Robert G. *College Teaching and the Development of Reasoning*. University of Nebraska, Lincoln, 1975–1977.

Griffiths, D. "Physics Teaching: Does It Hinder Intellectual Development?" *American Journal of Physics*, 44(1), 81–86, 1976.

Karplus, Elizabeth F. and Karplus, Robert. "Intellectual Development Beyond Elementary School I: Deductive Logic". *School Science and Mathematics*, 70(5), 398–406, 1970.

Karplus, Robert and Peterson, Rita W. "Intellectual Development Beyond Elementary School II: Ratio, A Survey". *School Science and Mathematics*, 70(9), 813–820, 1970.

— *Science Teaching and the Development of Reasoning*. University of California, Berkeley, California, 1977.

— "Formal Thought and Education—A Modest Proposal". Paper presented to the Eighth Annual Symposium of the Jean Piaget Society, Philadelphia, May, 1978.

Kolodyi, G. "The Cognitive Development of High School and College Science Student". *Journal of College Science Teaching*, 5, 20–22, 1975.

Lawson, Anton E. and Wollman, Warren T. "Encouraging the Transition from Concrete Formal Cognitive Functioning: An Experiment". *Journal of Research in Science*, 13, 413–430, 1976.

— *Biology Teaching and the Development of Reasoning*. University of California, Berkeley, California, 1976.

Piaget, Jean. "Intellectual Evolution from Adolescence to Adulthood". *Human Development*, 15, 1–2, 1972.

Renner, John W. and Lawson, Anton E. "Piagetian Theory and Instruction In Physics". *The Physics Teacher*, 11(3), 165–169, 1973.

Schatz, Dennis. *Astronomy Teaching and the Development of Reasoning*. American Astronomical Society, 1978.

Shayer, M. "Some Aspects of the Strengths and Limitations of the Application of Piaget's Developmental Psychology to the Planning of Secondary School Science Courses". Master's Dissertation, University of Leicester, 1972.

Piaget's Theory in a Nutshell

(original version written by Robert Karplus for the workshop on Physics Teaching and the Development of Reasoning)

In reading the student responses to the puzzles in Module 1, you undoubtedly recognized that type A answers were more complete, more consistent, and more systematic, in short, were better than type B answers. In fact, you may have been somewhat surprised to learn that many college students gave type B answers.

We suggest that each of the two types of answers demonstrates the use of either concrete or formal reasoning as described by the Swiss psychologist and epistemologist, Jean Piaget, in this theory of intellectual development. We shall, therefore, give you some general background regarding Piaget's theory and then apply it to the problem-solving and reasoning patterns used by students who responded to the puzzles in Module 1.

Dr. Piaget began his inquiry into the origins of human knowledge early in the 20th century. He sought to understand how knowledge develops in the human minds, i.e. to understand the genesis of knowledge. He called himself a genetic epistemologist to emphasize his interest in both the development of knowledge in the human species and the development of knowledge by an individual. Dr. Piaget's life long work had several distinct phases as shown in Figure 1.

From the large collection of Piaget's work we are only selecting a few concepts.

The fundamental units of knowing, for Piaget, are schemes. A scheme is a class of physical or mental actions you can perform on the world. Notice, that in the Piagetian sense knowledge is better described as knowing, as an active process. Hence, we will often use the

---

Robert Karplus, Physics Teaching and the Development of Reasoning, AAPT © 1975, (subsequent additions were made by R. G. Fuller and D. Moshman, UNL).

Jean Piaget (1895–1980)
Geneva, Switzerland
Three Periods of Work
1922–29—Started at Binet's Lab
—Began Semi-clinical interviews
—Discovered and described "Children's
Philosophies" e.g. "Sun Follows Me," Egocentrism
1929–40—Studied His Own Three Children
—Traced Origins of Child's Spontaneous
Mental Growth to Infant Behavior
e.g. Peek-a-Boo, Conservation reasoning
1940–80—Development of Logical Thought in Children and
Adolescents
—Child's Construction of His World.
Mind is not a passive mirror
—Child can reason about things but not about propositions.

Figure 1.

term reasoning to indicate the active, systematic process by which you come to know, or solve, something.

Two concepts of Piaget that we believe are most helpful to college teachers are: (1) sequences or stages in the development of schemes and (2) self-regulation (equilibration). Schemes develop gradually and sequentially and always from less effective to more effective levels. We shall discuss important schemes below.

The second key idea, self-regulation, refers to a process whereby an individual's reasoning advances from one level to the next. This advance in reasoning is always from a less to a more integrated and better adapted level. Piaget views this process of intellectual development as analogous to the differentiation and integration one sees in embryonic development. It is also seen as an adaptation analogous to the adaptation of evolving species. The process of self-regulation is discussed in a later module.

Piaget characterized human intellectual development in terms of four, sequential stages of reasoning. (See Figure 2)

The first two, called sensory-motor and pre-operational, are usually passed by the time a child is 7 or 8 years old. The last two, however, are of particular interest to college teachers; they are called the stages of concrete operational reasoning and of formal operational reasoning. What follows are some schemes that constitute important aspects of concrete reasoning and formal reasoning.

Concrete Schemes.

C1 Class inclusion. An individual uses simple classifications and generalizations (e.g. all dogs are animals, only some animals are dogs.)

C2 Conservation. An individual applies conservation reasoning (e.g. if nothing is added or taken away, the amount, number, length, weight, etc. remains the same even though the appearance differs).

C3 Serial Ordering. An individual arranges a set of objects or data in serial order and establishes a one-to-one correspondence (e.g. the youngest plants have the smallest leaves). These basic reasoning patterns enable an individual to:

(a) use concepts and simple hypotheses that make a direct reference to familiar actions and objects, and can be explained in terms of simple association (e.g. the plants in this container are taller because they get more fertilizer);

Stage	Logical Knowledge Stages of Cognitive Development (Jean Piaget) Characteristics	Approximate Age Range (Years)
Sensory—Motor	Pre-verbal Reasoning	0–2
Pre-operational	No cause and effect reasoning Uses verbal symbols, simple classifications, lacks conservation reasoning	1–8
Concrete Operational	Reasoning is logical but concrete rather than abstract	8–?
Formal Operational	Hypothetical-deductive reasoning	11–(?)

Figure 2.

(b) follow step-by-step instructions as in a recipe, provided each step is completely specified (e.g. can identify organisms with the use of a taxonomic key, or find an element in a chemical solution using a standard procedure);

(c) relate one's own viewpoint to that of another in a simple situation (e.g. a girl is aware that she is her sister's sister).

However, individuals whose schemes have not developed beyond the concrete stage have certain limitations in reasoning ability. These limitations are demonstrated as the individual:

(d) searches for and identifies some variables influencing a phenomenon, but does so unsystematically (e.g. investigates the effects of one variable but does not necessarily hold the others constant);

(e) makes observations and draws inferences from them, but does not consider all possibilities;

(f) responds to difficult problems by applying a related but not necessarily correct algorithm;

(g) processes information but is not spontaneously aware of his own reasoning (e.g. does not check his/her own conclusions against the given data or other experience).

The above characteristics typify concrete operational reasoning.

Formal Schemes:

F1 Combinatorial Reasoning: An individual systematically considers all possible relations of experimental or theoretical conditions, even though some may not be realized in nature (recall the Treasure Hunt Puzzle or Algae Puzzles).

F2 Separation and Control of Variables. In testing the validity of a relationship, an individual recognizes the necessity of taking into consideration all the known variables and designing a test that controls all variables but the one being investigated (e.g. in the Mealworm Puzzle, recognizes the inadequacy of the setup using Box 1).

F3 Proportional Reasoning. The individual recognizes and interprets relationships in situations described by observable or abstract variables (e.g. the rate of diffusion of a molecule through a semi-permeable membrane is inversely proportional to the square root of its molecular weight. Mr. Tall was six buttons tall and Mr. Short was 4 buttons tall, therefore, Mr. Tall must be one and a half times bigger than Mr. Short in any system of measurement.)

F4 Probabilistic Reasoning. An individual recognizes the fact that natural phenomena themselves are probabilistic in character, that any conclusions or explanatory model must involve probabilistic considerations, and that useful quantitative relationships can be derived, for example, the ratio of actual events to the total number possible (e.g. in the Frog Puzzle the ability to assess the probability of certain assumptions holding true such as: the frogs mingled thoroughly, no new frogs were born, the bands did not increase the death or predation rate of the banded frogs, and use of the ratio of 1 to 6).

F5 Correlational Reasoning. In spite of random fluctuations, an individual is able to recognize causes or relations in the phenomenon under study by comparing the number of confirming and disconfirming cases (e.g. to establish a correlation of say, blond hair with blue eyes and brunette hair with brown eyes, the number of blue-eyed blonds and brown-eyed brunettes minus the number of brown-eyed blonds and blue-eyed brunettes is compared to the total number of subjects).

These schemes, taken in concert, enable an individual to accept a hypothesized statement (assumptions) as the starting point for reasoning about a situation. One is able to reason hypothetical-deductively. In other words, one is able to imagine all possible relations of factors, deduce the consequences of these relations, then empirically verify which of those consequences, in fact occurs. For example, in the Island Puzzle, such an individual could explain "If there were a plane route between Island A and C, then people could get from A to B but that is forbidden."

At the concrete operational stage, some formal schemes may be absent or they are only intuitively understood. Hence they are applied only in familiar situations and only partially and unsystematically. One can be said to be reasoning at the formal level when formal schemes have become explicit and useful as general problem-solving procedures. We consider the concrete/formal dichotomy a useful heuristic to guide us in our classroom activities. It is NOT a new system of pigeon holes into which you place students. They can serve as another perspective by which you can more clearly view the reasoning used by your students.

In the table below, we summarize some differences between concrete and formal reasoning.

CHARACTERISTICS OF CONCRETE AND FORMAL REASONING

CONCRETE REASONING	FORMAL REASONING
Needs reference to familiar actions, objects, and observable properties.	Can reason with concepts, relationships, abstract properties, axioms, and theories; uses symbols to express ideas.
Uses concrete schemes C1–C3.	Uses formal schemes F1–F5 as well as C1–C3.
Schemes F1–F5 are either not used, or used only partially, unsystematically, and only in familiar contexts.	
Needs step-by-step instructions in a lengthy procedure.	Can plan a lengthy procedure given certain overall goals and resources.
Limited awareness of one's own reasoning. May be oblivious to inconsistencies among various statements one makes, or contradictions with other known facts.	Is aware and critical of one's own reasoning; actively checks conclusions by appealing to other known information.

Teachers who are interested in applying these ideas in their teaching should be aware that many theoretical and experimental issues relating to Piaget's work are still being investigated. Piaget's original notion was that all persons progress through the major stages in the same, invariant sequence, though not necessarily at the same rate. Recent studies suggest strongly that, although almost everyone becomes able to use concrete schemes, many people do not come to use the same formal schemes effectively.

Piaget's research has been a very rich resource for ideas about the construction of knowledge. A number of scholars around the world, known by the label "constructivists", are continuing to study the implications of Piaget's epistemology for education and learning. For example, the original version of this essay was written by Dr. Robert Karplus, a physicist and science educator at the University of California-Berkeley, who developed an elementary school (K-6) science curriculum based on Piaget's ideas.

Since the above patterns of reasoning that have been described as formal represent extremely worthwhile educational aims and indeed are fundamental to developing meaningful understanding of theoretical and complex disciplines, the finding that many college students in this country do not effectively employ formal schemes on a great many content tasks presents a real challenge.

In addition to this finding, five further points regarding concrete and formal reasoning should be kept in mind by teachers:

- First, formal reasoning is more than this or that specific behavior. It is also an orientation towards approaching and attempting to solve problems. For this reason, a person who is confident and experienced in one area may reason hypothetico-deductively (formally) in that area, but may be unwilling or unable to generate hypotheses and reason flexibly in a threatening or unfamiliar area.
- Second, a person's ability to effectively deal with problems using formal knowing is really open-ended in that one may deepen and broaden one's understanding in a particular domain, and/or add new intellectual areas within which one can reason formally.
- Third, many persons demonstrate the use of reasoning patterns which seem to be a mixture of concrete and formal schemes when solving particular problems. This type of reasoning can perhaps best be termed transitional.
- Fourth, a person develops formal schemes from concrete schemes through the process of self-regulation. Concrete schemes involving class inclusion, serial ordering, and conservation about real objects, events, and situations are the valuable *prerequisites* for the development of formal schemes.
- Fifth, sometimes by applying memorized formulae, words or phrases, students can appear to be using formal schemes and/or be comprehending formal subject matter, when they are in fact not.

Although this essay has not touched on many aspects of Piagetian theory, we will briefly mention its major implications for college teaching. These ideas will be expanded upon in later modules.

The theory's main implications for physics teaching are:

1. Reasoning is an active, constructing process that must engage your students in developing more adequate schemes.
2. Be aware that some of your students may sometimes use predominantly concrete schemes.
3. Be aware that many of the topics and concepts you teach require formal reasoning. You should figure out which topics these are.
4. Try to arrange your subject matter so it follows the developmental progression of familiar, concrete, real to less familiar, less concrete, and more theoretical.

5. Demonstrate to your students a questioning, dynamic, and active attitude towards the course you teach. Generate hypotheses, discuss alternative explanations and encourage your students to do the same. Turn your classroom into a laboratory where real problems are investigated and knowing is derived from acting on evidence that is produced. Rewarding this type of activity by your students helps students (i) realize that many hypotheses are constructed, (ii) reflect upon the meaning of hypotheses, (iii) examine alternative hypotheses, (iv) examine evidence and its meaning, and (v) construct formal schemes.

CHAPTER 9

The Scientist's Science Educator: How One Young Physicist's Life Was Changed by the Example of Robert Karplus

Alan J. Friedman*

Meeting Robert Karplus and working with him for a decade changed my career and my life. This happened not so much because of what Bob and I did together at the Lawrence Hall of Science, but because he demonstrated to me a possibility I had not known existed: a scientist could treat education as a science, rather than as an art form.

In the early 1970's I was an assistant professor of physics at a Hiram College, a small liberal arts college in Ohio. My Ph.D. in physics was recently hung on the wall, and I was aiming straight for a career as an academic, doing teaching and experimental physics. I really enjoyed teaching undergraduates, and assumed that at a small college one could do both teaching and research reasonably well. Good teaching, I believed, was a gift. In my undergraduate years and through graduate school I had some excellent teachers, some poor

* Director, New York Hall of Science.

A Love of Discovery: Science Education—The Second Career of Robert Karplus,
Edited by Robert G. Fuller, Kluwer Academic / Plenum Publishers, New York, 2002.

ones, and some in the middle. There was a rough correlation between the quality of an individual as a scientist and as a teacher, but they all *worked* at being good scientists, while few worked (at least visibly) at being good teachers. Whether they were good at teaching or not seemed more of an inherited trait. Some had it, some didn't. My professors all taught by lecture, demonstration, office hour, and graduate student assistant. So did I, but we didn't have graduate students at Hiram.

Several things happened at Hiram to unsettle my career. First, I discovered that unless there were sufficient numbers of students taking physics courses, my position as the most recent member of the three-person physics faculty would be in jeopardy. In an effort to build up enrollments, I tried teaching some "physics for poets" courses. One of them I taught with my wife, a former English major. It was called "Physics in Twentieth Century British and American Fiction," and was a lot of fun. There wasn't a great deal of physics in the class, of course, and the students were mostly in art and literature, unlikely to enroll in any further classes in physics. I doubted that any real physicists would take any interest in this course. But enrollment went up, fortunately.

In 1971 there was a dinner party at which a College administrator casually mentioned that since the college was going to reduce the size of the physics department, despite the recent modest increase in enrollment, had I made any plans yet? After confirming during the next few days the fact that my position was indeed to be phased out, at least temporarily, I invented Plan B. There were fellowships available from the National Endowment for the Humanities for scientists who wanted to spend a year studying the Humanities. That course in science and literature had been fun, so maybe I could spend a year playing with that while looking for another assistant professorship position, or seeing if the finances improved at Hiram so I could return there. I'd given a paper on the course at an AAPT meeting, and perhaps that would help with the fellowship application.

I applied, and amazingly enough, got one of those fellowships. I could go any place to study that would accept me, and it would cost them nothing. I applied to the English departments at Berkeley, the University of Michigan at Ann Arbor, and a few other campuses with strong literature programs as well as strong physics departments.

Then came a surprise letter from Berkeley; not from the English department, but from some place I'd never heard of, the Lawrence Hall of Science. Robert Karplus wrote on March 3, 1972, to ask if I had plans to publish a description of my work on that science and literature course. It seemed that the Council on Physics in Education was hoping to develop a resource letter and possibly a reprint volume on teaching physics to non-specialists. Karplus was also "very interested in this problem and consider it an important but somewhat neglected area of teaching."

I didn't know who Karplus was at the time, but I looked him up. My notes on his letter read, "He has written several physics books, many articles, etc." A real physicist, from Berkeley no less, was interested in my work as a teacher, not as a physicist. By this time I had received letters expressing a willingness to host my visit from the English departments at both Berkeley and Michigan. But the weather was said to be better in Berkeley, and there was at least one physicist there, Robert Karplus, who was interested in what I was doing.

I wrote back promptly, saying that yes, I was submitting a paper to journals, and by the way, I was thinking of coming to Berkeley for a year. Did he have any advice on finding a good apartment? Could I meet him during my stay? He was gracious and encouraging, and we began corresponding. I decided to go to Berkeley.

Shortly after arriving I went to visit Karplus at the Lawrence Hall. I had a hard time figuring out what he was doing. His office had no physics apparatus, no stacks of computer printouts, but there were these little plastic figures ("Mr. O" I learned later) and plastic tubs of rubber tubing, string, and soda straws. This, I discovered, was what Bob Karplus did when he was at the Lawrence Hall of Science. He was studying the ways children learned science, rather than science itself.

I had never imagined that real physicists did such things seriously. But Bob was indeed serious. He invited me to sit in on seminars at the Lawrence Hall and in the physics department, and I soon learned about Jean Piaget, the Karplus Learning Cycle, and SCIS.

My official work was studying literature. I had a great time doing that. But I kept visiting the Lawrence Hall of Science, trying to understand how the work of Bob Karplus and the Lawrence Hall fit into my picture of the world, in which the practice of science and the practice of education were two well-separated realms.

Bob Karplus and his colleagues were doing science, complete with experiments, theories, critical analysis, and publications. People at those seminars were just as sharp, and just as cutting, as they were in American Physical Society sessions presenting research results in physics. Bob and his colleagues didn't seem to be teachers, as I understood the term, because there was not much lecturing. But they certainly were educators. It finally dawned on me that one could be doing science and doing education *simultaneously*, instead of having to oscillate between the two professions as I had always seen my professors do and had always done myself.

What Bob did was just as challenging, just as rigorous, and just as much fun as doing experiments in x-ray crystallography or solid state physics. It was the learning process which was being investigated, rather than the behavior of low-temperature anti-ferromagnets (my dissertation topic). But it was still science, and as a career, had some advantages besides.

Fewer than a dozen people in the world were interested in what I had learned about the phase transition in manganese bromide, while millions of students were learning science with tools Bob and his colleagues had developed. There were also thousands of good scientists in the field of solid-state physics; but how many good scientists were working, *as scientists*, on how to teach physics? Aside from a handful, including Bob, W. M. Laetsch and Fred Reif at Berkeley, I didn't know of any. Finally, it was hard to find important experiments that could be done with modest equipment in my area of physics. The obvious ones were already taken. But in education—it was like physics must have been pre-Galileo—just about everything interesting was still up for grabs.

So it seemed to me I had a better chance of contributing, and becoming recognized, by becoming a frog in that relatively small pond around Bob Karplus.

I later learned that there were a lot more scientists than I thought who took education very seriously, but the number worldwide was still only a tiny fraction of the number of people in most branches of science itself. As my year of leave from Hiram College and the NEH fellowship ran out, I began to look for ways to stay at Berkeley, perhaps at the Lawrence Hall of Science, to see if this career option might really pan out. Then Bob invited me to a dinner party at his home and in the course of the evening, mentioned that Mauri Gould at the Lawrence Hall of Science was about to obtain an old, obsolete planetarium projector from UCLA. Mauri was thinking about ways to make use of it at the Lawrence Hall of Science, but didn't have any particular ideas yet.

I had enjoyed developing student activities in a small planetarium near Hiram College. Now, thanks to those seminars Bob invited me to, I knew there was even a theoretical justification for what I had enjoyed: those activities were *concrete* learning experiences for students, rather than the *formal* experiences of the traditional lecture-under-the-stars. Thank you, Bob Karplus; thank you, Jean Piaget.

The next morning I was up at the Lawrence Hall of Science meeting with Mauri. I wrote a few pages describing some education experiments that could be done using the recycled planetarium. Mauri and Bob were encouraging, and recommended me to W. M. (Mac) Laetsch, LHS Director. I spent the next 11 years at the Lawrence Hall, working with Bob, Mac, Mauri, and a remarkable group of people at what I still think of as the center of the science of science learning.

Bob Karplus became one of my mentors, but he also undertook one of the less rewarding roles for a faculty member, serving as the "Principal Investigator" for nearly all of my grant funded projects. The University required that a member of the faculty senate be listed as the PI on all proposals, and Bob graciously consented to do this for my projects even when they were far from his center of interest. For example, he served as the PI for a National

Endowment for the Humanities grant that helped me finish a book with Carol Donley, an English professor at Hiram College, on Einstein's impact on contemporary literature. There was a lot of paperwork, but also the expectation that he would regularly review and help improve those projects.

Even with the NEH grant, Bob's insights as a scientist/educator proved invaluable. He read the manuscript for the book (*Einstein as Myth and Muse*, Cambridge University Press, 1985), and while he didn't have much to say about our literary criticism, he paid great attention to our attempts to explain physics for our audience of English majors and cultural historians. He insisted that the science had to be right, even if greatly simplified. He made many suggestions for improvements, nearly all of which we adopted. He kept pressing us to go back and explain to our readers what was really important in the development of science. It wasn't just that a new equation worked better, it was that a new form of thinking worked better. In our description of the Newtonian revolution, Bob taught us to communicate that it was not just that Newton's equations of motions worked well, but rather that *interaction* replaced *predisposition* as the dominant mode of interpreting motion.

I moved more and more into work outside the formal education system, involving planetariums, exhibitions, "sports/science tip cards," and afterschool programs. Bob continued to provide theoretical underpinnings and critical reviews of nearly all my projects. One aspect of his thinking which I still try to incorporate in my work is constant attention to the situation of the learner; when this is going on, what is the learner going to make of it?

In my pre-Karplus days of college classroom lecturing about physics, I concentrated on telling the story. It was my responsibility to get the chain of logic all together, not skip any key steps, and arrive at the end before the bell rang. I did try to make it fun, with lots of demonstrations, occasional humor, and some passionate words at the end about how wonderful all this was. But Bob kept coming back to what the learner would be thinking at each stage of the story. Was I sure the learners knew what a particular word meant, in the context I was using it? How could a remark be interpreted differently from the way I intended? What counted was not that I got the story told by the end of the hour, but that the learner heard a story which made sense at the end of the hour.

Looking back at *Introductory Physics: A Model Approach*, Bob's 1969 textbook, I can see all the signs of the approach he taught me. Chapter 2, "Reference Frames," begins:

> You may associate the word "relativity" with mathematical mystery and scientific complexity, yet the basic concept, which we will try to explain in these pages, is simple.

The chapter begins by addressing the reader directly. Use of the second person noun seemed very strange to me at the time, but Bob helped me to realize that this could help engage the terrified non-science major. Even more important, however, is that the critical opening sentence takes direct cognizance of the state of mind of the reader. The word "relativity" does indeed come with a lot of baggage, and defusing some of the heavy burden in the reader's mind is a valuable first step towards getting the story across. Some of the physics explanations in *Einstein as Myth and Muse* have second person presentations ("If you toss a baseball you'll have given it about one joule of energy, in the form of motion."), and I think those passages work well.

Even today, I find myself looking back to *Introductory Physics: A Model Approach* when I am trying to figure out how to explain some point in science. The Introduction and Section 1.1 of Chapter 1, Section 1.3 on "Theories and models in science," and the first few sections of Chapter 3, on the concept of interaction, are remarkable and unusual examples of science presentations, rich with the attention to the learner's needs that characterize Bob's approach for me.

For the past 28 years I have been working as a science educator, trying to apply the lessons I learned from Bob Karplus, Mac Laetsch, Herb Thier, Larry Lowery, and all of my colleagues at the Lawrence Hall of Science. Bob continues to represent, for me, the model for how to combine two professional passions into one. From him I learned that it is possible, if not always easy, to do science and science education simultaneously.

Chapter 1
The nature of science,
sections 1.1 to 1.3

Have you ever sorted the books in your library according to their subject matter, only to find a few remaining that "didn't fit"? In a way, this problem is similar to problems that face a scientist. He makes observations on crystals or atomic particles or orbiting planets and must face the fact that some of his observations do not fit his expectations. Such an experience can be unsettling, but it can also lead to new understanding and insight.

One of the primary objectives of this text is to introduce you to a few of the powerful interpretations of natural phenomena used by the physicist to help him organize his experience. The text discusses some of these phenomena and the patterns of behavior they exhibit. You, in turn, are asked to examine your own experience for additional data to support or contradict these ideas. Occasionally, an unexpected outcome may compel you to reorganize your thinking. A critical approach to all aspects of the text is in order.

Unfortunately, modern culture has become fragmented into specialties. Science was once a branch of philosophy. In modern times, however, science, especially physics, is no longer an intellectual discipline with which every educated person is familiar. There are many reasons for this state of affairs (Figure 1.1). Probably the most important is that many individuals do not feel a need for a formal study of nature. They develop a commonsense "natural philosophy" as a result of their everyday experiences with hot

R. Karplus, Chapter 1, The nature of science, Section 1.1 to 1.3, pp. 3–14, *Introductory Physics: A Model Approach*. W. A. Benjamin, Inc. New York, © 1969.

Figure 1.1. What do you think of these reasons?
(a) The familiar surface of the moon. Should you know more about this silvery disk where people may someday live?
(b) Albert Einstein, popular symbol of theoretical, abstruse physics. Should grants for pure research be justifiable in terms of contemporary social needs?
(c) Nuclear explosion, Nevada Proving Grounds, 1957. Physics has become deeply involved in war and peace.
(d) A steel mill's waste gases bring with them one of the hidden costs of our technological civilization. Compare with Figure 2(c).

and cold objects, moving objects, electrical equipment, and so on. For most people, this seems quite adequate.

A second reason is that many of the questions with which modern physicists are concerned seem remote from everyday life. They deal with subnuclear phenomena, ultra-low or extremely high temperatures, cosmic dimensions, and other extraordinary conditions. The physics that is accessible to the beginning student has a cut-and-dried aspect that lacks the excitement of a quest into the unknown. Therefore, many students tend to think of physics as a finished story that must be memorized and imitated, rather than as a challenge to the creative imagination.

A third reason is the frequently indirect nature of the evidence on which physicists base their conclusions. As a result of this indirect evidence, experimental observations are related to theoretical predictions only through long and complicated chains of reasoning, often of a highly mathematical kind.

A fourth reason, of relatively recent origin, is that science has become identified with the invention of destructive weapons (the atomic bomb and biological warfare) and technological advances whose by-products (smog, detergents) threaten our natural environment. Many individuals reject science, and especially physics, as alien to sensitive, imaginative, and compassionate human beings.

"Why does this magnificent... science, which saves work and makes life easier, bring us so little happiness? The simple answer runs—because we have not yet learned to make a sensible use of it."

Albert Einstein

In this text we will try to overcome these difficulties. We will limit the diversity of topics treated, make frequent reference to the phenomena of everyday experience, and examine carefully the ways in which observations can be interpreted as evidence to support various scientific theories. The goal is to develop your understanding of how physical concepts are interrelated, how they can be used to analyze experience, and that they are employed only as long as there are no better, more powerful alternatives.

The reasons why an educated person should have some understanding of physics have been stated many times (Figure 1.2). Physics is a part of our culture and has had an enormous impact on technological developments. Many issues of public concern, such as air and water pollution, industrial energy sources, disarmament, nuclear power plants, and space exploration, involve physical principles and require an acquaintance with the nature of scientific evidence. Only a wider public understanding of science will ensure that its potential is developed for the benefit of mankind rather than devoted to the destruction of civilization. More personally, your life as an individual can be enriched by greater familiarity with your natural environment and by your ability to recognize the operation of general principles of physics even in small everyday things, such as children swinging or hot coffee getting cold.

1.1 THE SCIENTIFIC PROCESS

The present formulation of science consists of concepts and relationships that mankind has abstracted from the observation of natural phenomena over the centuries. This overall evolutionary process has been marked by occasional major and minor "scientific revolutions" that reoriented entire fields of endeavor. Examples are the Copernican revolution in astronomy, the Newtonian revolution in the study of moving objects, and the introduction of quantum theory into atomic physics by Bohr. The net result has been the development of the conceptual structure and point of view with which the modern scientist approaches his work.

An investigation. Let us briefly and in an oversimplified way look at the way a scientist might proceed with an investigation. For instance, consider a ball that falls to the ground when you release it. After additional similar observations (other objects, such as pieces of wood, a feather, and a glass bowl, all fall to the ground when released), you are ready to formulate a hypothesis: all objects fall to the ground when released. You continue to experiment. Eventually, you release a helium-filled balloon and find that instead of falling, it rises. That is the end of the original hypothesis. Can you modify it successfully? You could say, "All objects fall to the ground when released in a vacuum." This statement is more widely applicable, but it is still limited to regions near the earth or another large heavenly body where there is a "ground." In space, far from the earth, "falling to the ground" is meaningless because there is no ground.

What happens to an object released in space, far from the earth or another body?

This simple description has glossed over two important decisions that were made. First was the judgment as to what constituted "similar" observations. For instance, in the example you included the balloon along with the ball, wood, feather, and so on. Yet you might have considered the balloon to be very different from the other objects observed. Then its rising rather than falling would not have been considerd pertinent to the hypothesis of falling objects. Even for some time after Galileo's telescopic observations of the moon more than 300 years ago, there was controversy as to whether it and other heavenly bodies were material objects to which the hypothesis of falling objects should apply.

The second decision was the judgment about what aspects of the observations were to be compared. You decided to compare the motion of the bodies when they were released. Aristotle, who also thought about falling bodies, was more concerned with their ultimate state of rest on the ground and therefore reached conclusions very different from yours.

*"...from my observations,...
often repeated, I have been led
to that opinion which I have
expressed, namely, that I feel
sure that the surface of the
Moon is not perfectly smooth,
free from inequalities and
exactly spherical, as a large
school of philosophers
consider with regard to the
Moon and the other heavenly
bodies, but that, on the
contrary, it is full of
inequalities, uneven, full of
hollows and protuberances, just
like the surface of the Earth
itself, which is varied
everywhere by lofty mountains
and deep valleys."
        Galileo Galilei
    Sidereus Nuncius, 1610*

The scientific point of view. Usually the answers to these two kinds of questions are tacitly agreed upon by the

*Matter includes all solid, liquid,
and gaseous materials in the*

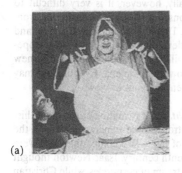

Figure 1.2. Is physics relevant?
(a) Do you base your actions on a crystal ball or on scientific evidence and reasoning?
(b) What clues enable you to identify the vertical direction?
(c) The waste gases from a steel mill are cleaned by the action of "precipitators," which make use of electric fields. Compare with Figure 1(d).
(d) Medical x-ray photography after an unlucky fall on a skiing trip.

universe. We will treat "matter" as an undefined term whose meaning must be grasped intuitively. Properties of matter, to be described later in this text, include mass, extent in space, permanence over time, ability to store energy, elasticity, and so on.

members of the scientific community and constitute what we may call the "scientific point of view." One aspect of this point of view is that there is a real physical universe, that *matter* exists, and that it participates in natural phenomena. A second is the assumption that natural phenomena are *reproducible*: that is, under the same set of conditions the same behavior will ensue. Other aspects of the point of view have to do with the form of an acceptable explanation of a phenomenon. Occasionally, however, it is very difficult to interpret new observations when they are considered from the current point of view. Then there is the need for bold and imaginative thinking to develop a new point of view. Hopefully, this new approach will be better able to explain the new observations and the known phenomena. Eventually it may become the accepted scientific point of view.

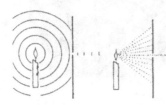

The work of Christian Huygens (1624–1695) and Isaac Newton (1642–1727) on the nature of light will be discussed in Chapters 5, 6, and 7.

The theory of light. A fascinating story in the history of physics that illustrates these remarks deals with the nature and interactions of light. Two competing ideas were advanced in the seventeenth century. Isaac Newton thought that light consisted of a stream of corpuscles, while Christian Huygens believed that light was a wave motion. Up to that time, experiments and observations on light rays had apparently been made without questioning further the nature of the rays.

In spite of contradictory evidence, Newton's corpuscular theory of light was preferred, largely because of the success of Newton's laws of the motion of material bodies subject to forces. Small bodies (corpuscles) probably provided a more acceptable explanation to Newton's contemporaries and followers than did the waves proposed by Huygens.

During the nineteenth century, however, new experimental data on the passage of light near obstacles and through transparent materials contradicted Newton's corpuscular theory conclusively and supported the wave theory. Waves and their motion became the accepted way to explain the observed properties of light.

This point of view flourished until the beginning of the twentieth century, when results of further experiments on the absorption and emission of light by matter conflicted with the wave theory and led to the presently accepted quantum theory of light. Already, however, there are contradictions within this theory, so that it, too, will have to be modified. This is one field of currently active research, and several proposals for new theories are being studied intensively to determine which holds the most promise.

Scientific "truth." Science is, therefore, never complete; there are always some unanswered questions, some unexpected phenomena. These may eventually be resolved within the accepted structure of science, or they may force a revision of the fundamental viewpoint from which the phenomena were interpreted. Progress in science comes from two sources: the discovery of new phenomena and the invention of novel interpretations that illuminate both the new and the well-known phenomena in a new way. Scientific truth is therefore not absolute and permanent; rather, it means agreement with the facts as currently known. Without this qualification, the statement that scientists seek the truth is misleading. It is better to say that scientists seek understanding.

1.2 DOMAINS OF MAGNITUDE

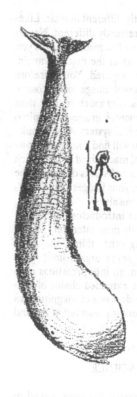

When and how does a person's experience of space and time originate? Probably the foundations are laid before birth, but the most rapid and important development takes place during an infant's early exploration of his environment. By crawling around, touching objects, looking at objects, throwing objects, hiding behind objects, and so on, he forms simple notions of space. By getting hungry and feeling lonely, by enjoying entertainment and playing, by watching things move and by moving himself, he forms notions of time. Even though the adult commands more effective skills with which to estimate, discriminate, and record space-time relations, the infant's need to relate his environment to himself is never really outgrown.

Size. As you look about and observe nature, you first recognize objects, such as other people, trees, insects, furniture, and houses, that are very roughly your own size. We will call the domain of magnitude of these objects the *macro domain.* It is very broadly defined and spans living creatures from tiny mites to giant whales. All objects to which you relate easily are in this domain.

All other natural phenomena can be divided into two additional domains, depending on whether their scale is much larger or much smaller than the macro domain. The former includes astronomical objects and happenings, such as the planet earth, the solar system, and galaxies. We will call this the *cosmic domain.* Much smaller in scale than the macro domain is the one that includes bacteria, molecules, atoms, and subatomic units of matter; we will call it the *micro domain.*

The phrase geologic times *is sometimes used to denote very long time intervals because geologic processes are extremely slow.*

Time. It is useful to introduce the concept of domains into time scales as well as into physical size. Thus times from seconds or minutes up to years are macro times in the sense that they correspond to the life-spans of organisms. Beyond centuries and millenia are *cosmic times,* whereas micro times are very small fractions of a second. As with physical sizes, the mental images you make for processes of change always represent in seconds or minutes what really may require cosmic times or only micro times to occur.

Applications. In order of size, then, the three domains are the micro, macro, and cosmic. The division is a very broad one, in that the earth and a galaxy, both in the cosmic domain, are themselves vastly different in scale. Likewise, bacteria and atomic nuclei are vastly different. Nevertheless, the division is useful because the mental images you make of physical systems are always in the macro domain, where your sense experience was acquired. You therefore have to remember that your mental image of a cosmic system, such as the solar system, is very much smaller than the real system. Similarly, your mental image of a micro system is very much larger than the real system. As you make mental images of these systems, you will find yourself endowing them with physical properties of macrosized objects, such as marbles, ball bearings, and rubber balls. This device can be very misleading because, of course, your images are in a different domain from the objects themselves.

When we pointed out in the introductory section to this chapter that physicists frequently must interpret indirect evidence, we had in mind, among other things, the three domains of magnitude. Since our sense organs limit us to observations in the macro domain, all interpretations concerning the other domains require extended chains of reasoning. An illustration relating the domains of magnitude to units of space and time measurement is presented at the end of this chapter in Figure 1.11.

1.3 THEORIES AND MODELS IN SCIENCE

In the preceding section we contrasted the roles played in science by observation and interpretation. Observations of experimental outcomes provide the raw data of science. Interpretations of the data relate them to one another in a logical fashion, fit them into larger patterns, raise new questions for investigation, and lead to predictions that can be tested.

Scientific theories are systematically organized interpretations. Examples are Dalton's atomic theory of chemical reactions, Newton's theory of universal gravitation, Einstein's theory of relativity, and Skinner's conditioning theory of learning in animals and man. Within the framework of a scientific theory, observations can be interpreted in much more far-reaching ways than are possible without a theory. In Newton's theory of gravitation, for instance, data on the orbital motion of the moon lead to a numerical value for the total mass of the earth! In Dalton's theory, the volumes of chemically reacting gases lead to the chemical formulas for the compounds produced. All theories interrelate and extend the significance of the facts that fall within their compass.

Working models. Theories frequently make use of simplified mental images for physical systems. These images are called *working models* for the system. One example is the sphere model for the earth, in which the planet is represented as a uniform spherical body and its topographic and structural complexities are neglected. Another example is the particle model for the sun and planets in the solar system; in this model each of these bodies is represented as a simple massive point in space, and its size as well as its structure is ignored. Still another example is the "rigid body model" for any solid object (a table, a chair) that has a definite shape, but may bend or break under a great stress.

Unlike other kinds of models (Figure 3), a working model is an abstraction from reality. Our thoughts can never comprehend the full complexity of all the details of an actual system. Working models are always simplified or idealized representations, as we have already pointed out. Working models, therefore, and the theories of which they are a part, have limitations that must be remembered when their theoretical predictions fail to agree with observations.

The scientist's relationship to the models he constructs is ambivalent. On the one hand, the invention of a model engages his creative talent and his desire to represent the operation of the system he has studied. On the other hand, once the model is made, he seeks to uncover its limitations and weaknesses, because it is from the model's failures that he gains new understanding and the stimulus to construct more effective models. Both creative and critical faculties are involved in the scientist's work with models.

One feature of working models is frequently disturbing to non-scientists: you never know whether a model is right. In fact, the concepts "right" and "wrong" do not apply to models. Instead, a model may be more or less adequate, depending on how well it represents the functioning of its system. Even an inadequate model is better than none at all,

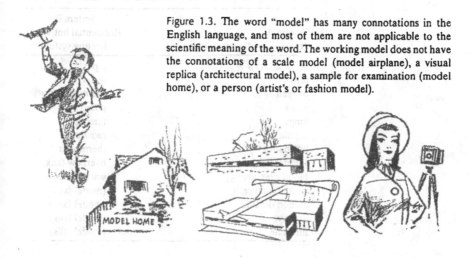

Figure 1.3. The word "model" has many connotations in the English language, and most of them are not applicable to the scientific meaning of the word. The working model does not have the connotations of a scale model (model airplane), a visual replica (architectural model), a sample for examination (model home), or a person (artist's or fashion model).

and even a very adequate model is often replaced by a still more adequate one. The investigator has to determine whether a particular model is good enough for his purposes, or whether he should seek a better one with which to replace his earlier effort.

Analogue models. Before a scientist constructs a theory, he often realizes that the system he is studying operates in a way similar to another system with which he is more familiar, or on which he can conduct experiments more easily. This other system is called an *analogue model* for the first system. You may, for instance, liken the spreading out of sound from a violin to the spreading out of ripples from a piece of wood bobbing on a water surface.

The analogue model for one physical System A is another, more familiar, System B, whose parts and functions can be put into a simple correspondence with the parts and functions of System A.

An analogy may be drawn between the human circulatory system and a residential hot water heating system (Table 1). It is clear that the human circulatory system fulfills several functions, whereas the heating system fulfills only one. The analogue model is, therefore, not complete, but it is nevertheless instructive.

The virtue of an analogue model is that System B is more familiar than System A. This familiarity can have several advantages:

Table 1. Analogue Model for the Human Circulatory System.

System A: Human circulatory system	System B: Residential hot water heating system
veins, arteries	pipes
blood	water
oxygen	thermal energy
heart	pump
lungs	furnace
capillaries	radiators
hormones	thermostat
(model fails)	overflow tank
blood pressure	water pressure
white blood cells	(model fails)
carbon dioxide	(model fails)
kidneys	(model fails)
intestine	(model fails)

1. Features of the analogue model can call attention to overlooked features of the original system. (Had you overlooked the role of hormones in the circulatory system, the room thermostat would have reminded you.)
2. Relationships in the analogue model suggest similar relationships in the original system. (Furnace capacity must be adequate to heat the house on a cold day; lung capacity must be adequate to supply oxygen needs during heavy exercise.)
3. Predictions about the original system can be made from known properties of the more familiar analogue model. (Water pressure is high at the inflow to the radiators, low at the outflow; therefore, blood pressure is high in the arteries, low in the veins.)

The limitations of the analogue model can lead to erroneous conclusions, however. On a cold day, for instance, the water temperature is higher in the radiators; therefore, you might predict that the oxygen concentration in the blood will be higher during heavy exercise. Actually, the heartbeat and the rate of blood flow increase to supply more oxygen—the oxygen concentration does not change greatly.

"There are two methods in which we acquire knowledge— argument and experiment." Roger Bacon (1214–1294)

Thought experiments. In a thought experiment, a model is operated mentally and the consequences of its operation are deduced from the properties of the model. A thought experiment differs from a laboratory experiment in that the latter serves to provide new information about what really happens in nature, whereas the former seeks new deductions from previous knowledge or from assumptions. By comparing the deductions with observations in real experiments, you can find evidence to support or contradict the properties or assumptions of the model.

A simple example of a mystery system (Figure 4) can be used to illustrate these ideas. Two working models for what might be under the cover in Figure 4 are shown in Figs. 1.4b and c. If you conduct simple thought experiments with these models, you quickly find out how satisfactory they are. In the first thought experiment, you imagine turning handle A clockwise. In model G, handle B will turn somewhat faster, because the second gear is smaller than the first, but it will turn counterclockwise. This prediction is in disagreement with the properties of the mystery system. In the second thought experiment, you turn handle A in model S. What can you infer from this second experiment? Can you suggest a satisfactory working model?

Thought experiments are important tools of the theoretical scientist because they enable him to make deductions

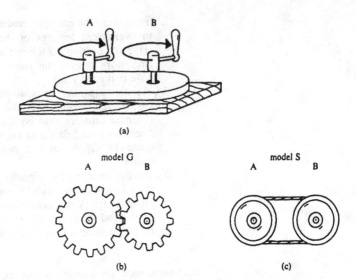

Figure 1.4. A mystery system. (a) When handle A is turned one revolution clockwise, handle B makes $2\frac{1}{2}$ revolutions clockwise. Make models for what is under the cover. (b) Large and small gear model. (c) Two pulley and string model.

from a working model or a theory. These deductions can then be compared with observation. The usefulness of a theory or model is determined by the agreement between the deduction and observation. Some very general theories, such as the theory of relativity, lead to consequences that appear to apply universally. Some models, such as the corpuscular model for light, are useful only in a very limited domain of phenomena.

Mathematical models and variable factors. Scientific theories are especially valuable if they lead to successful quantitative predictions. Working models G and S for the mystery system in Figure 4 both lead to quantitative predictions for the relationship between the number of turns of handles A and B. The relationship deduced from model S (that the handles turn equally) can be represented by the formula in Eq. 1.1. We will call such relationships *mathematical models*; the formula in Eq. 1.1 is an algebraic way of describing the relationship, which we have also described in words, and which can be described by means of a graph (Figure 1.5).

A familiar example of a mathematical model, applicable to an automobile trip, is the relation of the distance traveled, time on the road, and speed of the car (Eq. 1.2). The

Equation 1.1

Mathematical model (algebraic form)
number of turns of handle A
N_A
number of turns of handle B
N_B

$$N_A = N_B$$

Equation 1.2

Mathematical model (algebraic form)
distance = s, speed = v, time = t

$$s = vt$$

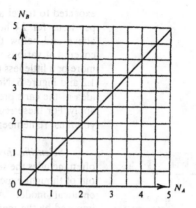

Figure 1.5. Mathematical model (graphical form). Number of turns of handle A, N_A; Number of turns of handle B, N_B.

distance is equal to the speed times the time. At 50 miles per hour, for example, the car covers 125 miles in $2\frac{1}{2}$ hours (Figure 1.6).

The physical quantities related by a mathematical model are called *variable factors* or *variables*, for short. The numbers of turns of handles A and B are two variable factors in Eq. 1.1 and Figure 1.5. The distance and elapsed time are two variable factors in Eq. 1.2 and Figure 6. The speed in this mathematical model is called a *constant*, because it does not vary. Under different conditions, as in heavy traffic, the speed might be a variable factor.

Like the working model for a system, the mathematical model for a relationship is not an exact reproduction of a real happening. No real car, for instance, should be

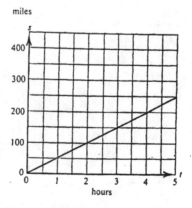

Figure 1.6. Mathematical model (graphical form). Distance = s, time = t, speed = 50 miles per hour.

expected to travel at the perfectly steady speed of 50 miles per hour for $2\frac{1}{2}$ hours. The actual speed would fluctuate above and below the 50-mile figure. The actual distances covered at various elapsed times, therefore, might be a little more or a little less than those predicted by the model in Eq. 1.2 and Figure 6. Nevertheless, the model gives a very good idea of the car's progress on its trip and it is very simple to apply. For these reasons the model is extremely useful, but you must remember its limitations.

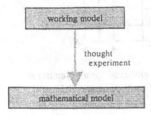

Scientific theories. The making of a physical theory often includes the selection of a working model, the carrying out of thought experiments, and the construction of a mathematical model. All physical theories have limitations imposed by the inadequacies of the working model and the conditions of the thought experiments. Occasionally a theory has to be abandoned because it ceases to be in satisfactory agreement with observations. Nevertheless, physical theories are extremely useful. It is probably the power of the theory-building process we have described that lies behind the rapid progress of science and technology in the last 150 years.

Chapter 3
The interaction concept

The interaction concept is being used more and more widely to explain social and scientific phenomena. At conferences, strong interaction may be evident among some participants, weak interaction among others. At the ocean shore, erosion is caused by the interaction of wind and water with rock. In the laboratory, magnets interact even when they are not touching.

A dictionary provides the following definitions.

interact *vi*: to act upon each other . . . ;
interaction *n*: action upon or influence on each other.

To say that objects interact, therefore, is to say that they have a relationship wherein they jointly produce an effect, which is the result of their action upon each other. In the examples cited above, anger may be the effect caused by strong (and irritating) interaction among the conference participants; crumbling and wearing away is the effect of the interaction of wind and water with rocks; and movement toward one another followed by sticking together is the effect of the interaction of the magnets.

3.1 EVIDENCE OF INTERACTION

We take the point of view that influence and interaction are abstractions that cannot be observed directly. What is observable are their effects. Congressional passage of unpopular legislation requested by the President would be an observable effect of the President's influence and therefore

R. Karplus, Chapter 3, The interaction concept, pp. 55–76, *Introductory Physics: A Model Approach.*
W. A. Benjamin, Inc. New York, © 1969.

would be called evidence of his influence. The change in direction of motion of a struck baseball is an observable effect of its interaction with the bat and therefore can be called *evidence of interaction*.

You may believe that you can sometimes observe the interaction itself, as when the bat hits the baseball or your typewriter prints a letter on a piece of paper. These examples, which include physical contact and easily recognized effects, seem different from those where magnets interact without contact or where erosion is so slow that the effects are imperceptible. This apparent difference, however, is an illusion. You *observe* only the close proximity of bat and ball, a sound, and the change of the ball's direction of motion, all of which are so closely correlated and so familiar that you instantly interpret them as evidence of interaction between bat and ball. The interaction of bat and ball is, however, merely the relationship whereby the observable effects are brought about, and relationships are abstractions which cannot be observed directly.

Indirect evidence of interaction. An example where the evidence is very indirect was described in Section 2.2. In his analysis of the films of President Kennedy's assassination, Alvarez interpreted a blurred photograph as evidence of interaction between the photographer and a rifle being fired. No one can question the blurring in the photograph, which is directly observable. But it is clear that not everyone may agree with Alvarez's interpretation that there is a relationship between the blur and a rifle shot.

Another example of interpretation of indirect evidence is the relationship that has been posited to exist between cigarette smoking and lung cancer. Lung cancer and cigarette smoking are separately observable and appear sufficiently correlated to warrant the conclusion that there is a pathological interaction between cigarette smoke and lung tissue. Yet most smokers do not take this interpretation of the evidence seriously enough to believe that they are slowly committing suicide.

Alternate interpretations. The critical problem in interpreting evidence of interaction is that any one of several different interactions might be responsible for the same observed effect. The typed letter in the example of the typewriter and the paper does not furnish conclusive evidence as to which typewriter made the letter, a question that sometimes arises in detective stories. A direct way to overcome this weakness is to find evidence that supports one hypothesis. For instance, the paper might be beside a typewriter, the ribbon on one typewriter might match the shade of the typed letter, or a defect in the machine's type might match one that appears in the typed letter. If so, the original identification of the typewriter is supported.

An indirect way to support one hypothesis is to eliminate alternatives. By checking many typewriters and finding how poorly they match the ribbon color and type impression, the detective may be able to eliminate them from further consideration. Supporting evidence for one alternative and/or evidence against other alternatives will enable you to establish one hypothesis conclusively or may only lead you to decide that one of them is more likely than the others. A procedure for finding such evidence by means of control experiments is described in Section 3.4.

When you suspect that there might be interaction, you should make a comparison between what you observe and what you would expect to observe in the absence of interaction. If there is a difference, you can interpret your observation as evidence of interaction and seek to identify the interacting objects; if there is no difference, you conclude that there was no interaction or that you have not observed carefully enough.

3.2 HISTORIC BACKGROUND

Mankind has not always interpreted observed changes or discrepancies as evidence of interaction. In ancient times, some philosophers took the view that changes were brought about by a fate or destiny that was inherent in every object. In primitive societies, events are frequently ascribed to supernatural forces. These forms of explanation and the modern interaction concept, however, do share a common feature: they are man's attempts to explain regular patterns in nature so as to enable him to anticipate the future and possibly to influence and control future events.

Cause and effect. When you observe two happenings closely correlated in space and time, you tend to associate them as cause and effect. Such a conclusion is reinforced if the correlation of the happenings persists in a regular pattern. The person who strikes a match and observes it bursting into flame infers that the striking caused the fire. The primitive man who performs a rain dance infers that the dance causes the ensuing rain. Even the laboratory pigeon that receives a pellet of grain when it pecks a yellow card becomes conditioned to peck that card when hungry. These individuals will repeat their actions—striking the match, dancing, or pecking the card—if they wish to bring about the same consequences again. After a sufficient number of successful experiences, all three will persist in their established behavior, even though some failures accompany their future efforts.

The interaction viewpoint. We may state the distinction between the modern scientific approach and the ways man formerly devised to explain happenings in the following way: the scientist ascribes happenings to interactions among two or more objects rather than to something internal to any one object. Thus, the falling of an apple is ascribed to its gravitational interaction with the earth and not to the heaviness inherent in the apple. The slowing down of a block sliding on a table is ascribed to friction between the block and the table and not to the power of the block to come to rest by itself. Fire is the manifestation of combustion, that is, the interaction of fuel and oxygen, and is not itself an element. The rain dance and the rain, however, cannot be put into this framework; therefore, this association is nowadays considered a superstition.

At any one time period, however, science cannot provide explanations for all possible happenings. When a new phenomenon is discovered, the interacting objects responsible for it must be identified, and this may be difficult. The origin of some of the recently discovered radiation reaching the earth, for example, is yet to be found. Cancer-producing agents and the structures in the body with which they interact to produce the disease also have yet to be identified.

3.3 SYSTEMS

The word "system" has entered our daily lives. Communication systems, electronic systems, and systems analysis are discussed in newspapers and magazines and on television. In all these discussions, and in this text as well, the word "system" refers to a whole made of parts.

The systems concept is applied whenever a whole, its parts, and their interrelationships must all be kept clearly in mind, as illustrated in the following two examples. Traffic safety studies take into account an entire driver-car system and do not confine themselves merely to the engineering of the car or the health of the driver. A physician realizes that the human heart, though a single organ, is really a complex system composed of muscles, chambers, valves, blood vessels, and so on. The system is physically or mentally separated from everything else so that the relations among the parts may be studied closely.

To simplify our terminology, we will often refer to the whole as "system" and to the parts as "objects." Thus, the car and the driver are the objects in the driver-car system, and the muscles, chambers, and so on are objects in the system called "the heart." By using the word "object" to refer to any piece of matter (animate or inanimate, solid, liquid, or gaseous), we are giving it a broader meaning than it has in everyday usage.

Sometimes one of the parts of a system is itself a system made of parts, such as the car (in the driver-car system), which has an engine, body, wheels, and so forth. In this case, we should call the part a *subsystem*, which is a system entirely included in another system.

In a way, everyone uses the systems concept informally, without giving it a name. Everyone focuses his attention temporarily on parts of his environment and ignores or neglects other parts because the totality of impressions reaching him at one moment is too complex and confusing to be grasped at once. The system may have a common name, such as 'atmosphere' or "solar system," or it may not, as in the example of the jet fuel and liquid oxygen that propel a rocket. The systems concept is particularly useful when the system does not have a common name, because then the group of objects under consideration acquires an individual identity and can be referred to as "the system including car and driver" or more briefly as "the driver-car system" once the parts have been designated.

Conservation of systems. Once a system is identified, it is useful to have criteria according to which the same system can be identified at later times and in spite of changes that have occurred. For the chemist, the conservation of matter is very important. One criterion for defining the identity of a system through changes in time is that no matter be added to or removed from the matter originally included in the system. When jet fuel burns, the fuel and oxygen become carbon dioxide and water. Therefore, the chemist thinks of the carbon dioxide and water as being the same system as the jet fuel and oxygen, even though the chemical composition and temperature have changed.

The psychotherapist and the economist do not use the same criterion as the chemist for following the identity of a system over time. The psychotherapist focuses his attention on a particular individual with a personality, intellectual aptitudes, and emotions. He, therefore, selects this individual as a system that is influenced by its interaction with other individuals and by its internal development. It retains its identity even though it exchanges matter with its environment (breathing, food consumption, waste elimination). For the economist, all the production, marketing, and consuming units in a certain region constitute an economic system that retains its identity even though persons may immigrate or emigrate and new materials and products may be shipped in or out.

The physicist studying macro-domain phenomena finds the matter-conserving system most useful. This is, therefore, the sense in which we will use the systems concept in this text. In the micro domain, however, the concepts of matter and energy have acquired new meanings during the last few decades, and other criteria for defining systems are used.

You can apply conservation of matter to the selection of systems in two ways. By watching closely, you can

"... in all the operations of art and nature, nothing is created; an equal quantity of matter exists both before and after the experiment, ... and nothing takes place beyond changes and modifications in the combinations of these elements. Upon this principle, the whole art of performing chemical experiments depends."
Antoine Lavoisier
Traité Elémentaire de Chimie, *1789*

Place two identical pieces of clean writing paper in front of you. Pick up one piece and call it System P. (1) Wrinkle the paper in your hand into a ball. Is what you now hold in your hand System P? (2) Is the paper lying on the table System P? (3) Tear the wrinkled paper in half, and hold both pieces. Is what you now hold in your hand System P? (4) Put down one of the two torn pieces. Is what you now hold in your hand System P?

determine whether you see the same system before and after an event. For instance, when a bottle of ginger ale is opened, some of the carbon dioxide gas escapes rapidly. The contents of the sealed bottle (we may call it System A), therefore, are not the same system as the contents of the opened bottle, which may be called System B. The escaping bubbles are evidence of the loss of material from the bottle.

In the second kind of application, you seek to keep track of the system even though its parts move from one location to another. Thus, after the bottle is opened, System A consists of System B plus the escaped gas; the latter, however, is now mixed with the room air and can be conveniently separated from the room air only in your mind. For this reason we stated at the beginning of this section that a system of objects need only be separated mentally from everything else; sometimes the physical separation is difficult or impossible to achieve, but that is immaterial for purposes of considering a system.

State of a system. To encompass the continuity of the matter in the system as well as the changes in form, it is valuable to distinguish the *identity* of the system from the *state* of the system. The identity refers to the material ingredients, while the state refers to the form or condition of all the material ingredients (Figure 3.1). Variable factors, such as the distance between objects in the system, its volume, its temperature, and the speeds of moving objects, are used to describe the state. In Chapter 4 we will relate matter and energy, which are of central concern to the physical scientist, to changes in the state of a system. There we will describe the ways in which a system may store energy and how energy may be transferred as changes occur in the state of a system. From an understanding of energy storage and transfer has come the extensive utilization of energy that is at the base of modern technology and current civilization.

Investigations of interacting objects. In their research work, physicists study systems of interacting objects in order to classify or measure as many properties of the

Initial state of system T   Final state of system T

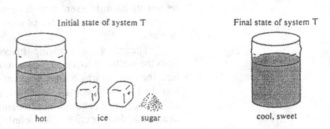

hot   ice   sugar   cool, sweet

Figure 3.1. Change in the state of a system.

interactions as they can. They try to determine which objects are capable of interacting in certain ways, and which are not (e.g., magnetic versus nonmagnetic materials). They try to determine the conditions under which interaction is possible (a very hot wire emits visible light but a cold wire does not). They try to determine the strength of interaction and how it is related to the condition and spatial arrangement of the objects (a spaceship close to the earth interacts more strongly with the earth than does one that is far away from the earth). Physicists try to explain all physical phenomena in terms of systems of interacting objects or interacting subsystems.

Working models for systems and the structure of matter. There are some happenings, however, such as the contraction of a stretched rubber band, that involve only a single object and appear to have no external causes. In such cases, the scientist makes a working model in which the object is made of discrete parts. A working model for the rubber band is made of parts called "rubber molecules." The properties of the entire system are then ascribed to the motion and the interaction of the parts. Some models are very successful in accounting for the observed behavior of the system and even suggest new possibilities that had not been known but that are eventually confirmed. Such a model may become generally accepted as reality: for instance, everyone now agrees that rubber bands are systems made of rubber molecules. Also, further model building may represent the rubber molecules as subsystems composed of parts called "atoms," and explain the behavior of the molecules in terms of the motion and interaction of atoms.

This kind of model building is called the search for the structure of matter—how ordinary matter in the macro domain is composed of interacting parts, and these parts in turn are composed of interacting parts, and so on into the micro domain. One of the frontiers of science is the search for ultimate constituents, if such exist. Since man will always find more questions to ask, it is unlikely that he will ever accept the concept of an "ultimate constituent."

3.4 COLLECTING EVIDENCE OF INTERACTION

Interactions are recognized by their effects, that is, by the difference between what is actually observed and what would have been observed in the absence of interaction. Such a difference is evidence of interaction. The systems concept is of great value here because it enables you to designate and set apart (at least mentally) the objects that are being compared as you look for a difference. One approach is to compare a system before an event (in its so-called final state) with the same system after an event (in its so-called initial state). For

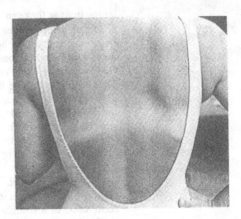

Figure 3.2. The skin shows evidence of interaction with the sun only where it was exposed to sunlight. It can be compared to the unexposed areas.

example, you compare a section of bare skin on the morning and the evening of a day at the beach (Figure 3.2). The section of skin is the system. In this experiment you assume on the basis of past experience that the skin color would not have changed in the absence of interaction. The observed change in skin color is therefore evidence of interaction with the sun.

As another example, take some sugar and let it dissolve in water in a glass beaker to form a solution. At the beginning of the experiment, the water-sugar system consists of dry crystals and colorless, tasteless water. At the end, there are no crystals and the liquid tastes sweet. The change in the state of this system is evidence of interaction between sugar and water.

Control experiment. Consider now an experiment in which you put yeast into a sugar solution in a glass and let this system stand in a warm place for several days. You will observe bubbles, an odor, and a new taste—that of ethyl alcohol. These changes can be interpreted as evidence of interaction within the water–sugar–yeast system. Can you narrow down the interacting objects more precisely or are all three parts necessary?

For comparison, suppose you can conduct experiments in which one ingredient is omitted. You dissolve sugar in water without yeast, you dissolve yeast in water without sugar, and you mix sugar and yeast. Each of these is called a *control experiment;* from their outcomes, you can answer the question above. By designing other control experiments, you can try to determine whether the glass container was

Figure 3.3. When you try to determine which electric circuit breaker supplies power to a particular light fixture, you turn on the switch of the fixture and then turn off the circuit breakers one at a time. In one of these "experiments" the bulb darkens, in the others it does not. Each turning off serves as a control experiment to be compared to the situation in which all circuits are turned on.

necessary, and whether the temperature of the environment made any difference.

By carrying out control experiments, you try to identify those objects in the system that interact and those whose presence is only incidental (Figure 3.3).

Inertia. One other important concept in the gathering of evidence of interaction is the concept of *inertia.* Inertia is the property of objects or systems to continue as they are in the absence of interaction, and to show a gradually increasing change with the elapse of time in the presence of interaction. For example, you expected the pale skin on the girl's back (Figure 3.2) to remain pale as long as it was not exposed to the sun. You expect a rocket to remain on the launching pad unless it is fired. You expect sugar crystals to retain their appearance if they are not heated, brought into contact with water, or subjected to other interactions. You expect an ice cube to take some time to melt even when it is put into a hot oven.

Your everyday experience has taught you a great deal about inertia of the objects and systems in your environment. When you compare the final state with the initial state of a system and interpret a difference as evidence of interaction, you are really using your commonsense background regarding the inertia of the system. You must be careful, however, because occasionally your commonsense background can be misleading.

Inertia of motion. The motion of bodies also exhibits inertia. Curiously, motion is one of the most difficult subjects to treat scientifically because of commonsense experience. When you see a block gliding slowly on an air track (Figure 3.4), you almost think it must contain a motor because you expect such slowly moving objects to come to rest after a very short time. As a matter of fact, the block is only exhibiting its inertia of motion because the frictional interaction with the supporting surface is very small. You must, therefore, extend your concept of inertia to cover objects in motion (such as the block), which tend to remain in motion and only gradually slow down if subject to a frictional interaction. You must also extend it to objects at rest (such as the rocket), which tend to remain at rest and only gradually acquire speed if subject to an interaction. Change in speed from one value to another—where the state of rest is considered to have "zero" speed—is therefore evidence of interaction. Galileo already identified inertia of motion even though he did not give it a name. Isaac Newton framed a theory for moving bodies in which he related their changes in speed and direction of motion to their interactions. The "laws of motion," as Newton's theory is called, will be described in Chapter 14.

One of the important concepts in the laws of motion is that of the *inertial mass*, a quantitative measure of the inertia of moving bodies. An operational definition of inertial mass, in contrast to that of mass (often called

> "... we may remark that any velocity once imparted to a moving body will be rigidly maintained as long as the external causes of acceleration or retardation are removed ..."
> Galileo Galilei
> Dialoghi della Nuove
> Science, 1638

OPERATIONAL DEFINITION
Inertial mass is measured by the number of standard units of mass required to give the same rate of oscillation of the inertial balance.

(a)   (b)

Figure 3.4. An air track. (a) Small holes in the track emit tiny jets of air. When a close-fitting metal piece passes over an opening, the air is trapped and forms a thin film over which the metal piece can slide with very little friction. (b) The closeness of fit can be seen in this end view.

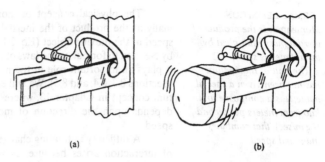

(a)   (b)

Figure 3.5. The inertial balance. (a) The inertial balance consists of an elastic steel strip, which oscillates back and forth after the free end is pulled to the side and released. (b) When objects are attached to the end of the strip, the oscillations take place more slowly. The inertial mass of a stone is equal to the number of standard objects required to give the same count of oscillations per minute. To measure the inertial mass of the stone, the stone is attached to the end of the steel strip and set into oscillation. The number of oscillations in 1 minute are counted. Then the stone is taken off, a number of standard objects are attached, and their number adjusted until the count of oscillations is equal to the count obtained with the stone.

gravitational mass, because gravity is decisive in the operation of the equal-arm balance) given in Section 1.5, makes use of a mechanical device called the *inertial balance* (Figure 3.5) to compare an object to the standard units of inertial mass. The operation of the inertial balance occurs in a horizontal plane, to eliminate the effects of gravity. The body attached to the end of the steel strip is repeatedly speeded up and slowed down by the oscillation of the strip. The inertia of the body, therefore, strongly influences the rate of oscillation of the strip: large inertia (resistance to change of speed) means slow oscillations, small inertia means rapid oscillations.

The generally accepted standard unit of inertial mass is the kilogram, represented by the same platinum-iridium cylinder as the unit of gravitational mass. Even though inertial and gravitational masses are measured in the same units, they are different concepts and have different operational definitions.

A second important concept in the laws of motion is the *momentum* of a moving body. The word is commonly applied to a moving object that is difficult to stop. A heavy trailer truck rolling down a long hill may, for instance, acquire so much momentum that it cannot be brought to a stop at an intersection at the bottom. By contrast, a bicyclist coasting down the same hill at the same speed has much less momentum because he is less massive than the truck.

Momentum is the product of inertial mass multiplied by instantaneous speed.

The unit of momentum has not been given a special name; it is a composite unit, kilogram meters per second (kg m/sec), that combines mass and speed.

Equation 3.1

momentum $\mathcal{M}$
speed v
inertial mass M_I

$$\mathcal{M} = M_I v$$

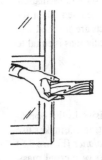

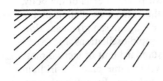

The physical concept of momentum is defined formally as the product of the inertial mass multiplied by the speed of the moving object (Eq. 3.1). This concept was used by Newton to formulate the laws of motion (Chapter 14) and it plays an important role in the modern models for atoms (Sections 8.3, 8.4, and 8.5). We will elaborate on the momentum concept in Chapter 13, where we will describe how it depends upon the direction of motion as well as on the speed.

A difficulty in treating changes of motion as evidence of interaction arises because, as we have pointed out in Chapter 2, motion must be defined relative to a reference frame. An object moving relative to one reference frame may be at rest relative to another. Evidence of interaction obtained from observation of moving objects, therefore, will depend on the reference frame. We will ordinarily use a reference frame attached to a massive body such as the earth (for terrestrial phenomena) or the sun (for the solar system).

Combined interaction. A block held in your hand does not show evidence of interaction (i.e., it remains at rest), yet it is clearly subject to interaction with the hand and with the earth. This is an example of what your conceptual framework is forced to describe as two interactions combining in such a way that they compensate for one another and give the net effect of no interaction. Situations such as this raise the question of the strength of interaction; how can you compare two interactions to determine whether they can compensate exactly or not, other than to observe their combined effect on the body? This is a question we will take up in Chapter 11.

Radiation. A situation that contains a different element of mystery is illustrated in Figure 3.6. Two mirrors are facing each other at a separation of several meters. There is no mechanical connection between them. At a central point near one mirror is a device called a detector, which is connected to a dial. If a lighted match is placed at a central point in front of the opposite mirror, you see a deflection on the dial (Figure 3.6a). After a little experimentation, you recognize that the placement of the match and the dial deflection are definitely correlated. This is the evidence of interaction between the match and the detector. If, now, the match is held in position and a cardboard shield is placed in various positions in the apparatus, the deflection falls to zero (Figure 3.6b, c, and d). Without anyone touching any of the visible objects used for the experiment, an effect was produced. The inference is that something was passing from the match to the detector by way of the mirrors, and that this passage was interrupted by the shield.

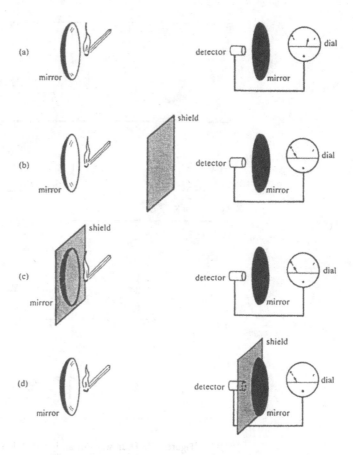

Figure 3.6. Four steps in the investigation of the interaction of a match flame with a detector show the effects of a shield placed in various locations.

We, therefore, construct a working model that is just like the experimental system but includes in addition an "object" that passes from the match to the first mirror, the second mirror, and the detector. The scientist calls this "model object" *radiation*. In terms of this model, he can describe the effect of the shield on the dial reading as evidence of interaction between the shield and the radiation, he can describe the path of the radiation, he can describe the match as a radiation source, and he can describe the detector as a radiation detector.

Another experiment, with a rocklike material X and a detector with a dial, is illustrated in Figure 3.7. From the evidence you may conclude that material X is not an ordinary

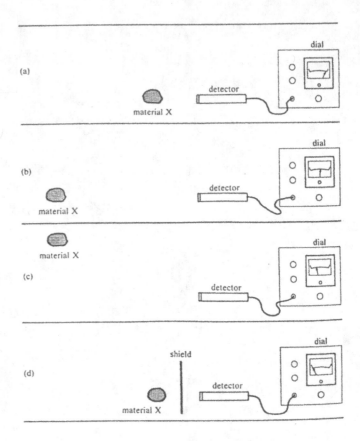

Figure 3.7. Four steps in an investigation of material X show
evidence of interaction between the material and the detector.

inert rock but is a source of radiation, and you make a
working model that includes an "object," again called radiation, that passes between material X and the detector. After
this discovery, you can study the spatial distribution of the
radiation by holding the detector in various directions and at
various distances from the rock, you can study the interaction of the radiation with various shields (cardboard, glass,
iron, and aluminum) placed to intercept it, and so on. From
this kind of investigation you become more familiar with the
radiation from material X and may, eventually, think of it as
a real object and not only as part of a model.

The discovery of evidence of interaction is a challenge
to identify the interacting objects and to learn more about
the interaction: the conditions under which it occurs, the kind
of objects that participate, the strength and speed with which

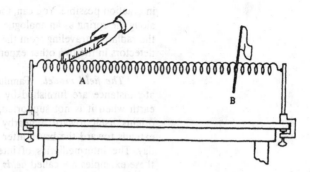

Figure 3.8. The ruler interacts with the flag by way of the long spring. Is this an example of interaction-at-a-distance?

the evidence appears, and so on. It can be the beginning of a scientific investigation.

3.5 INTERACTION-AT-A-DISTANCE

Consider now a common feature of the two experiments with radiation. In both cases, you observed evidence of interaction between objects that were not in physical contact. We speak of this condition as *interaction-at-a-distance* because of the distance separating the interacting objects. The human mind seems to resist this concept and, therefore, constructs working models that include radiation to make interaction possible between the two objects. The usefulness of these models is confirmed by the effects of the shields that intercept the radiation.

An experiment that significantly resembles the radiation experiments can be carried out with the system shown in Figure 3.8. A spring is supported at the ends by rigid rods. If a ruler strikes the spring at point A, you see a disturbance in the spring, which is one evidence of interaction, and then movement of the flag at B, another evidence of interaction. The first movement is evidence of interaction between the ruler and the spring. The second is evidence of interaction of the spring with the flag.

The experiment with the spring and the flag becomes another example of interaction-at-a-distance, however, if you choose to focus on the system including only the ruler and the small flag. The motion of the flag correlated with the motion of the ruler is evidence of interaction-at-a-distance between these two objects. Of course, in this experiment you can see the spring and a disturbance traveling from the ruler along the spring to the flag. You do not need to construct a working model with a "model object" to make the interaction possible. You can, therefore, use the disturbance along the spring as an analogue model to help you visualize the radiation traveling from the match or material X to the detectors in the two other experiments.

The field model. Familiar examples of interaction-at-a-distance are furnished by a block falling toward the earth when it is not supported, by a compass needle that orients itself toward a nearby magnet, and by hair that extends toward the brush after it is brushed on a very dry day. The intermediaries of interaction-at-a-distance in all these examples are called *fields*, with special names, such as *gravitational field* for the block–earth interaction, *magnetic field* for the compass needle–magnet interaction, and *electric field* for the brush-hair interaction. We may call this approach the *field model* for interaction-at-a-distance.

Radiation and fields. Do radiation and fields really exist, or are they merely "theoretical objects" in a working model? As we explained in Section 1.3, the answer to this question depends on how familiar you are with radiation and fields. Since radiation carries energy from a source to a detector, while the field does not accomplish anything so concrete, radiation may seem more real to you than fields. Sunlight, the radiation from the sun to green plants or to the unwary bather, is so well known and accepted that it has had a name for much longer than has interaction-at-a-distance. Nevertheless, as you become more familiar with the gravitational, magnetic, and electric fields, they also may become more real to you.

For the scientist, both radiation and fields are quite real. In fact, the two have become closely related through the field theory of radiation, in which the fields we have mentioned are used to explain the production, propagation, and absorption of radiation. More on this subject is included in Chapter 7.

Gravitational field. Two fields, the gravitational and the magnetic, are particularly familiar parts of our environment. At the surface of the earth the gravitational field is responsible for the falling of objects and for our own sense of up and down. The plumb line (Section 1.4) and the equal-arm balance (Section 1.5) function because of the gravitational interaction between the plumb bob or the weights and the earth. We, therefore, use a plumb line to define the direction of the gravitational field at any location. Because the earth is a sphere, the direction of the gravitational field varies from place to place as seen by an observer at some distance from the earth (Figure 3.9). More about the gravitational field will be described in Chapter 11.

"The physicist . . . accumulates experiences and fits and strings them together by artificial experiments . . . but we must meet the bold claim that this is nature with . . . a good-humored smile and some measure of doubt."
    Goethe
    Contemplations of Nature

OPERATIONAL DEFINITION
The direction of the gravitational field is the direction of a plumb line hanging freely and at rest.

Figure 3.9. The gravitational field near the earth is directed as indicated by plumb lines. The field appears to converge on the center of the earth.

gravitational field
directions

Magnetic field. The magnetic field is explored conveniently with the aid of a *magnetic compass*, which consists of a small magnetized needle or pointer that is free to rotate on a pivot (Fig. 3.10). When the compass is placed near the magnet, the needle swings back and forth and finally comes to rest in a certain direction. Because of its interaction with other magnets, the compass needle functions as detector of a magnetic field at the point in space where the compass is located. It is most commonly used to identify the direction of the magnetic field at the surface of the earth, which lies close to the geographic north–south direction (Figure 17). Since the compass needle has two ends, we must decide which end indicates the direction of the magnetic field. The accepted direction of the magnetic field is that of the geographic north-seeking end of the needle (henceforth called the "direction of the needle"), as shown by the arrows in Figure 3.11.

Strength of the magnetic field. When you place a compass near a magnet, you notice that the needle swings back and forth rather slowly if it is far from the magnet and

(a) (b) (c)

Figure 3.10. Examples of magnetic compasses. (a) The compass needle is often enclosed in a case for better protection. (b) The pivot may permit the needle to rotate in a horizontal plane. (c) The pivot may permit the needle to rotate in a vertical plane.

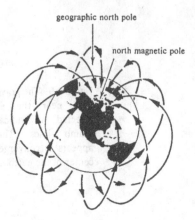

geographic north pole

north magnetic pole

Figure 3.11. The magnetic field of the earth, represented by the arrows, lies close to the geographic north-south direction, but does not coincide with it. In the magnetic dipole model for the earth, the two magnetic poles lie near the center of the earth on a line through northern Canada and the part of Antarctica nearest Australia.

quite rapidly if it is close to the magnet. You can use this observation as a rough measure of interaction strength or magnetic field strength: rapid oscillations are associated with a strong field, slow oscillations with a weak field. You thereby discover that the magnetic field surrounding a magnet has a strength that differs from point to point; the field strength at any one point depends on the position of the point relative to the magnet.

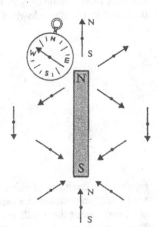

Figure 3.12. The arrows represent the compass needles that indicate the magnetic field near the bar magnet.

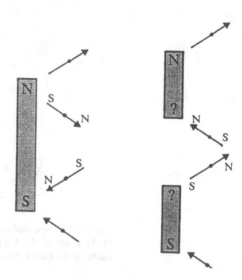

Figure 3.13. A bar magnet is cut in half in an effort to separate the north pole from the south pole. Arrows represent compass needles. Each broken part still exhibits two poles, the original pole and a new one of the opposite kind.

William Gilbert (1544–1603) an Elizabethan physician and scientist, wrote the first modern treatise on magnetism, De Magnete. *Gilbert worked with natural magnets (loadstones). In one chapter of this work, Gilbert introduced the term* electric *(from the Greek* electron *for amber).*

"... thus do we find two natural poles of excelling importance even in our terrestrial globe... In like manner the loadstone has from nature its two poles, a northern and a southern... whether its shape is due to design or to chance... whether it be rough, broken-off, or unpolished: the loadstone ever has and ever shows its poles."
            William Gilbert
            De Magnete, 1600

Magnetic pole model. When you explore the magnetic field near a bar magnet, you find that there are two regions or places near the ends of the magnet where the magnetic field appears to originate. This common observation has led to the *magnetic pole* model for magnets already used by William Gilbert. In this model, a magnetic pole is a region where the magnetic field appears to originate. The magnetic field is directed away from north poles and toward south poles according to the accepted convention (Figure 18). All magnets have at least one north pole and one south pole. Opposite poles of two magnets attract one another, like poles repel. If you apply these findings to the compass needle itself, you conclude that the north-seeking end of the needle contains a north pole (it is attracted to a magnetic south pole, Figure 3.12).

An obvious question now suggests itself: can a magnetic pole be isolated? So far, physicists have failed in all their attempts to isolate magnetic poles (Figure 3.13), in that they have not been able to narrow down the regions inside magnets where the magnetic field originates. They have found instead that the magnetic field appears to continue along lines that have no beginning or end but loop back upon themselves (Figure 3.14). Thus, magnetic poles appear to be useful in a working model for magnets when the magnetic

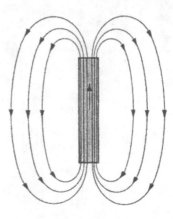

Figure 3.14. So-called magnetic field lines indicate the direction of the magnetic field. Lines inside the magnet close the loop made by the lines outside the magnet.

field outside magnets is described, but they fail to account for the field inside magnets.

Display of the magnetic field.   Another technique for exploring a magnetic field is to sprinkle iron filings in it (Figure 3.15). The filings become small magnets and, like the compass needle, tend to arrange themselves along the direction of the magnetic field. They produce a more visual picture of the magnetic field. This method is less sensitive than the compass, because the filings are not so free to pivot.

Electromagnetism.   Not quite 150 years ago, Hans Christian Oersted, while preparing for a lecture to his students, accidentally found evidence of interaction between a compass needle and a metal wire connected to a battery. Such a wire carries an electric current (see Chapter 12). One of the most startling properties of the interaction was the tendency of the compass to orient itself at right angles to the wire carrying the electric current. Oersted's discovery is the basis of the *electromagnet*, a magnet consisting of a current-carrying coil of wire which creates a magnetic field. The distributions of iron filings near current-carrying wires are shown in Figure 3.16.

Electric field.   Somewhat less familiar than gravitational or magnetic fields is the electric field, which is the intermediary in the interaction of the brush and the brushed hair. Electric fields also are intermediaries in the interaction of thunderclouds that lead to lightning, the interaction of phonograph records with dust, and the interaction of wool

Hans Christian Oersted (1777–1851) inherited his experimental acumen from his father, an apothecary. His famous discovery of the magnetic field accompanying an electric current took place as he was preparing for a lecture demonstration for his students. It initiated the fusion of two hitherto separate disciplines into the branch of physics called electromagnetism.

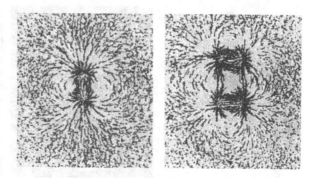

Figure 3.15. Iron filings were sprinkled over a piece of paper that concealed one or two bar magnets.

skirts with nylon stockings. Objects that are capable of this kind of interaction are called *electrically charged*.

Electrically charged objects. Objects may be charged electrically by rubbing. Many modern plastic materials, especially vinyl (phonograph records), acetate sheets, and spun plastics (nylon, dacron, orlon in fabrics) can be charged very easily. Electric fields originate in electrically charged objects and are intermediaries in their interaction with one another.

Display of the electric field. Individual grass seeds, which are long and slender in shape, like iron filings, orient themselves when they are placed near charged objects (Figure 3.17). Their ends point toward the charged objects. The patterns formed by the seeds re very similar to the iron filing patterns in a magnetic field. Using the more familiar

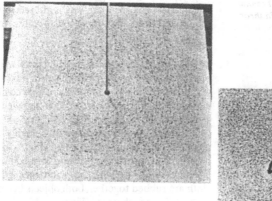

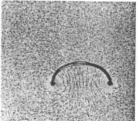

Figure 3.16. Iron filings near current-carrying wires clearly show the closed loops of the magnetic field lines.

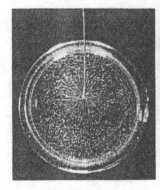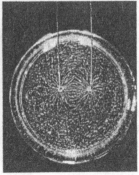

Figure 3.17. Grass seeds suspended in a viscous liquid indicate
the direction of the electric field near charged objects.

*Benjamin Franklin
(1706–1790) Born in Boston,
Massachusetts, the son of an
impoverished candle maker,
Ben was apprenticed to an
older brother in the printing
trade. When his apprenticeship
was terminated, he left for
Philadelphia where he
supported himself as a printer,
and eventually earned the
fortune that freed him for
public service. His political
career is, of course, celebrated,
but it is less widely known that
he was one of the foremost
scientists of his time. Franklin
proposed the one-fluid theory
of electricity, and introduced
the terms "positive electricity"
and "negative electricity."*

magnetic field as an analogue model for the electric field, we
define the direction of the electric field to be the direction of
the grass seeds.

Early experiments with electrically charged objects.
Benjamin Franklin and earlier workers conducted many
experiments with electrically charged objects. Gilbert had
already found that almost all materials would interact with
charged objects. One important puzzle was the ability of
charged objects to attract some charged objects (light seeds,
dry leaves, etc.) but to repel certain others. It was found, for
instance, that two rods of ebonite (a form of black hard
rubber used for combs and buttons) rubbed with fur repelled
one another. The same was true of two glass rods rubbed with
silk. But the glass and ebonite rods attracted one another
(Figure 3.18). Since a glass rod interacted differently with a
second glass rod from the way it interacted with an ebonite
rod, it followed that glass and ebonite must have been
charged differently.

Two-fluid model for electric charge. Two working
models for electrically charged objects were proposed. In
one, there were two kinds of electric fluids or "charges" (a
word for electrical matter) that could be combined with ordinary matter. Charges of one kind repelled charges of the
same kind and attracted charges of the other kind.

Franklin's experiment. In a highly original experiment, Franklin found that the two kinds of charges could not
be produced separately, but were formed in association with
one another. Thus, when an uncharged glass rod and silk
cloth are rubbed together, both objects become charged, but
with different charges (Figure 3.19). When the silk is
wrapped around the rod, however, the rod–silk system does
not produce any electric field, even though the two objects
separately do. Franklin therefore concluded that the two

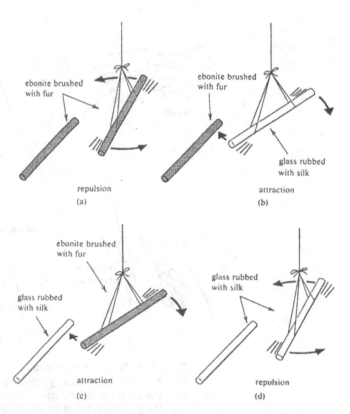

Figure 3.18. Hard rubber rods brushed with fur and glass rods rubbed with silk are permitted to interact. One rod is suspended by a silk thread and is free to rotate. The other rod is brought near until movement gives evidence of interaction.

kinds of charges were opposites, in that they could neutralize each other. Accordingly, he called them "positive" and "negative," the former on the glass rod, the latter on the silk (and on the ebonite). Combined in equal amounts in one object, positive and negative charges add to zero charge (an object with zero charge is uncharged or electrically neutral).

One-fluid model for electric charge. Franklin's model, to explain this observation, provided for only a single "electric fluid." Uncharged objects have a certain amount of this fluid. Positively charged objects have an excess of the fluid, whereas negatively charged objects have a deficiency. When two uncharged objects are charged by being rubbed together, fluid passes from the one (which becomes negative) to the other (which becomes positive). The fluid is conserved (neither created nor destroyed), so the two objects *together*

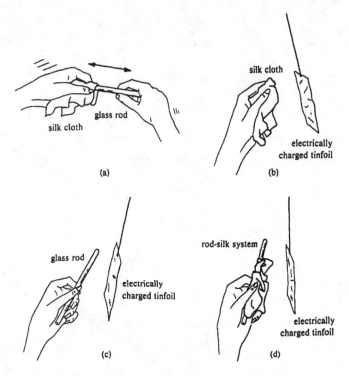

Figure 3.19. Franklin's experiment.
(a) A glass rod is charged by rubbing a piece of silk.
(b) The silk is tested for electric charge by interaction.
(c) The glass rod is tested for electric charge.
(d) The rod–silk system is tested for electric charge.

have just as much fluid as at the beginning of the experiment; hence they form an electrically neutral system. Clearly the isolation of one or two electric fluids would be an exciting success of these models. We will discuss the subject further in Chapters 8 and 12.

SUMMARY

Pieces of matter (objects) that influence or act upon one another are said to interact. The changes that occur in their form, temperature, arrangement, and so on, as a result of the influence or action are evidence of interaction. For the study of interaction, pieces of matter are mentally grouped into systems to help the investigator focus his attention on their identity. As he gathers evidence of interaction, the investigator compares the changes he observes with what would have happened in the absence of interaction. Sometimes he may carry out control experiments to discover this; at other times he may draw on his experience or he may make assumptions.

Pieces of matter that interact without physical contact are interacting-at-a-distance. Radiation and fields have been introduced as working model intermediaries for interaction-at-a-distance. The gravitational field, the magnetic field, and the electric field are the fields important in the macro domain. All three fields have associated with them a direction in space.

Robert Karplus
1970

CHAPTER 10

The First Career of Robert Karplus—Theoretical Physics

Robert G. Fuller*

Robert Karplus was a brilliant physicist. He showed his tremendous intellectual ability at an early age. He and his younger brother, Martin, were both to have extremely successful scientific careers.

Robert was born in 1927 into an intellectual and successful secular Jewish family in Vienna, Austria. The Karplus family had been in Vienna for many years. They were prominent citizens. Many of Robert's uncles had been in medical practice in Vienna. There is today a Karplus Street in Vienna.

One can only imagine the stress on such families in Austria around 1930. In retrospect it seems that the Karplus parents were preparing their sons for lives outside Austria. The boys began the study of English at a very early age. They were tutored weekly by their aunt, who had lived for a short time in England. Bob was extremely quick and progressed rapidly in his study of English.

At the age of nine, Bob had an accident that significantly shaped the rest of his life. He fell out of a tree, landed on his head, and was knocked unconscious. He remained in a coma for several days. When he regained consciousness, he had lost the hearing in one ear and his sense of balance was impaired.

*University of Nebraska-Lincoln.

A Love of Discovery: Science Education—The Second Career of Robert Karplus,
Edited by Robert G. Fuller, Kluwer Academic / Plenum Publishers, New York, 2002.

(As a result, he was classified 4F and not eligible for U.S. military service in World War II.) His behavior changed from one that bordered on hyperactivity to one that was more cerebral. The kind of reckless energy that he had previously put into physical activities went into mental activities.

Two days after the Nazis conquered Vienna in March of 1938, Bob's mother took him and Martin on a skiing holiday to Switzerland. They returned to Austria only many years after the end of World War II.

As Martin was to recall: "I think one of the major things that my mother had the responsibility for was the trip to Switzerland. I remember the whole time from leaving Austria until getting to the U.S. as kind of pleasure trip; it was always a very pleasant time—we went to the beach in the summer time. In retrospect it must have been very hard times, but somehow my mother managed. My father was in jail in Vienna at the time. I can't explain in detail what they did. I remember our having a great time. We went to school there and we started learning Swiss Deutsche. We were quite fluent in it at the time. We used to speak it with each other so that our mother would not know what we were saying. It was just a nice experience." (Endnote 1)

Their father was able to arrange to be released from jail in Vienna and was able to join them just in time to sail for the U.S. Bob's uncle, Eduard, had come to the U.S. to study at MIT. He had married an American woman and he was working at Raytheon. He had posted a $5000 bond for the release of Bob's father from jail in Vienna. After the Karplus family reached New York City in 1939 they went to Boston to join Bob's uncle. After a short time they moved to Newton, Massachusetts, because they believed that the public schools in Newton were better than those in New York City. Bob was graduated from Newton High School in 1943 and was number one in his class. He became a U.S. citizen in 1944.

Bob enrolled at Harvard and began to pursue an academic major in chemical physics. During the war years, Harvard offered a very condensed curriculum, which was well suited to Bob's exceptional intellectual skills. He earned his Bachelor's and Master's degrees by 1946 and completed his Ph.D. in 1948 when he was just 21 years old! His research in chemical physics, was directed by E. Bright Wilson, Jr. His thesis combined experimental work with theoretical work directed by Julian Schwinger, who was later to win a Nobel prize in Physics. Professor Wilson was to remark that Robert Karplus was the best Ph.D. student that he had ever directed.

Bob enjoyed dancing, especially folk dancing. It was at a folk dancing event at Harvard that he met Elizabeth Frazier, a young woman who had a bachelor's degree in Physics from Oberlin and was enrolled in Radcliffe College's graduate program in teaching.

After completing his Ph.D. in Chemical Physics in 1948, Robert Karplus went to the Institute for Advanced Study in Princeton, NJ, where he learned quantum electrodynamics (QED). The field of QED was very hot in theoretical physics in the late 1940's. Richard P. Feynman and Julian Schwinger were later awarded Nobel prizes for their work on QED. Freeman Dyson, who was also at the Institute for Advanced Study, had proposed a systematic scheme whereby the results of QED theory could be calculated. Two bright young theorists at the Institute, Robert Karplus and Norman Kroll, decided to use Dyson's suggestions and QED theory to calculate an actual physical observable, some property of a real system where QED would give a result that would be measurably different from the results of classical physics. They chose the fourth-order corrections in QED to the magnetic moment of the electron. One can only imagine the excitement that drove them. Here would be the first detailed calculation, based upon this new theory in physics, that would provide a test of the validity of that theory and its usefulness in predicting, in detail, the behavior of a physical system. Karplus and Kroll worked for about a year, carrying out very difficult mathematical calculations by hand. Usually they calculated separately and only compared their work when they had obtained an intermediate result. (Remember that this was around 1949, long before computers were available.) Their collaboration resulted in publication in the 1950 Physics Review of the famous Karplus and Kroll paper (Karplus and Kroll, 1950). This paper was a major success for QED. It offered theoretical physicists a technique that brought the results of QED into the laboratory for verification. Their results were in agreement with the then accepted experimental results. It was a triumph, not only for Feynman, Schwinger and Dyson, but also for Karplus and Kroll. In addition to his famous paper with Kroll, Karplus published in the early 50s other papers that have stood the test of time, such as his paper with Neuman on the scattering of light by light (Karplus and Neuman, 1951) and his papers on positronium with Klein (Karplus and Klein, 1952).

His post-doctoral colleagues at the Institute reported that Bob was a very warm, outgoing, and friendly person. He was very joyful. He had a quick wit and a good sense of humor and he enjoyed making word-game type jokes. This latter attribute, it was reported, drove Professor Schwinger to distraction. Bob's friendships knew no boundaries of class or rank; again and again his post-doctoral Ph.D. students remarked on his open and friendly way of treating them.

In the Summer of 1950, Robert Karplus returned to Harvard as an Assistant Professor and continued to do research in theoretical physics as well as teach upper level physics courses. He began to attract graduate students and broaden his theoretical interests. The broad expertise of Bob Karplus, in both theory and experiment, lead to a variety of "superhero" stories that may still survive at Harvard.

Since Harvard was unlikely to give tenure to one of its own graduates, Karplus began to look for a position elsewhere. In those days, Harvard gave a half-year of sabbatical leave to faculty who had been in residence for three and a half years. In March, 1953, Dr. Edward Teller from the University of California, Berkeley visited Harvard and had a long interview with Karplus. He was quite impressed. The Physics Department Chairman at Berkeley wrote to the Dean of the College of Letters and Science in May of 1953 saying, "Dr. Karplus was one of the most promising of the young theoretical physicists of the country. We definitely hope to add him permanently to our staff." (Endnote 2) Karplus was offered and accepted a position as Visiting Lecturer with rank equivalent to Associate Professor. In the summer of 1953, Karplus went to the University of California, Berkeley, for the summer and fall. There he was a fantastic success! The senior theoretical physicists at Berkeley began in October of 1953 to urge the leadership of Berkeley to offer Karplus a permanent faculty appointment. They wrote such things as "I am most enthusiastic about Karplus as a theoretical physicist, as a teacher and as a man ... He has done something to the atmosphere of the place that makes it very reminiscent of the pre-war days when Robert Oppenheimer was at his best." (Endnote 3) And "It gives me great pleasure to write in support of the prospective appointment of Robert Karplus to be Associate Professor of Physics at Berkeley. I believe firmly that Karplus is the best all-around theoretical physicist in the country at the rank proposed." (Endnote 4) To make an even stronger impression on the Dean of the College, the Department had solicited comments about Karplus from internationally famous physicists Hans Bethe and Robert Oppenheimer. They also wrote extraordinary comments about Robert Karplus. "In my opinion, Dr. Karplus is one of the best, possibly the best, theoretical physicist of his generation. (Endnote 5) "He was a member of the Institute; and I got to know him quite well as a physicist, for his work was of great interest to me, and reasonably well as a man. On both counts I rate him very highly ... I now write this to you in confidence that this appointment will be a source of pride and pleasure to you and to your colleagues." (Endnote 6) With such glowing letters of support Berkeley could hardly do otherwise than to offer Karplus a tenure appointment. Robert Karplus went to Berkeley as Associate Professor of Physics in Summer 1954, when he was only 27 years old.

Robert Karplus did not disappoint the Berkeley Physics Department He began to attract students to his work and he published with a variety of collaborators on a wide range of theoretical topics. He worked with J. M. Luttinger and published a paper on the anomalous Hall effect in ferromagnetic materials. (Karplus and Luttinger, 1954) Karplus and M. A. Ruderman published the first applications of dispersion relations to particles other than protons (Karplus and Ruderman, 1955). With D. A. Hamlin, Karplus wrote a paper that is still widely referenced on the drift of particles in the earth's magnetic field. (Hamlin, Karplus, Vik, and Watson, 1961)

In 1958, the department was ready to urge his promotion to full Professor. As Karplus would be only 31 years old, the department compared him with other U.S. theoretical physicists, about his age, who were already full Professors: T. D. Lee, age 31, at Columbia {Eds. note—a 1957 Nobel prize winner in Physics}; M. Gell-Mann, age 28, at Cal. Tech. {Eds. note—a 1969 Nobel prize winner in Physics}; C. N. Yang, age 35, at the Institute for Advanced Study {Eds. note—a 1957 Nobel prize winner in Physics}; and F. Low, age 36, at M.I.T. By this time, Bob had made other contributions to the Physics Department beyond his research. He had become respected as a good teacher:

> He is unusually interested in teaching and I have heard numerous favorable accounts of his classroom work. I have sat in several of his classes... and I thought his approach was clear, challenging and quite original. He also has a strong experimental and chemical background for a theoretical physicist, thus giving a satisfying breadth to his teaching. (Endnote 7)

Karplus had also worked to build up the theoretical physics group at Berkeley.

> I should particularly emphasize the contributions Dr. Karplus has made to building up our Physics Department. His influence was most important in securing for the University the services of such outstanding men as Professors Geoffrey Chew and Kenneth Watson. Due greatly to his initiative we can now boast of a theoretical physics staff which is second to no other one in the United States." (Endnote 8)

Thus it was in the summer of 1958, that Robert Karplus became a full Professor of Physics at the University of California, Berkeley, just ten years after completing his Ph.D. at Harvard.

The career of Robert Karplus as a theoretical physicist was exceptional. He published 49 refereed papers in physics from 1946 to 1967, more than two per year. He was the senior author of 32 of them. His coauthors numbered 32 physicists, more than 90 percent of them went on to become Fellows of the American Physical Society and two were Nobel prizewinners in Physics. Sixteen of the articles of which he was the senior author have received more than 1750 citations by other physics authors! He was brilliant, prolific, outgoing, energetic and well liked and admired by his colleagues.

On the personal side of his life, he was also prolific. Robert Karplus and Elizabeth Frazier were married in December 1948, and lived in an apartment near the Institute for Advanced Study. They started their family in 1950 with the birth of their daughter, Beverly. They moved back to Watertown, MA, in the Summer of 1950. Their second daughter, Margaret, was born in Cambridge, MA, in 1952. They moved again in 1953 to a rental apartment in Berkeley while Bob was a visiting Lecturer at UCB. Their third child, Richard, was born in Berkeley, CA, in 1953. After Bob had returned from Harvard to Berkeley as an Associate Professor, they bought property in Orinda, a suburb just east of Berkeley and built a house there. They added four more children to their family, all born in Oakland, CA: Barbara in 1955, Andrew in 1957, David in 1960 and Peter in 1962.

Perhaps, being the father of seven children may explain the change that took place in the career of Robert Karplus. The transition to his second career occurred between 1957 and 1962 and it was as exceptional as his first career had been.

ENDNOTES

1. Martin Karplus, telephone interview, March, 1999.
2. R. T. Birge, letter to Dean Davis, May, 1953.
3. L. W. Alvarez, letter to Chairman Birge, December, 1953.
4. C. Kittel, letter to Chairman Birge, December, 1953.
5. H. A. Bethe, letter to Chairman Birge, December, 1953.
6. Robert Oppenheimer, letter to Chairman Birge, January, 1954.
7. C. Kittel, letter to Chairman Helmholtz, October, 1957.
8. Edward Teller, letter to Chairman Helmholtz, October, 1957.

REFERENCES

Hamlin, D. A., Karplus, R., Vik, R. E., and Watson, K. M., 1961, Mirror and azimuthal drift frequencies for geomagnetically trapped particles, *J. Geophys. Res. (USA)*, **66**(1), 1–4.

Karplus, R., and Klein, A., 1952, Electrodynamic displacement of atomic energy levels I. Hyperfine structure, *Phys. Rev.*, **85**, 972–984. Electrodynamic corrections to the fine structure of positronium, Letter in *Phys. Rev.*, **86**, 257. Electrodynamic displacement of atomic energy levels III. The hyperfine structure of positronium, *Phys. Rev.*, **87**, 848–858.

Karplus, R., and Kroll, N. M., 1950, Fourth-Order Corrections in Quantum Electrodynamics and the Magnetic Moment of the Electron, *Phys. Rev.*, **77**, 536–549.

Karplus, R., and Luttinger, J. M., 1954, Hall effect in ferromagnetics, *Phys. Rev.*, **95**, 1154–1160.

Karplus, R., and Neuman, M., 1951, The Scattering of Light by Light, *Phys. Rev.*, **83**, 776–784.

Karplus, R., and Ruderman, M. A., 1955, Applications of Causality to Scattering, *Phys. Rev.*, **98**, 771.

CHAPTER 11

The Transition Years
What a Shame, What a Waste—To Give All That Up to Work with Children and Teachers

Robert G. Fuller*

A co-worker of Robert Karplus completed her work at Berkeley and went to another university. The news that she had worked with Karplus preceded her. When she arrived at her new campus, she was told that she should have a conversation about Karplus with a famous physicist on that campus. So she did. The famous physicist happened to be a fairly taciturn person with little interest in education.

He said, "Well, you worked with Karplus, I understand." "Yes," was the reply. "What a shame. What a waste." he said. "What do you mean?" she asked. "He was a very gifted theoretical physicist and to give up all that to work with children and teachers."

By 1958 Robert Karplus was held in high regard for his work in theoretical physics. He was already a full Professor of theoretical physics in a department that thought of itself as second to none. What happened for him to change his career?

*University of Nebraska-Lincoln.

A Love of Discovery: Science Education—The Second Career of Robert Karplus,
Edited by Robert G. Fuller, Kluwer Academic / Plenum Publishers, New York, 2002.

Personal motives are always mixed. None of us does anything for a single reason. People who worked with Karplus offered a variety of explanations for his career change. Almost none of the physicists read his papers in education and nearly all of his education co-workers would have found the content of his physics papers beyond their comprehension. His change in careers surely had many facets. Here I offer a brief exploration of some of them.

In the late 1950s the excitement of theoretical physics may have started to wane for Robert Karplus. He had been doing theoretical physics for a decade. There is some indication that he was feeling a sense of malaise about his career in theoretical physics.

When asked by one of his co-workers what caused him to change from theoretical physics? He said, "Well, it was, I suppose, the loneliness of it. I would go to work and close the door and it would take me an hour or longer to get into the intellectual space before I could dig into the work on a theoretical piece; and I would ponder and visualize and ponder it for four or five hours. Then I was exhausted and I would move out of that space and close down for the day. That, and the fact that there were only four or five people in the world who understood what I was trying to do, meant that I had very little communication and it was a fairly lonely enterprise."

While Karplus was brilliant, he was not number one in theoretical physics. Others had intellectual gifts that exceeded his. He had always been number one in his schools and in his family. Did he look around for another place to be number one?

Was it the challenge of doing science education that attracted Robert Karplus? As Richard Hake recently posited, "Science education is not rocket science—it is much harder." Was it the incredible challenge of trying to teach science to children in a time of national need that attracted his attention? Educating children may not have the glamour and prestige of theoretical physics, but it certainly offers intellectual challenges that are at least as great as those facing physicists.

Robert Karplus was already the father of five children by the time he became a full Professor of Physics and his first-born child, Beverly, was eight years old. One of the roles as father that he took extremely seriously was to be the intellectual mentor of his children. In those days in Orinda, teachers let the parents of their pupils enter the classroom and Beverly volunteered her dad. He came in and gave a little physics lecture and it was a disaster! One can almost see the angst on his face. He just felt awful about it. He was painfully aware that he did not connect with any of the children. He was really shaken by that. He wanted for them to share in these ideas. It was very personal for him. He really cared about those kids. He was passionately interested in science. He was so incredibly intense and caring.

Why did Robert Karplus change careers? Two physicist co-workers who now live thousands of miles apart, tell the following story:

Robert Karplus placed the toy truck in front of a child.

He rolled the truck slowly across the desk.

"Did the truck move?" he asked.

"No." replied the child.

(It is difficult to learn the fundamental concepts of motion when an object that goes from one location to another does not move. Perhaps he had misunderstood. He moved the truck back to its starting position. Again, he slowly rolled the toy truck across the desk to a new location.)

"Did the truck move?" he asked again.

"No." the child replied once again.

"Can you explain to me why you say the truck did not move?" Karplus asked.

"It did not move;" responded the child triumphantly. "You moved it!"

In that moment of puzzlement Robert Karplus was hooked. The physics that he knew and loved had not prepared him for such an experience. He discovered the importance of one's mental state in the shaping of learning and reasoning.

He began his journey of discovery through the works of Whorf, Vygotsky and Piaget. He had been captured by a challenge worthy of his brilliant mind. The path from outstanding theoretical physicist to elementary science had been started. Robert Karplus never looked back.

The Robert Karplus clan, circa, summer 1980

Left to right: Front row: Elizabeth Karplus, Nathan Hellweg held by Sue Karplus (pregnant with Valerie), Robert Karplus, Karen Karplus, Peter Karplus, Beverly Karplus Hartline holding infant Jeffrey Harltine. Back row: Richard Karplus with Ben Hellweg on shoulders, Margaret Karplus Hellweg holding infant Miriam Hellweg, Barbara Karplus, Andy Karplus, David Karplus, Jason Hartline on shoulders of Fred Hartline. The grandchildren at that time were: Ben Hellweg (b. Dec 1974), Jason Hartline (b. May 1976), Nathan Hellweg (b. Nov 1976), Miriam Hellweg (b. Dec 1979) and Jeff Hartline (b. April 1980).

The only missing spouse was Christian Hellweg.

CHAPTER 12

The Second Career—
Science Education

Robert G. Fuller*

By 1963, Robert Karplus had turned his professional interests to the Science Curriculum Improvement Study (SCIS). Together with Herb Thier, he worked almost exclusively on elementary school science materials for the rest of the 1960's. He did maintain his professional activity in physics as a member of the physics faculty of the University of California, Berkeley, but his last productive year in theoretical physics was 1962 and he co-authored just two theoretical physics papers after that. From 1963 until his cardiac arrest in 1982, Karplus worked exclusively in science education.

In this book, I have tried to capture the essence of Robert Karplus' enduring contributions to science education. This has been done by assembling reprints of some of his papers around common themes. In this final chapter, I want to highlight again his significant work, much of which serves as the basis for the national conversations about science education in the U.S. today—many years after he completed his work.

As Robert Karplus began his work in science education, he found the work of Jean Piaget well suited to his own understanding of scientific reasoning. As a result he took Piaget's work seriously. Karplus replicated the results

*University of Nebraska-Lincoln.

A Love of Discovery: Science Education—The Second Career of Robert Karplus,
Edited by Robert G. Fuller, Kluwer Academic / Plenum Publishers, New York, 2002.

of Piaget's semi-clinical interviews using American children and later he conducted an international study. His admiration for the work of Piaget encouraged him to concentrate part of his attention on student reasoning. The reader will find these aspects of his work detailed in his "Piaget in a Nutshell" reprint in Chapter 8, the "Can Physics Develop Reasoning?" paper in Chapter 7, and all of the reasoning beyond elementary school reprints in Chapter 6.

Robert Karplus realized the importance of converting the SCIS elementary science materials into a systematic teaching process that would enable teachers to successfully use these materials while enabling students to learn and enjoy science. His early attempts at this are discussed in Chapter 5. He, along with others, developed the learning cycle instructional strategy. The learning cycle is the topic of Chapter 4.

Physicist Karplus brought to his curriculum development work the same approach that he had used in doing physics. He, and his co-workers, developed a scientific process of curriculum development. They followed a feedback loop model of *develop-field test-revise*. Only the very best lessons survived. This process is emphasized in the reprint in Chapter 2 and some specific lessons are discussed in Chapter 3.

Very early in their work together Robert Karplus and Herb Thier recognized the necessity for the SCIS teachers to enjoy science. After the curriculum had been completed, Karplus continued to emphasize teacher development. He, along with others, created a series of teacher workshops on the topic of science teaching and the development of reasoning. These workshops offered teachers a first hand experience of the results of the work of Piaget and Karplus and how these results were related to what happens in the science classroom. Furthermore, Karplus and Peterson made a film of students performing various reasoning tasks, e.g. Formal Reasoning Patterns, to help teachers realize the importance of thinking about student reasoning. The teacher development aspects of Karplus' work are discussed in Chapter 8.

Robert Karplus never thought of school science as an elite subject for special students. In his very first papers on science education he emphasized science for everyone. Now "Science for All" is a part of the national science goals for the U.S. Karplus and his co-workers tried to make sure that their science materials were, indeed, for all students. They tried the SCIS lessons on students of all kinds. Only lessons that worked for all students made it into the final set of SCIS materials.

Robert Karplus struck his co-workers as an infectiously happy person. This part of him is well documented in the essays written especially for this book. Over and over again the authors of these essays testify to the influence of Karplus on their lives and their careers. He loved the act of discovering new things himself. He wanted all people to know the joy of discovery. "Don't tell me, let me find out" is an essential part of the legacy of Robert Karplus for all of us.

Appendices

Robert G. Fuller*

CONTRIBUTOR'S BIOSKETCHES

Chapter 2—by Rita W. Peterson

Rita W. Peterson, Senior Lecturer in Education, University of California, Irvine

Rita Peterson is currently a senior lecturer in education at the University of California, Irvine. Peterson did her undergraduate work in education with minors in biology and art at the University of California, Hayward. She received her master and Ph.D. degrees in education from the University of California, Berkeley. Throughout her research career, Dr. Peterson has studied factors that influence the understanding of science—at the individual level and at the national and international levels. At the individual level, she studies the cognitive development of children and adolescents and ways that cognitive development influences their understanding of science. From studying children's natural expression of curiosity in school and non-school settings, she discovered changes in curiosity from childhood to adolescence that led to her receiving the highest award in her field, the Outstanding Research Award from the National Association for Research in Science Teaching. Her studies of the development of problem solving abilities from childhood to adolescence, with physicist, Robert Karplus, led to their making the first American film on Piaget's theory of logical thought as it relates to solving problems in science,

*University of Nebraska-Lincoln.

A Love of Discovery: Science Education—The Second Career of Robert Karplus,
Edited by Robert G. Fuller, Kluwer Academic / Plenum Publishers, New York, 2002.

a film that is still used in universities throughout the world. Presently, Dr. Peterson is studying the role that neuro-developmental (brain-mind) variations play in students' academic success and failure. In collaboration with Melvin Levine, M.D., at the University of North Carolina, she works with groups of elementary, middle/high school and college students, tutors, and teachers to explore strategies that will allow teachers, tutors, and parents to assist students who are failing in science and other subjects, based on understanding individual student profiles of neuro-developmental strengths and weaknesses. At the national and international levels, Dr. Peterson has analyzed major changes in science education programs and movements in the U.S. and Pacific Rim countries. Her research has been supported by the National Academy of Education, the National Science Foundation, and the Japan Society for the Promotion of Science.

Among her best-known books are: *The Brain, Cognition, and Education*, with Friedman and Klivington (Academic Press, 1986) and *Science and Society*, with Bowyer, Butts, and Bybee (Merrill, 1984). Her research has been described in over 100 publications, invited papers and keynote addresses. Presently Dr. Peterson serves as Retiring President (2000–2001) of the Pacific Division of the American Association for the Advancement of Science.

Chapter 3—by Herbert D. Thier
Herbert D. Thier, Academic Administrator Emeritus, Lawrence Hall of Science, University of California, Berkeley.

Herb Thier is Director of the Science Education for Public Understanding Program [SEPUP] and a number of other grants at the University. Thier received his B.A. in Physics and Biology from the State University of New York, Albany in 1953 and his M.A. in School Administration in 1954. He received his Ed.D. in Curriculum and Administration from New York University in 1962. Thier was a science teacher, science coordinator and school administrator between 1954 and 1963. He was Assistant Superintendent of Schools for Instruction when he met and began to work with Bob Karplus on the start up of the Science Curriculum Improvement Study. Since 1963 he has been leading Instructional Materials Development and Teacher Enhancement projects in science at the Lawrence Hall of Science. He received (with M. Linn), the JRST *Research in Science Teaching Award*, of the National Association for Research in Science Teaching in 1975. Thier received the Distinguished Service to Science Education Award, of the National Science Teachers Association in 1994 and the Distinguished Service to Science Education Award, of the Connecticut Science Supervisors Association in 1996. He is a Fellow of the American Association for the Advancement of Science and is listed in Marquis' *Who's Who in America*. His new book titled: Developing Inquiry-Based Science Materials: A Guide for Educators will be published in June, 2001 by Teachers College Press.

Chapter 4—by Anton E. Lawson
Anton E. Lawson, Professor of Biology, Arizona State University

Anton Lawson is currently a professor in the Biology Department at Arizona State University. Lawson did his undergraduate work at the University of Arizona from 1963 to 1967. He received a masters degree from the University of Oregon in 1968 and a Ph.D. from the University of Oklahoma in 1973. After spending one year at Purdue University, Lawson joined the Karplus research group at the Lawrence Hall of Science in Berkeley. From 1974–1977, Karplus, Lawson, Warren Wollman, Helen Adi, and others collaborated on several projects researching the relationships between students' intellectual development and learning science and mathematics. In 1977 Lawson moved to Arizona State where he continues to conduct research and teach. In 1981, the Association for the Education of Teachers of Science honored Lawson with its Outstanding Science Educator of the Year Award. Lawson has three times been awarded the National Association for Research in Science Teaching's JRST for the most significant research contribution of the year. He has also been awarded that organization's Distinguished Contributions Award.

Chapter 5—by J. Myron Atkin
J. Myron Atkin, Professor of Education, Stanford University

Mike Atkin taught science for seven years in New York elementary and secondary schools. He joined the faculty of the University of Illinois at Urbana-Champaign in science education in 1955 and moved to the Stanford University faculty in 1979. At both universities, he also was Dean of Education—from 1970 to 1979 at Illinois and from 1979 to 1986 at Stanford. He has chaired the Education Section of the American Association for the Advancement of Science and serves as a frequent consultant on education to the Organization for Economic Cooperation and Development (OECD) in Paris. At the National Research Council of the National Academies of Science, he was a member of the Mathematical Sciences Education Board and the National Committee on Science Education Standards and Assessment—and is now chair of the Committee on Science Education K-12. He chaired the Committee on Science and Engineering Education of Sigma Xi, the honorary scientific research society, and was vice-chair (1985–86) of the Advisory Committee for Science and Engineering Education at the National Science Foundation. During 1986–1987, he served as Senior Advisor to the Education Directorate at NSF. In the 1960s, he directed one of the first two NSF-supported curriculum projects for children below the high school level, the University of Illinois Astronomy Project. His recent and current major projects include (1) principal investigator for an NSF-supported research project on formative assessment in science classrooms, (2) co-chair of the International Steering Committee of a 13-country study of innovations in science, mathematics, and technology education conducted under the auspices of the OECD, (3) principal investigator for two American case studies (Project 2061 and California science reform) that were part of the OECD project, (4) ongoing study of local, inter-institutional alliances to improve science education that bring school districts into collaborative relationships with government laboratories, universities, industry, and museums, (5) evaluation of a five-year NSF-supported Mills College program to improve science teaching in the elementary schools of Oakland, California, and (6) evaluation of an NSF-supported gender-equity project linking the University of California at San Francisco with the San Francisco Unified School District.

Chapter 6—by Helen Adi Khoury
Helen Adi Khoury, Associate Professor of Mathematics Education, Northern Illinois University

Helen Adi Khoury is an Associate Professor of mathematics education in the Department of Mathematical Sciences at Northern Illinois University, in DeKalb, Illinois. After receiving her doctorate from Florida State University, Helen worked with Robert Karplus, at the Lawrence Hall of Science, in UC Berkeley, on his NSF-funded projects for one year as a post- doctoral research associate during 1976–1977. Professionally, it was a highly rewarding year for Helen to work, to do research, to gain insights, and to discuss with Bob and other colleagues issues related to students' intellectual development and the reasoning strategies they use in various mathematical situations. With Robert Karplus as a mentor, Helen learned early in her career how to focus on trying to understand students' mathematical thinking in her research, teaching, and teacher-education efforts.

Helen Adi Khoury is an author of several studies, published in leading research journals. These studies tend to focus on understanding and promoting students' reasoning in several multiplicative situations. Over the years, she directed several research-based Eisenhower teacher-enhancement projects. On these projects, she works with mathematics teachers to help them implement the vision of reform in their mathematics classrooms as they teach mathematics for sense-making. Helen has also been fortunate to direct the work of an excellent group of doctoral students in mathematics education at Northern Illinois University. The doctoral dissertations she directed have also focused on understanding students' thinking in various problem-solving proportion and function situations.

Chapter 8—by Jane Bowyer
Jane Bowyer, Abbie Valley Professor of Education, Mills College

Jane Bowyer has been a professor at Mills College in Oakland, California for the past 25 years. She is currently the Abbie Valley Professor of Education and Department Head at Mills. She received her Ph.D. and M.A. in science education from the University of California, Berkeley and her B.A. in education at Miami University in Oxford, Ohio. Professor Bowyer has co-authored many publications for the Science Curriculum Improvement Study, published numerous articles on science education and co-authored the book entitled Science and Society. She has been the recipient of numerous National Science Foundation grants focusing on science teacher development. The most recent five-year award for four million dollars was used to increase the quantity and quality of elementary school science for Oakland's 35,000 children by working with their teachers on science content, pedagogy and school reform leadership skills. Dr. Bowyer is an active member of NSTA, NARST, AAAS, and AERA and has received the NSTA Award for the Outstanding Research Paper that has Applications for Classroom Teaching, the Outstanding Research Paper Award of NARST, and a Fulbright Senior Research Fellowship.

Jane's relationship with Professor Karplus began in 1968 in Orinda, California. She was a neighbor of the Karplus' and a teacher in their neighborhood school where she became acquainted with the elementary science materials that Karplus and his colleagues were developing. At Karplus' suggestion she joined the SCIS development team and simultaneously pursued her doctorate. Karplus was the dissertation advisor for her thesis, Cummulative Effects of the Science Curriculum Improvement Study and the Development of Scientific Literacy. Until Karplus' death he and Jane collaborated on projects involving science teacher development.

Chapter 9—by Alan J. Friedman
Alan J. Friedman, Director, New Your Hall of Science
Alan J. Friedman has been the Director of the New York Hall of Science, New York City's public science-technology center, since 1984. Before coming to New York Dr. Friedman served as Conseiller Scientifique et Muséologique for the Cité des Sciences et de l'Industrie, Paris, and was the Director of Astronomy and Physics at the Lawrence Hall of Science, University of California, Berkeley for 12 years. He is the co-author, with Carol C. Donley, of Einstein as Myth and Muse (Cambridge University Press, 1985). The American Association for the Advancement of Science recognized Dr. Friedman's work in developing the New York Hall of Science by naming him the 1996 winner of the AAAS Award for Public Understanding of Science and Technology.

Dr. Friedman received his Ph.D. in Physics from Florida State University and his B.S. in Physics from the Georgia Institute of Technology. He is a Fellow of the American Association for the Advancement of Science and the New York Academy of Sciences. His interests include museums, science education, and the relations between science and the broader culture.

Chapters 1, 7, 10, 11 and 12—by Robert G. Fuller
Robert G. Fuller, Professor of Physics, University of Nebraska-Lincoln

Robert Fuller is currently a professor in the Department of Physics and Astronomy of the University of Nebraska-Lincoln (UNL). He did his undergraduate work in physics at the Missouri School of Mines and Metallurgy from 1953 to 1957. He received his masters and Ph.D. degrees in physics from the University of Illinois. He has taught high school physics at the Methodist English High School in Rangoon, Burma as well as working as a research physicist at the U.S. Naval Research Laboratory before joining the faculty of the UNL in 1969. He began collaboration with Dr. Karplus in 1973 and worked with him on producing the physics teaching and development of reasoning workshop for the American Association of Physics Teachers(AAPT) in 1974. Dr. Fuller joined Karplus and his team as visiting professor of physics at the University of California, Berkeley in 1976–77. Fuller was president of the AAPT in 1980. He directed the Piagetian-based, multidisciplinary program for college freshmen, the ADAPT program, at UNL from 1975–97. In 1992, he received the Robert A. Millikan prize, from the AAPT, for his outstanding contributions to physics education.

Let's Look to the Future

AAPT President Robert Karplus

In three years our Association will be celebrating its 50th birthday. We can look back with satisfaction at many accomplishments that include our widely-read journals, national and local meeting programs that have made possible the exchange of ideas among us, the recognition of distinguished contributions to physics teaching through our awards, and the establishment of an Executive Office that provides us with a variety of ongoing services to make our teaching more effective.

As we look to the future and the approach of the 21st century, our goals may require us to find new ways of orienting our efforts. Let me first remind you of Article II of our constitution, which states "The objectives of the Association will be the advancement of the teaching of physics and the furtherance of the appreciation of the role of physics in our culture." That's a very broad mandate, especially once you realize that the general public hardly distinguishes physics from science. Fortunately we are not alone in this endeavor, though we are unique in our concern with physics. Clearly we must join with other organizations concerned with science teaching, science-related government agencies, and private foundations involved in education. Furthermore, we have to realize that the objective can be promoted not only through teaching in a school or college, but also through activities in a science museum, planetarium, or other informal learning environment, through leadership in a community youth group or adult education program, and through knowledgable conversations with family members, friends, and professional associates who do not have a science background.

It is easy to overlook the rapid social changes in our culture, social changes that are partly fueled by the impact of knowledge and technology based on physical principles. Many decisions faced by our government and by us as citizens have ever more complex scientific aspects. And the need on the part of everyone to understand the strengths and limitations of physics as a discipline become ever more urgent.

AAPT President Robert Karplus

Our Association has already begun to respond to these challenges. Several years ago we established a *Committee on Science Education for the General Public* (jointly with APS) which has offered sessions at our meetings, conducted a "Store-Front Physics Competition," and is considering use of mass media. Starting this year, we have a *Committee on Professional Concerns* to attend to the conditions under which we teach, rather than emphasizing the content of instruction that has been our more traditional interest.

Our meetings have changed their character somewhat. Perhaps you had a chance to attend the Chicago meeting, or at least looked over the many offerings described in the December *Announcer*. Our meetings now present many more occasions that invite your active participation in laboratory activities, equipment construction workshops, short courses on various topics, and poster sessions where you may converse with the author of each paper. Perhaps this trend has been stimulated by more emphasis on active student involvement in many physics courses that we teach—after all, a meeting is a teaching opportunity in which our colleagues play the role of students. Perhaps this style of meeting will help us attract more students as we find out how to involve them actively in their own learning.

One great source of strength in science education during the past twenty years has been the National Science Foundation. The Foundation has helped us and individual members work towards our objectives through course development, teacher education, the Commission on College Physics, international activities, and fellowships. New programs such as the Comprehensive Assistance to Undergraduate Science Education program (CAUSE) and the Research Initiation and Support program (RIAS) are having their first impact this year, and "Science for the Citizen" programs are being planned.

Now the Foundation needs our help. NSF's education programs have been put on the defensive by attacks channeled through the Congress and emanating from certain parent groups and publishers. They have criticized some specific course developments and the alleged use of teacher education for promotion of specific curricula. As a result, research, development, and teacher education for precollege science programs were curtailed abruptly. When new guidelines are issued in final form, the prospects of their leading to further significant innovation can be appraised. It is clear that broad support for NSF's education programs from the scientific community and the public are necessary if the new guidelines are to be applied with imagination and wisdom.

The same process that led to the present conditions allows you to express your views to President Carter, your representatives and senators, members of the National Science Board, officials of the National Science Foundation, and scientists in your community. President Carter will appoint a new Director of the National Science Foundation—a director who will, I hope, sincerely support the educational mission of the Foundation in a broad sense. The entire community of scientists must recognize the great importance of public understanding of science as it relates to energy conservation and resource development, management of raw materials, production of food, the processing of information, the exploration of space, and the maintenance of a strong scientific research establishment.

You can also help within our Association. Our 34 Regional Sections carry our programs and activities to many members and nonmembers who are unable to come to national meetings. Some sections have found unique ways of serving their area. If you are taking part in a special section program, please submit descriptive material to the executive office for publication in the *Announcer*. If you do not know your section's programs, obtain names of the officers from the March, 1976, *Announcer* and contact them. If your section is not as active as you would like it to be, enliven its programs by helping to plan them, by being a candidate for office, contributing to a meeting, or by suggesting a novel activity that would contribute to the membership. If you live in one of the areas that does not have a section (Alaska, Hawaii, Idaho, Montana, Nevada, North Dakota, Ohio, and Utah) think about organizing one! As few as ten AAPT members may apply to the secretary for authority to form a section with the power to hold meetings.

Are you a member of AAPT? If not, won't you consider joining the Association to work with us in realizing our objectives? If you are a member, then please let us know more insistently just how you might contribute to our common purpose and how the Association could better serve you. Even though you may think of AAPT as a somewhat impersonal organization, it is really made up of people like you and me. We officers and the staff of the executive office at Stony Brook appreciate your thoughtful suggestions and will act on them (see the "Letters" column in the *Announcer*) or forward them to the appropriate group.

Let's plan exciting new programs in physics and physics teaching for our students, the general public, and ourselves!

Meet Your New AAPT President

W. M. Laetsch

His entry in "Who's Who in America" lists research interests ranging from analytical chemistry through quantum field theory to developmental psychology. He has published papers on "A New Method of Determination of Manganese," "Fourth Order Corrections in. Quantum Electrodynamics and The Magnetic Moment of the Electron," "Radiation of Hydromagnetic Waves," "Opportunities for Concrete and Formal Thinking on Science Tasks," and he has produced a series of films illustrating Piaget's development theory of intellectual development.

AAPT President Robert Karplus is obviously a man of many capabilities. His multifaceted life began in Vienna, Austria, and he attended elementary school in Vienna. He arrived in the United States in 1938, and enrolled in the sixth grade at the Boston Latin School. Following graduation from high school in Newton, Massachusetts at the age of 16, he entered Harvard where he obtained a B.S. in chemistry. Bob remained at Harvard for a Ph.D. in chemistry. He then plunged into theoretical physics at the Institute of Advanced Study at Princeton, where he was an F. B. Jewett Fellow from 1948 to 1950. His work during this period was concerned with quantum electrodynamics. He returned to Harvard as an assistant professor of physics in 1950. The fact that his career would later take an interesting course became apparent during that period. In the words of a graduate student who attended his lectures, and who is now a physics colleague at Berkeley, his lectures "reflected an unconventional point of view."

Bob's westward migration resumed in 1954 when he came to Berkeley as an associate professor of physics. He earned his professorship in 1958, and this year also marked his first experiments in elementary school education. What began as volunteer science instruction for his children's classes, developed into a major science curriculum development

---

W. M. Laetsch, Lawrence Hall of Science, University of California, Berkeley, California 94720.

Reprinted with permission from Laetsch, W. M., Meet Your New AAPT President, *The Physics Teacher* 70, pp 126. Copyright, American Institute of Physics [for *Physics Today*], 1977.

Figure 1. Elementary school students find President Karplus can get down to their level.

project with international impact, and initiated an additional career in science education and developmental psychology.

In the post-Sputnik era when many university scientists were becoming involved in secondary school curriculum development programs, Bob decided that the real challenge was at the elementary and junior high level. We forget that this was an unconventional approach for that time, and Bob likes to recall that a number of university colleagues with interest in school science had cautioned against working at the elementary school level because of the problems with children's understanding and the teachers' lack of preparation in regular science disciplines. Bob's interests were in developing the conceptual base for the development of curricular materials at this level, and his interest led to the National Science Foundation's first grant at the elementary school level in 1959. This modest beginning grew into the Science Curriculum Improvement Study Project (SCIS), which is a comprehensive kindergarten through sixth grade science program currently used by approximately four million elementary school children in the United States. It is also widely used abroad, and has been translated into French, Swedish, Danish, Italian, and Japanese.

The project has been strongly influenced by the ideas of the Swiss psychologist, Jean Piaget, on intellectual development in children. A primary concern of SCIS was the development of instructional materials appropriate to the intellectual level of children of different ages. As a result of his interest in intellectual development, Bob initiated a research program at the Lawrence Hall of Science a number of years ago on intellectual development, particularly at the post-elementary school level. In fact, this project has even been concerned with this question at the undergraduate level.

Bob is primarily identified by many educators with elementary school science, but he also has a very high reputation on the Berkeley campus for his excellence as an instructor of physics. He has been a leader on the campus in developing personalized systems of instruction for his course, and he is author of two elementary physics texts. In addition to his contributions to elementary and undergraduate science instruction, Bob was an initiator and continues to be an active member of the Graduate Group in Science and Math Education on the Berkeley campus.

Bob's wife, Betty, also has a physics background, and is currently teaching in special education in a local high school. They have collaborated on a number of studies on intellectual development. As the father of seven children, Bob had ample reason to be concerned with elementary school science education. The children are now distinguishing themselves in a variety of fields, including science. Bob is now a grandfather three times over, so watch out for a plunge into preschool education.

This man has navigated with considerable ease through the waters of theoretical chemistry, theoretical physics, and developmental psychology, so he obviously has wide interests. Bob built the kitchen cabinets and installed the wiring of his hilltop home in Orinda, and he enjoys gardening, cooking, and photography. He is an early morning jogger, and a later-night telephone caller. One of his notable, if short-lived, enthusiasms was an attempt to start a shampoo business with the family washing machine as the center of production. While he did not become a shampoo king, it is quite likely that this experience has or will surface in his science teaching.

The new AAPT President is an outstanding scholar in several fields, an inspired teacher, and a congenial and always cooperative colleague. At the moment, he is also a harassed administrator since he is serving a stint as Acting Director of the Lawrence Hall of Science. He still retains his "unconventional point of view," and this combined with his multiple talents will certainly provide creative leadership for the AAPT.

Memorial to Robert Karplus
1927–1990

Eyvind H. Wichmann, Professor of Physics

Robert Karplus, Professor Emeritus of Physics, died on March 20, 1990, after a long illness following an incapacitating heart attack eight years earlier. He is survived by his wife Elizabeth, seven children and fourteen grandchildren, and by his brother Martin.

Robert Karplus was born in Vienna, in 1927. He was educated at Harvard University. He obtained his B.S. in Chemistry and Physics in 1945; his M.A. in Chemistry in 1946, and his Ph.D. in Chemical Physics in 1948. He was an F.B. Jewitt Fellow at the Institute for Advanced Study, Princeton, in 1948–50, and an Assistant Professor of Physics at Harvard University in 1950–54. In 1954 he joined the faculty of the Department of Physics here at Berkeley, as an associate professor. He became a Professor of Physics in 1958, and remained an active member in the Department until his illness.

The early scientific publications of Karplus were in the field of chemical physics, with a strong physics orientation. The wartime developments in microwave technology had provided important new tools for research both in chemistry and physics, and some of Karplus's early papers concern (theoretical) issues in microwave spectroscopy. Around 1948 his interest shifted to quantum electrodynamics and what is today called particle physics. In the later half of the decade of 1940 there occurred some remarkable breakthroughs in quantum electrodynamics and quantum field theory. In the new approaches by Tomonaga, Schwinger and Feynman, it seemed possible to overcome the famous "divergence difficulties", which had presented seemingly unsurmountable obstacles to any real progress in the past, and there was now hope that computations could be carried out to test the theory against the new experimental facts which accumulated at the same time. In later years quantum electrodynamics has been simplified in many ways, but at that time it was a very difficult subject, and very much unknown territory, calling for brave explorers. For many years Robert Karplus was at the forefront of the development of the theory, and his name is associated with the theoretical predictions of many of the experimental results by which the theory was tested. He studied the magnetic moment of the electron (with Norman Kroll); scattering of light by light (with Maurice Neuman); the fine structure of positronium (with Abraham Klein), and the general theory of electrodynamic displacements of atomic energy levels, with applications to the Lamb shift and to hyperfine structure, in a sequence of three fundamental papers with Abraham Klein and Julian Schwinger. His collaboration with Norman Kroll on the computation of the fourth order radiative correction to the magnetic moment of the electron is particularly famous. This was the first *really* difficult problem attacked, which involved all the intricacies of renormalization theory in quantum electrodynamics, and its successful solution did much to inspire confidence in the theory. This was a golden time, when the theoretical predictions agreed with experimental results to an astonishing degree, and Karplus was one among a very small group of enthusiastic pioneers who established the results which are now the classical results in the textbooks.

Later Karplus turned this attention to strong-interaction physics (then called meson-nucleon physics). Even today this field is not as well understood as quantum electrodynamics. Progress in the theory has been very slow, and the present experimental picture of this field is also very different from what it was in 1950. During the 1950's Karplus made a number of interesting contributions to partial solutions of the burning issues at the time. The papers from that time show a wide range of interest within a large field. A characteristic feature of his research was that the topics selected always had some definite connection with experimental problems, and one can say that the selections showed good taste and physical insight. Actually his interests were not exclusively confined to particle physics: among his publications there are papers on such diverse subjects as the Hall effect, van Allen radiation, and magnetohydrodynamic waves. His last publication in "pure physics," dated 1962, concerns magnetohydrodynamic waves.

Around 1960 Karplus became increasingly interested in science education at the elementary level. He once told me that the original motivation for this came from visits to his children's classrooms. Eventually this became his principal interest, although he continued to supervise graduate students in theoretical physics for several years. The first task which he took on was the Elementary School Science Project in the early sixties, which involved cooperative efforts with groups at other institutions, and which eventually led to the 15-year, NSF-funded Science Curriculum Improvement Study (SCIS), centered at the Lawrence Hall of Science. He served as Associate Director, and Acting Director of LHS in the 1970's. His interests in science teaching and science learning broadened steadily, and came to encompass science teaching at all levels, and the psychology of learning. Karplus published a large number of papers on these issues from 1962 to 1983, which show his broad range of interests. One particular concern of his, which was like a red thread in his work, is worth mentioning. He wanted to make science, and in particular physics, tangible and "hands on."

Karplus's work in pure physics was widely recognized and acclaimed. He was a Fellow of the American Physical Society, and the recipient of Guggenheim and Fulbright fellowships. His contributions in his "second career" were also widely appreciated. He received Distinguished Service Citations from the American Association of Physics Teachers (and served as president in 1977), and from the National Science Teachers Association. In 1980 he was awarded the Oersted Medal, as well as an Honorary Doctorate by the University of Gothenburg, Sweden.

Bob Karplus was highly regarded and respected by his colleagues. He had an open and generous personality, always willing to help and listen to others. When I first came to Berkeley in 1957, Bob did everything he could to make me and other newcomers feel at home, and to integrate us into the activities of the theoretical group. It was all done in an unpretentious manner and with a good deal of humor. I soon noticed that he also interacted with his students in the same manner. One notable feature of Bob was his cheerful attitude and optimistic approach to seemingly intractable problems in physics.

His illness and his passing away was a great loss for the Department and for his many friends.

It Was a Great Time . . . In Memoriam . . . Let Us Continue

Robert Karplus, 1927–1990

On March 20th, 1990 Robert Karplus, long time member and friend of the NSTA, died of a heart attack after eight years of severe disability due to a previous heart attack. In 1978, the NSTA recognized Bob Karplus's extraordinary contributions as a scientist and science educator by awarding him their Citation for Distinguished Service to Science Education. In part that citation stated:

Born in Vienna, Austria, and educated in chemistry and chemical physics at Harvard, Professor Robert Karplus was a leading U.C. Berkeley elementary particle theorist when he began dabbling in elementary school science education in the mid-1950s. What started as occasional visits to his children's classrooms to involve them and their classmates in hands-on and intellectually challenging science activities, gave Karplus insight into difficulties his college physics students were having exploring, observing, and understanding natural phenomena and theories. His experiences indicated textbook-based science teaching alone did not give students of any age the integration of the conceptual understanding and process skills he liked to refer to as scientific reasoning which is necessary to become a scientist or understand science as a citizen. What Karplus and his colleagues learned and formalized during his three decades as a science educator and student of science education is used today in essentially all experience-based elementary science curriculum development.

His contributions were based on his growing understanding in three areas: children's cognitive learning strategies and levels; varied conceptual organizations of science; and a learning-cycle, guided-discovery system to make that complex integration of concepts and process, referred to as scientific reasoning, available to students of varying ages and levels of expertise.

Through Bob's leadership this "scientific reasoning" approach to science teaching and learning was introduced to both secondary school and college level learners and their teachers through the development and publication of the "Science Reasoning Workshops" in Biology, Chemistry, Physics and General Science. As in all his other instructional development efforts Bob not only gave leadership to the development but was an active planner of and participant in the local leadership development efforts necessary to make the "Science Reasoning Workshops" an integral part of the teaching and learning of science in schools and colleges nationwide.

Beginning with Elementary School Science Project, continuing with early '60s science projects in Massachusetts, Maryland, and Minnesota, and culminating with the 15-year, NSF-funded Science Curriculum Improvement Study (SCIS) at UC Berkeley's Lawrence Hall of Science, Karplus expanded his activities and contributions in elementary science education. The trademark of his approach was to try some activity in the classroom, see and understand what works and what does not, and continually evaluate and improve the activities and programs on all fronts. He recognized that good programs resulted only as a result of cooperation among practicing teachers and science educators who had expertise working with children and scientists who had subject matter expertise. Together they could construct an effective, experience-based program for all children; separately, they could not.

Bob and I met and began working together in 1962 on what was to become the Science Curriculum Improvement Study (SCIS). Our close professional interaction extended over twenty years until Bob's first disabling heart attack in 1982. Looking back at nearly thirty years of a close personal relationship with the most unique individual I have ever known I realize Bob Karplus's legacies to me personally and our profession generally go far beyond the "Learning Cycle", SCIS, the research on children's cognitive development and the many other specific contributions he made to our profession. His first legacy was establishing by example the validity and the value of the participation of the successful research scientist in the science instructional development process. Not as a critic but as a co-worker, confident about the unique contribution he could make and highly interested in and respectful of the unique contributions others could make. Second he truly believed in the importance of the American public school as a social institution responsible for protecting and enhancing the quality of life for all individuals in our democratic society. That is why all of his work focused on enhancing scientific literacy for all individuals while encouraging interest in science as a career. Coupled with his deep personal belief in the value, uniqueness and importance of each individual learner this led to the emphasis in all of his work on learning as the basis for teaching rather than the opposite emphasized so often these days.

As part of his presentation of the 1990 Robert Karplus lecture at the Atlanta NSTA Convention, Professor Jerome Bruner, the noted learning psychologist, summed up his interactions with Bob as follows:

> His ideas about how to teach science were not only elegant but from the heart. He knew what it felt like "not to know," what it was like to be a "beginner." As a matter of temperament and principle, he knew that not knowing was the chronic condition not only of a student but of a real scientist. That is what made him a true teacher, a truly courteous teacher.
>
> What he knew was that science is not something that exists out there in nature, but that it is a tool in the mind of the knower—teacher and student alike. Getting to know something is an adventure in how to account for a great many things that you encounter in as simple and elegant a way as possible.

In conclusion, let his own words describe the curiosity, the enthusiasm, the delight he had found in problem solving cooperatively with other scientists, teachers, and children as the Science Curriculum Improvement Study's funding from the National Science Foundation ended in 1977:

> Elementary school science education has changed during the almost twenty years since I first became aware of the challenges and opportunities in this curriculum area. Thanks to the efforts of many thousands of teachers, enlight-

ened planning by hundreds of school boards and school administrators, and encouragement from millions of parents, the Science Curriculum Improvement Study's materials and methods have gained a firm place in many elementary schools.

Little did I realize in 1958, when I was visiting my daughter Beverly's second-grade class to teach a few science lessons, that my professional career would be redirected from theoretical physics to science education. My fate was sealed when the National Science Foundation's very first grant for course content improvement in elementary school science was awarded to three colleagues and me at the University of California, Berkeley, in 1959. At that time there began an intensive learning experience that introduced me to a new field and one that is still continuing. . . .

By 1963, the National Science Foundation had responded to the urging of many scientists, science educators, and teachers by announcing an extensive program of elementary science course development. Together with Herbert D. Thier, who became assistant director of SCIS, I began the work of producing and classroom-testing new teaching materials. We planned a ten-year development and publication program, which was completed successfully with the help of about two hundred associates in Berkeley and at the five trial centers in New York, Michigan, Oklahoma, Los Angeles, and Hawaii. . . .

In thinking back over this period of time, I am awed that all these people and institutions had to make positive contributions to accomplish the change in elementary school science that has come about. A gap anywhere in the chains leading from the development laboratories to the children in the classroom would have been fatal.

I am grateful for the opportunity to have a part in these activities and to work together with so many thoughtful, inventive, conscientious, and persistent individuals. . . .

I believe we have accomplished the original purpose of SCIS, to narrow the gap between the concept of science held by scientists and that experienced by boys and girls in their classrooms. The dynamic nature of science arising from the interplay of observations and ideas comes alive when pupils investigate the dissolving of salt, the life cycle of mealworms, the energy transfer from a rubber band to a propeller, and the feeding relations in a community of grass, crickets, and frogs. . . .

An additional outcome, which I did not anticipate at all in 1959, has been the success of many teachers and science educators in using the SCIS program with special pupil groups. Inner city children, rural youngsters, youngsters from non-English-speaking homes, and Native American boys and girls became more interested in science and in learning by [using] the . . . SCIS materials in their own ways to satisfy their own curiosity. More recently the growing concern with special education has led to successful use of the SCIS program or adaptations with visually impaired, emotionally disturbed, hard of hearing, and mentally retarded children. In all these situations, the benefits have included not only science learning but also social aspects arising from the children working together in small groups on their projects. . . .

What does the future hold? Much emphasis has been given very recently to the "basics," and one wonders whether schools will have money to offer anything more than a minimal program. Yet the very lack of funds makes it all the more urgent that children become active learners who can benefit from all their life experiences, not only from their lessons in school. And what better way is there of accomplishing this aim than by advancing the scientific attitudes of curiosity, inventiveness, critical thinking, and persistence through the SCIS program or a similar laboratory-based curriculum? In such an approach, already practiced in some communities, the science program is closely linked with language development, mathematics, and social studies. What do you think?

Herbert D. Thier

Obituaries

Robert Karplus

Robert Karplus

After an eight-year illness following a heart attack, theoretical physicist Robert Karplus died on 20 March 1990. Since 1954 he had been a faculty member of the physics department of the University of California, Berkeley.

Karplus, born in Vienna in 1927, received both his undergraduate and graduate education at Harvard University. His early interest was in chemical physics: His 1945 BS degree was in chemistry and physics, his 1946 MA in chemistry and his 1948 PhD thesis in microwave spectroscopy. During a two-year postdoctoral stint at the Institute for Advanced Study in Princeton, Karplus's interest shifted to high-precision quantum electrodynamics, whose development at that point lay at the frontier of science. Karplus's focus on QED continued throughout his next four years as a junior faculty member at Harvard. He collaborated with Norman Kroll, Abraham Klein and Julian Schwinger, among others, in a golden period of theoretical development. Karplus's contributions to precise QED predictions included Feynman-graph calculations of the electron's magnetic moment, scattering of light by light and the fine structure of positronium as well as of atoms. Especially attention getting was his graph-based work with

Reprinted with permission from Chew, G. F., Obituaries—Robert Karplus, *Physics Today* 45(3), pp 80. Copyright, American Institute of Physics [for Physics Today], 1992.

Kroll on the fourth-order correction to the electron's magnetic moment; the complexity of this calculation was then unprecedented at the frontiers of physics. In 1954 Karplus moved to the University of California as a tenured faculty member.

In the mid-1950s Karplus's interests broadened to include strong interactions, where even though perturbation theory was invalid, Feynman graphs had become recognized as somehow relevant. Landau eventually formalized the relevance in 1959 through his rules associating S-matrix singularities with graphs, from which work analytic S-matrix theory emerged. Karplus's work with collaborators such as Charles Sommerfield and Eyvind Wichmann was a precursor to Landau's insight.

Karplus entered a totally different career in the early 1960s. He became fascinated by the process of science learning in young children (Karplus himself had seven children), and he commenced research in this area, which had largely been ignored in the US. His Elementary School Science Project led to the 15-year, NSF-funded Science Curriculum Improvement Study. The SCIS was centered at the Lawrence Hall of Science in Berkeley, where Karplus served as associate and then acting director during the 1970s. What Karplus and colleagues uncovered through the SCIS is used today in developing almost all experience-based elementary-science curriculums.

Karplus was himself a talented teacher, an attribute related to his cheerful, generous personality and his sense of humor. He made important contributions to self-paced teaching at the college level, and he served a term as president of the American Association of Physics Teachers in 1977–78.

Geoffrey F. Chew
University of California, Berkeley

ROBERT KARPLUS

PERSONAL DATA

Born: February 23, 1927, Vienna, Austria
Naturalized U.S. Citizen: December 4, 1944
Married: December 27, 1948, to Elizabeth J. Frazier
Professional career ended with a cardiac arrest: June, 1982, Seattle, Washington
Died: March 20, 1990

EDUCATION

B. S. Harvard University, Chemistry and Physics, 1945
M. A. Harvard University, Chemistry, 1946
Ph. D. Harvard University, Chemical Physics, 1948

EMPLOYMENT

F. B. Jewett Fellow, Institute of Advanced Study, Princeton University, 1948–50
Assistant Professor of Physics, Harvard University, 1950–54
Associate Professor of Physics, University of California, Berkeley, 1954–58
Professor of Physics, University of California, Berkeley, 1958–1982
Guggenheim Fellow and Fulbright Research Grantee, 1960–61
Visiting Professor, University of Maryland, 1962–63
Associate Director, Lawrence Hall of Science, 1969–1982
Guggenheim Fellow and Visiting Professor, M. I. T., 1973–74
Director, Science Curriculum Improvement Study, 1961–77
Director, Intellectual Development Project, 1975–1982
Acting Director, Lawrence Hall of Science, 1976–77
Director, CAUSE Project, 1976–1979
Chairman, Graduate Group in Science and Mathematics Education, 1978–1980
Dean, School of Education, University of California, Berkeley, 1980

HONORS AND AWARDS

Fellow of American Physical Society
Distinguished Service Citation, American Association of Physics Teachers, 1972
National Science Teachers Association Award for Distinguished Service to Science Education, 1978
President, American Association of Physics Teachers, 1977
Honorary degree, Doctor of Philosophy, University of Gothenburg, 1980
Oersted Medal, American Association of Physics Teachers, 1981

SELECTED BIBLIOGRAPHY
Science Research Papers:

1. R. Karplus and J. J. Lingane, New Method of Determination of Manganese, 1946, *Industrial and Engineering Chemistry*, 18, 191–194.
2. R. Karplus and E. R. Blount and M. Fields, Absorption spectra VI. The infrared spectra of certain compounds containing conjugated double bonds, 1948, *J. Am Chem Soc.*, 70, 194.
3. R. Karplus and E. R. Blount, The infrared spectra of polyvinyl alcohol, 1948, *J. Am Chem Soc.*, 70, 862.
4. R. Karplus and J. Schwinger, A Note on Saturation in Microwave Spectroscopy, 1948, *Phys. Rev.*, 73, 1020–1026.
5. R. Karplus, Frequency Modulation in Microwave Spectroscopy, 1948, *Phys. Rev.*, 73, 1027–1034.

6. R. Karplus, Saturation effects in the microwave spectrum of ammonia, 1948, Letter in *Phys. Rev.* **73**, 1120.

7. R. Karplus, Saturation effects in the microwave spectroscopy, 1948, Letter in *Phys. Rev.* **74**, 223–224.

8. R. Karplus, Note on the energy of a rotating molecule, 1948, *J. Chem. Phys.*, **16**, 1170–1171.

9. R. Karplus and A. H. Sharbaugh, Second-order Stark Effect of methyl chloride, 1949, *Letter in Phys. Rev.* **75**, 889–890, (March); erratum **75**, 1449.

10. R. Karplus and N. M. Kroll, Fourth-Order Corrections in Quantum Electrodynamics and the Magnetic Moment of the Electron, 1949, Letter in *Phys. Rev.*, **76**, 846–847.

11. R. Karplus and N. M. Kroll, Fourth-Order Corrections in Quantum Electrodynamics and the Magnetic Moment of the Electron, 1950, *Phys. Rev.*, **77**, 536–549.

12. R. Karplus and R. S. Halford, Motions of Molecules in Condensed Systems, VI. The Infrared Spectra for Vapor, Liquid, and Two Solid Phases of Methyl Chloroform, *J.Chem. Phys.*, 1950, **18**, 910–912.

13. R. Karplus and M. Neuman, Non-Linear Interactions Between Electromagnetic Fields, 1951, *Phys. Rev.*, **80**, 380–385.

14. R. Karplus and M. Neuman, The Scattering of Light by Light, 1951, *Phys. Rev.*, **83**, 776–784. (8/15)

15. R. Karplus, A. Klein and J. Schwinger, Electrodynamic displacement of atomic energy levels, 1951, Letter in *Phys. Rev.* **84**, 597.

16. R. Karplus, A. Klein and J. Schwinger, Electrodynamic displacement of atomic energy levels II. Lamb shift, 1952, *Phys. Rev.*, **86**, 288–301.

17. R. Karplus and A. Klein, Electrodynamic displacement of atomic energy levels I. Hyperfine structure, 1952, *Phys. Rev.*, **85**, 972–984.

18. R. Karplus and A. Klein, Electrodynamic corrections to the fine structure of positronium, 1952, Letter in *Phys. Rev.* **86**, 257.

19. R. Karplus and A. Klein, Electrodynamic displacement of atomic energy levels III. The hyperfine structure of positronium, 1952, *Phys. Rev.*, **87**, 848–858.

20. R. Karplus, Quantum Electrodynamics, 1953, *J. Opt. Soc. Amer.*, **43**, 237–239.

21. R. Karplus, M. Kivelson and P. C. Martin, A Note of Meson-Nucleon Scattering, 1953, *Phys. Rev.*, **90**, 1072–1075.

22. T. Fulton and R. Karplus, Bound state corrections in two-body systems, 1954, *Phys. Rev.*, **93**, 1109–1118.

23. R. Karplus and J. M. Luttinger, Hall effect in ferromagnetics, 1954, *Phys. Rev.*, **95**, 1154–1160.

24. R. Karplus and M. A. Ruderman, Applications of Causality to Scattering, 1955, *Phys. Rev.*, **98**, 771.

25. C. J. Goebel, R. Karplus and M. A. Ruderman, Momentum dependence of phase shifts, 1955, *Phys. Rev.*, **100**, 240–241.

26. K. M. Case, R. Karplus and C. N. Yang, Experiments with slow K mesons in deuterium and hydrogen, 1956, *Phys. Rev.*, **101**, 358–359.

27. K. M. Case, R. Karplus and C. N. Yang, Strange particles and the conservation of isotopic spin, 1956, *Phys. Rev.*, **101**, 874–876.

27a. K. M. Case, C. N. Yang and R. Karplus, A reply to the criticism by Mr. A. Gambra, 1957, *Nuovo Cimento*, **5**, 1004.

28. M. Ruderman and R. Karplus, Spin of the Λ^0 Particle, 1956, *Phys. Rev.*, **102**, 247–250. (April 1)

29. R. Karplus and K. M. Watson, Structure of a many-particle quantum-mechanical medium, 1957, *Phys. Rev.*, **107**, 1205–1218.

30. R. Karplus and K. M. Watson, Two-nucleon forces and nuclear saturation, 1957, *Amer. J. Phys.*, **25**, 641–647.

31. G. F. Chew, R. Karplus, S. Gasiorowicz and F. Zachariasen, Electronmagnetic structure of the nucleon in local-field theory, 1958, *Phys. Rev.*, **110**, 265–276.

32. R. Karplus, C. M. Sommerfield and E. H. Wichmann, Spectral Representations in Perturbation Theory. I. Vertex function, 1958, *Phys. Rev.*, **111**(4), 1187–1190.

33. R. Karplus, C. M. Sommerfield and E. H. Wichmann, Spectral Representations in Perturbation Theory. II. Two particle scattering, 1959, *Phys. Rev.*, **114**(1), 376–382.

34. R. Karplus, L. Kerth and T. Kycia, K-meson nucleon Interaction, 1959, *Phys. Rev. Letters*, **2**(12) 510–513.

35. R. Karplus, and L. S. Rodenberg, Inelastic final state interactions: K^- absorption in deuterium, 1959. *Phys. Rev.*, **115**(4), 1058–1069.

36. A. J. Dessler and R. Karplus, Some properties of Van Allen radiation, 1960, *Phys. Rev. Letters*, 4(6) 271–274.
37. R. Karplus, Radiation of Hydromagnetic Waves, 1960, *Phys. of Fluids*, 3, 800–805.
38. W. E. Francis and R. Karplus, Hydromagnetic Waves in the ionosphere, 1960, *J. geophys. Res. (USA)*, 65(11), 3593–3600.
39. D. A. Hamlin, R. Karplus, R. E. Vik and K. M. Watson, Mirror and azimuthal drift frequencies for geomagnetically trapped particles, 1961, *J. geophys. Res. (USA)*, 66(1), 1–4.
40. P. G. O. Freund and R. Karplus, A Natural Boundary of the Scattering Amplitude and an Unphysical Sheet, 1961, *Nuovo Cimento*, 21(3), 519–523.
41. P. G. O. Freund and R. Karplus, Anomalous thresholds of reaction amplitudes, 1961, *Nuovo Cimento*, 21(3), 531–540.
42. R. Karplus and Y. Yamaguchi, Quasi-elastic nucleon-nucleus scattering at high energy, 1961, *Nuovo Cimento*, 22(3), 588–603.
43. A. J. Dressler and R. Karplus, Some effects of diamagnetic ring currents on Van Allen radiation, 1961, *J. geophys. Res. (USA)*, 66(8), 2289–2296.
44. P. G. O. Freund and R. Karplus, The analytic structure of scattering and reaction amplitudes in unphysical Riemann sheets, 1961, *El. Part. Conf. Aix-en-Provence*, Vol. 1, 327–330.
45. A. J. Dragt and R. Karplus, Analyticity and unitarity of generic transition amplicatues, 1962, *Nuovo Cimento*, 26(1), 168–176.
46. R. Karplus, W. E. Francis and A. J. Dragt, The attenuation of hydromagnetic waves in the ionosphere, 1962, *Planet. Space Sci. (GB)*, 9, 771–785.
47. A. J. Dragt and R. Karplus, Poles in Coupled Scattering Amplitudes, 1964, *J. Math. Phys.*, 5, 120–126.
48. G. Domokos and R. Karplus, Specualtions concerning large-angle meson-nucleon scattering at high energies, 1967, *Phys. Rev.*, 153(5), 1492–1496.

Science Education Papers:
49. R. Karplus, Beginning a Study in Elementary School Science, 1962, **Amer. J. of Phys.**, 30(1), 1–9.
50. J. M. Atkin and R. Karplus, Discovery or Invention? 1963, **The Science Teacher**, 29, 45–47.
51. R. Karplus and J. Cunningham, Free Fall Demonstration Experiment, 1962, **Amer. J. Phys. 30**, 656.
52. R. Karplus, The Science Curriculum—One Approach, 1962, **The Elementary School Journal**, 62(5), 243–252.
53. R. Karplus, H. D. Thier and C. A. Powell, A Concept of Matter for the First Grade, 1963, **J. of Res. in Sci. Teach.**, 1, 315–318.
54. R. Karplus and C. A. Powell, Objects Grab Bag, 1963, **Science and Children**, 2, 14–15.
55. R. Karplus, C. A. Powell and J. Reynolds, Using a Bathroom Scale, 1963, **Science and Children**, 2, 16–17.
56. R. Karplus, Meet Mr. O, 1963, **Science and Children**, 3, 19–23.
57. R. Karplus, One Physicist Looks at Science Education, 1964, Intellectual Development: Another Look. Washington, D. C. Association for Supervision and Curriculum Development.
58. R. Karplus, The Science Curriculum Improvement Study—Report to the Piaget Conference, 1964, **Journal of Research in Science Teaching**, 2, 236–240.
59. R. Karplus, The Science Curriculum Improvement Study, 1964, **Journal of Research in Science Teaching**, 2, 293–303.
60. R. Karplus, Teaching Physics in the Elementary Grades, 1964, **Physics Today**, 17(10), 34–38.
61. R. Karplus, One Physicist Experiments with Science Education, 1964, **American Journal of Physics**, 32(11), 837–839.
62. R. Karplus, Using A Bathroom Scale, 1964, **Science and Children**, 5, 12–13.
63. R. Karplus, Science, New Frontiers of Education. Edited by Fred Guggenheim and Corrine Guggenheim. New York: Gruen and Stratton, Inc., 1966, 114–135.
64. R. Karplus, Chemical Phenomena in Elementary School Science, 1966, **Journal of Chemical Education**, 43, 257–269.
65. R. Karplus, Conceptual Structure and Physics Teaching, 1966, **Amer. J. Phys.**, 34(8), 733.
66. C. Berger and R. Karplus, Models for Electric and Magnetic Interactions, 1968, **Science and Children**, 6, 43–49.
67. E. F. Karplus and R. Karplus, Intellectual Development Beyond Elementary School I: Deductive Logic, 1970, **School Science and Mathematics**, 70(5), 398–406.

68. R. Karplus and R. W. Peterson, Intellectual Development Beyond Elementary School II: Ratio, A Survey, 1970, **School Science and Mathematics, 70**(9), 813–820.

69. R. Karplus and E. F. Karplus, Intellectual Development Beyond Elementary School III: Ratio, A Longitudinal Study, 1972, **School Science and Mathematics, 72**(8), 735–742.

70. R. Karplus, Opportunities for Concrete and Formal Thinking on Science Tasks, A paper presented at the third annual meeting of the Jean Piaget Society, May 22, 1973 (published in their annual proceedings).

71. E. F. Karplus, R. Karplus and W. Wollman, Intellectual Development Beyond Elementary School IV: Ratio, The Influence of Cognitive Style, 1974, **School Science and Mathematics, 74**(7), 476–482.

72. W. Wollman and R. Karplus, Intellectual Development Beyond Elementary School V: Using Ratio in Differing Tasks, 1974, **School Science and Mathematics, 74**(8), 593–613.

73. R. Karplus and H. D. Thier, 1967, *A New Look at Elementary School Science*. Chicago: Rand McNally and Co.

74. R. Karplus, Physics for Beginners, 1972, **Physics Today, 25**(6), 36–47.

75. R. Karplus and J. Bowyer and J. Randle, Models: Electric and Magnetic Interactions Evaluation Supplement (Revised Edition). Berkeley: Science Curriculum Improvement Study, November 1974.

76. R. Karplus, Strategies in Curriculum Development: the SCIS Project, *Strategies for Curriculum Development*, Edited by Jon Schaffarzick and David H. Hampson. Berkeley, CA: McCutchan Publishing Corp., 1975.

77. R. Karplus, F. P. Collea, R. G. Fuller, J. W. Renner, and L. Paldy, Workshop on Physics Teaching and the Development of Reasoning, Stony Brook, NY: American Association of Physics Teachers, 1975.

78. R. Karplus, and J. R. Eakin, SCIS Final Report, Berkeley, CA: Lawrence Hall of Science, 1976.

79. A. Arons and R. Karplus, Implications of accumulating data on levels of intellectual development, 1976, **Amer. J. Phys., 44**(4), 396.

80. R. Karplus, A. E. Lawson, W. T. Wollman, M. Appel, R. Bernoff, A. Howe, J. J. Rusch, and F. Sullivan., Workshop on Science Teaching and the Development of Reasoning, Berkeley, CA: Lawrence Hall of Science, 1977.

81. R. Karplus, Science Teaching and the Development of Reasoning, 1977, **Journal of Research in Science Teaching, 14**(2), 169–175.

82. R. G. Fuller, R. Karplus, and A. E. Lawson, Can Physics Develop Reasoning? 1977, **Physics Today, 30**(2), 23–28.

83. E. W. Laetsch, Meet your new AAPT president, 1977, **The Physics Teacher, 15**(2), 70.

84. R. Karplus, Let's look to the future, 1977, **The Physics Teacher, 15**(2), 71.

85. R. Karplus, A Survey of Proportional Reasoning and Control of Variables in Seven Countries, 1977, **Journal of Research in Science Teaching, 14**(5), 411–417.

86. R. Karplus, The Lawrence Hall of Science, 1978, **Studies in Science Education, 5**, 107–110.

87. R. Karplus, The Intellectual Development Project at the Lawrence Hall of Science, 1978, **Studies in Science Education, 5**, 111–113.

88. R. Karplus, Science for Young Pupils, 1978, **Prospects, 8**(1), 48–57.

89. H. Adi, R. Karplus, A. E. Lawson, and S. Pulos, 1978, Intellectual Development Beyond Elementary School VI: Correlational Reasoning, **School Science and Mathematics, 78**(8), 675–683.

90. R. Karplus, A. E. Lawson and H. Adi, The Acquisition of Propositional Logic and Formal Operational Schemata During the Secondary School Years, **Journal of Research in Science Teaching**, 1978, **15**(6), 465–478.

91. R. Karplus and J. Bowyer, Inservice Staff Development: A Piagetian Model, **California Journal of Teacher Education**, 1978, **5**(3), 48–54.

92. R. Karplus, Proportional Reasoning in the People's Republic of China: A Pilot Study, **The Genetic Epistemologist**, 1978, **7**(3).

93. R. Karplus, E. F. Karplus, M. Formisano, and A. C. Paulsen, Proportional Reasoning and Control of Variables in Seven Countries, In J. Lockhead and J. Clement (eds.), Cognitive Process Instruction. Philadelphia, PA: The Franklin Institute Press, 1979.

94. B. Kurtz and R. Karplus, Intellectual Development beyond Elementary School VII: Teaching for Proportional Reasoning, **School Science and Mathematics**, 1979, **79**(5), 387–398.

95. R. Karplus, Authors' Reply to Comments on 'The Acquisition of Propositional Logic and Formal Operational Schemata During the Secondary School Years', **Journal of Research in Science Teaching**, 1979, **16**(4), 363–367.

96. A. E. Lawson, R. Karplus and H. Adi, Development of Correlational Reasoning in Secondary Schools: Do Biology Courses Make a Difference, **The American Biology Teacher**, 1979, **41**(7).

97. R. Karplus, Continuous Functions: Students' Viewpoints, **European Journal of Science Education**, 1979, **1**(4), 397–416.

98. R. Karplus, Teaching for the Development of Reasoning, A. E. Lawson (Ed.), Columbus, Ohio: ERIC-SMEAC, 1980. The Psychology of Teaching and Creativity.

99. R. Karplus, Early Adolescents' Structure of Proportional Reasoning, R. Karplus (Ed.). Berkeley, CA: Lawrence Hall of Science, 1980. Proceedings of the Fourth International Conference for the Psychology of Mathematics Education.

100. R. Karplus, Cognitive Correlates of Proportional Reasoning in Early Adolescence, R. Karplus (Ed.). Berkeley, CA: Lawrence Hall of Science, 1980. Proceedings of the Fourth International Conference for the Psychology of Mathematics Education,

101. R. Karplus, E. K. Stage and S. Pulos, Social context of Early Adolescents' Proportional Reasoning, Proceedings of the Fourth International Conference for the Psychology of Mathematics Education, R. Karplus (Ed.). Berkeley, CA: Lawrence Hall of Science, 1980.

102. R. Karplus, Educational Aspects of the Structure of Physics, **American Journal of Physics**, 1981, **49**(3), 238–241.

103. R. Karplus, H. Adi and A. E. Lawson, Intellectual Development Beyond Elementary School VIII: Proportional, Probabilistic, and Correlational Reasoning, **School Science and Mathematics**, 1980, **80**(8), 673–678.

104. R. Karplus, H. Adi and A. E. Lawson, Conditional Logic Abilities on the Four-Card Problem: Assessment of Behavioral and Reasoning Performances, **Educational Studies in Mathematics**, 1980, **11**, 479–496.

105. R. Karplus, Response by the Oersted Medalist: Autonomy and Input, **American Journal of Physics**, 1981, **49**(9), 811–814.

106. R. Karplus, S. Pulos and E. K. Stage, Proportional Reasoning of Early Adolescents: Comparison and Missing Value Problems in Three Schools,' Proceedings of Third Annual Conference for the Psychology of Mathematics Education, Tom Post (Ed.), Minneapolis, MN: University of Minnesota, 1981.

107. R. Karplus, E. K. Stage and S. Pulos, Early Adolescents' Attitudes Toward Mathematics: Findings from Urban Schools, Proceedings of Third Annual Conference for the Psychology of of Mathematics Education, Tom Post (Ed.). Minneapolis, MN: University of Minnesota, 1981.

108. R. Karplus, Education and formal thought: A modest proposal, New Directions in Piagetian Theory and Practice, I. E. Sigel, D. M. Brodzinaky, and R. M. Golinkoff (Eds.), Hillsdale, NJ: Lawrence Erlbaum Assoc., 1981.

109. R. Karplus, The Development of Reasoning, guest editorial in Piaget for Educators, by R. W. Bybee and R. B. Sund. Columbus, OH: Charles E. Merrill, 1982.

110. G.F. Chew, Obituaries Robert Karplus, 1992, **Physics Today**, **45**(3), 80.

Books:

R. Karplus and H. D. Thier., *A New Look at Elementary School Science*. Chicago: Rand McNally and Co., 1967.

R. Karplus, *Introductory Physics: A Model Approach*. W. A. Benjamin, Inc., New York, ©1969.

Films:

R. Karplus and C. S. Lavatelli, *CLASSIFICATION*, (1968) 16 minutes, Davidson Films, Inc., San Luis Obispo, CA.

R. Karplus and C. S. Lavatelli, *CONSERVATION*, (1968) 29 minutes, Davidson Films, Inc., San Luis Obispo, CA.

R. Karplus and R. Peterson, *FORMAL REASONING PATTERNS*, (1978) 32 minutes, Davidson Films, Inc., San Luis Obispo, CA.

Index